Restoration Ecology

To
Heleen, Arieke, Thibaud and Perrine

Restoration Ecology

The New Frontier

Edited by Jelte van Andel and James Aronson

Blackwell
Publishing

BLACKWELL PUBLISHING
350 Main Street, Malden, MA 02148-5020, USA
9600 Garsington Road, Oxford OX4 2DQ, UK
550 Swanston Street, Carlton, Victoria 3053, Australia

The right of Jelte van Andel and James Aronson to be identified as the Authors of the Editorial Material in this Work has been asserted in accordance with the UK Copyright, Designs, and Patents Act 1988.

First published 2006 by Blackwell Science Ltd

1 2006

Library of Congress Cataloging-in-Publication Data

Restoration ecology : the new frontier / Jelte van Andel and James Aronson (editors).
    p. cm.
  Includes bibliographical references and index.
  ISBN-13: 978-0-632-05834-1 (pbk. : alk. paper)
  ISBN-10: 0-632-05834-X (pbk. : alk. paper)
  1. Restoration ecology. I. Andel, Jelte van. II. Aronson, James, 1953–

  QH541.15.R45R517 2005
  333.7′153–dc22

                                    2005004142

A catalogue record for this title is available from the British Library.

Set in 9/11.5pt Rotis
by Graphicraft Limited, Hong Kong
Printed by TJ International, Padstow

The publisher's policy is to use permanent paper from mills that operate a sustainable forestry policy, and which has been manufactured from pulp processed using acid-free and elementary chlorine-free practices. Furthermore, the publisher ensures that the text paper and cover board used have met acceptable environmental accreditation standards.

For further information on
Blackwell Publishing, visit our website:
www.blackwellpublishing.com

Contents

v

Contributors

James Aronson
Centre National de la Recherche Scientifique
(CNRS–U.M.R. 5175), Centre d'Écologie
Fonctionnelle et Évolutive (CEFE), 1919 route de
Mende, 34293 Montpellier Cedex 5, France

Jan P. Bakker
University of Groningen, Community and
Conservation Ecology Group, P.O. Box 14, 9750 AA
Haren, The Netherlands

Jordi Cortina
University of Alicante, Department of Ecology, Ap.
Correus 99, 3080 Alacant, Spain

Anton Fischer
Technical University Munich, Department of
Ecology and School of Forest Science and Resource
Management, Am Hochanger 13, D-85354 Freising,
Germany

Holger Fischer
Technical University Dresden, Institut of Silviculture
and Forest Protection, P.O. Box 1117, D-01737
Tharandt, Germany

Ab P. Grootjans
University of Groningen, Community and
Conservation Ecology Group, P.O. Box 14, 9750 AA
Haren, The Netherlands

Ramesh D. Gulati
Netherlands Institute of Ecology, Centre for
Limnology, Rijksstraatweg 6, 3631 AC Nieuwersluis,
The Netherlands

Jim A. Harris
Cranfield University, Institute of Water and
Environment, Silsoe, Bedfordshire MK45 4DT, UK

Martin Janes
The River Restoration Centre, Silsoe Campus, Silsoe,
Bedfordshire MK45 4DT, UK

Bernhard Krautzer
Federal Research and Education Centre, Raumberg-
Gumpenstein, 8952 Irdning, Austria

Jenny Mant
The River Restoration Centre, Silsoe Campus, Silsoe,
Bedfordshire MK45 4DT, UK

Juli G. Pausas
Centro de Estudios Ambientales del Mediterráneo
(CEAM), C/ Charles R. Darwin 14, Parc Tecnologic,
46980 Paterna, Valencia, Spain

Theunis Piersma
University of Groningen, Animal Ecology Group,
P.O. Box 14, 9750 AA Haren, The Netherlands

Ramón Vallejo
Centro de Estudios Ambientales del Mediterráneo
(CEAM), C/ Charles R. Darwin 14, Parc Tecnologic,
46980 Paterna, Valencia, Spain

Jelte van Andel
University of Groningen, Community and
Conservation Ecology Group, P.O. Box 14
9750 AA Haren, The Netherlands

Rudy van Diggelen
University of Groningen, Community and
Conservation Ecology Group, P.O. Box 14
9750 AA Haren, The Netherlands

Ellen van Donk
Netherlands Institute of Ecology, Centre for
Limnology, Rijksstraatweg 6, 3631 AC Nieuwersluis,
The Netherlands

Sipke E. van Wieren
Wageningen University, Tropical Nature
Conservation and Vertebrate Ecology Group,
Bornsesteeg 69, 6708 PD Wageningen, The
Netherlands

Helmut Wittmann
Institute for Ecology, Johann-Herbst-Str. 23, A-5026
Elsbethen/Salzburg, Austria

Technical editing, references and index

Madelijn Marquerie
University Medical Centre Groningen (UMCG)
P.O. Box 30.001, 9700 RB Groningen,
The Netherlands

Design of figures

Dick Visser
University of Groningen, Biological Centre,
P.O. Box 14, 9750 AA Haren, The Netherlands

Foreword

Ecological degradation has a long history but was considered to be a problem only by a small group of nature conservationists, scientists and professional ecologists for much of this time. The awareness of the wider public, and the beginning of the general environmental movement, is usually traced back to the publication of Rachel Carson's *Silent Spring* in 1962. This global awareness was further enhanced, along with firm commitment by national governments, by the well-known conference on biodiversity in Rio de Janeiro in 1992. Since then, biodiversity has been recognized as a political goal and many countries have signed treaties and implemented programmes to prevent further losses. Societal activities that are detrimental to biodiversity and to environmental health have been identified, and policies are being developed to halt such activities or counteract their consequences. Issues like acid rain, desertification and global warming have drawn international attention and attempts are being made to eliminate these threats.

Nevertheless, many of the Earth's ecosystems still remain affected, some of them to a substantial degree. Ecological restoration as a tool to improve the quality of degraded systems is therefore extremely necessary and is – luckily – carried out on an ever-increasing scale. The Society for Ecological Restoration (now SER International, or SERI) was founded in 1987 to provide a focus for such activities and now has a global membership, drawn largely from academia, government and industry. Ecological restoration has even become a commercially viable activity. Existing consultancy and engineering companies have included restoration in their activities and new, specialized companies have emerged. Unfortunately, impressive as this increase in quantity may be in itself, it is not often matched by a similar increase in project quality. Much practical restoration is not backed up by sound scientific knowledge or, even worse, the knowledge does not exist. It is here that SERI plays a key role in facilitating information exchange and in stimulating research on restoration-related topics.

One of the continents where ecosystems are most affected is Europe, not least because of its long history of industrialization, high population density and high level of human pressure. However, this adverse situation also offers great opportunities to test all sorts of restoration techniques and theories. Much progress has been achieved in the last decade and made available, among others, through conferences from SERI Europe, the European chapter of SERI. The textbook on restoration ecology that you are reading is another way to exchange this knowledge. I am very pleased that Jelte van Andel and James Aronson and their team of authors have succeeded in creating such a thorough overview of the present science behind ecological restoration. I expect this book to become a landmark in the literature on restoration ecology for many years to come. It will certainly find its way to academics and practitioners alike and stimulate further cooperation. I am sure it will also serve as a stimulus for further research and improvement of ecological restoration.

Rudy van Diggelen
Chairman of SERI Europe

Preface

Restoration ecology is the field of study that provides the scientific background and underpinnings for practical ecological restoration, a field that is currently undergoing expansion. Together, ecologists, social scientists and managers are challenged to explore, test and apply current theories, models and concepts of academic ecology, even though many of them were not conceived or developed in view of direct applications. Throughout the book we, and our invited authors, explore to what extent currently available ecological concepts and theories can be made applicable in the specific, interventionist, trans-disciplinary context of ecological restoration.

Here are some sample questions. Can current models and insights into stability and disturbance be useful in coping with the task of judging when an ecosystem's trajectory can be considered healthy, so to speak, when it is seriously disturbed by human interventions, or when it is in or near to an anthropogenic threshold zone separating alternative steady states or basins of attraction? Can insight from the classic theory of island biogeography and the more recent metapopulation approach be applied to cope with the problem of rescuing species or species assemblages in increasingly fragmented habitats? Can current models be of use when constructing a reference system, for either a so-called natural system or a human-influenced one? Should the goal be to restore stable ecosystems, in terms of resistance and resilience, even though we know ecosystems are dynamic and changing in any case?

One intriguing issue for scientists and restoration practitioners is to determine the predictive value of a general model or theory in a very specific situation. In the real world we must find out not only what may happen, or is expected to happen in general, but also what will or does occur at a certain site, at a certain time, when steps are taken actively to repair and restore a damaged ecosystem or re-integrate a fragmented and humanized, or human-dominated, landscape. Through active involvement in real-world, large-scale projects, these insights can generate new information and stimulate new models, theories and concepts for all of ecology and related sciences.

The majority of chapters in the book are devoted to the science of restoration ecology, but the practice of ecological restoration is always present in the background. After the introductory chapters (Part 1), in Part 2 we explore the potential applicability of current ecological concepts and theories – at the levels of landscapes, ecosystems, communities and populations. Then, in Part 3, we focus on the ecological restoration of different ecosystems. We have assembled expert reviews, biome by biome, on ecosystems and habitats, special problems and restoration opportunities occurring in Europe. The case studies presented in the eight chapters of Part 3 aim at contributing to an ecological evaluation of the actual situation. In each of these chapters, we therefore start by setting the historical scene, describing the development from the original, near-natural ecosystems towards what these systems currently are as a result of land use. Then, after having identified the current threats and limitations for restoration or rehabilitation, referring to some kind of a reference situation or to a certain goal set, we explore the available ecological knowledge and models to help the reader understand the causes of successes and failures of ecological restoration efforts in the real world.

The reader should remember that the reconstruction of earlier-existing nature or ecosystems, or the development of 'new nature', or the design of 'new' ecosystems, cannot be realized in isolation from societal and political will and impact. The first chapter, therefore, delineates ecological restoration as a project for global society, and in the final part of the book we return at some length to the questions addressed at the broadest scales. In the two concluding chapters (Part 4), we define a number of reasons to restore ecosystems and reflect on broad notions of ecosystem health, emerging ecosystems and socioecological systems, and on the utilization of natural capital in the context of an emerging sustainability science. In addition, we discuss methods of diagnosis, evaluation and monitoring of whole systems, which is certainly one of the primary challenges for project managers, planners and of course researchers. We also review some of the challenges facing those who would practise ecological restoration on the landscape, regional, national and international scales, ideally in liaison or conjunction with efforts aimed at nature conservation and rational, sustainable management of land, water, vegetation and other natural resources.

Indeed, the ultimate goal of ecological restoration is to achieve sustainable, resilient and inter-connected ecosystems, and socio-ecological systems, providing goods and services to humans and habitat and well-being for non-human beings as well. But this is a lofty, very long-term goal to be tackled step by step. Scientists and other citizens in all societies need to address this challenge together, ensuring that ecology is used in all levels of decision-making, which in turn implies and requires a broadly inclusive approach to ecological research and applications, carried out in collaboration with social scientists and decision-makers from the local to the international level.

The present book aims at introducing MSc and PhD students, teachers, researchers and nature managers to the lively, volatile and stimulating interface between the science of restoration ecology and the practice of ecological restoration. It can be a useful document in workshops and master classes. Although we focus on the European context (entirely extra-tropical, and with few, if any, truly arid lands), we hope and intend to address a much larger, global audience. The structure we adopt is that of a textbook, but to make the most of the contents a firm background in basic ecology is recommended. We hope that some readers at least will find that we have been somewhat helpful in paving the way to the future. Today's students, after all, are tomorrow's decision-makers.

Acknowledgements

The editors are indebted to the team of enthusiastic authors, all international experts in their fields, who have made available a huge amount of expertise, while agreeing to and respecting a certain format and basic terminology agreed upon early on. We are very grateful to Madelijn Marquenie for her careful fine-tuning of the editorial work, in particular through the verification and quality-assessment for the numerous bibliographic references. We are also indebted to Dick Visser for the professional design of all the figures, which strongly contributed to the visual and educational quality of our book.

In the spirit of sustainability – one of the leitmotifs for this volume and, we hope, for the 21st century – we dedicate this book to our children, as representatives of the next generation to inherit this beautiful and precious planet in peril.

Jelte van Andel *and* James Aronson
Groningen and Montpellier

PART 1

Introduction to restoration ecology

1

Ecological restoration as a project for global society

Jim A. Harris and Rudy van Diggelen

1.1 Restoration ecology and ecological restoration

1.1.1 Is there a problem?

We rely on the integrity of the Earth system for our continued existence as a species, along with many others. **Sustainability** has therefore become an increasingly important issue in thinking about the future. As early as the 1970s Meadows *et al.* (1972) realized that there are limits to (economic) growth but it was not until 1987 that the so-called Brundtland report (WCED 1987) put sustainability firmly on the political agenda. The idea was accepted as a political goal at the conferences of Rio de Janeiro (1992) and its successor in Johannesburg in 2002 (Arrow *et al.* 1995). Central to the idea of sustainability is the notion that long-term thinking is essential to ensuring that the world remains a suitable place to live for future generations. This is not restricted to environmental conditions alone but also has social and economical components. Jepma and Munasinghe (1998) showed the relations between these three fields in a triangle diagram (Fig. 1.1).

Economic sustainability is based on the concept of maximizing income while at least maintaining the capital. The **social** component of sustainability seeks to maintain the resilience of social and cultural systems and their capacity to withstand shocks. The **environmental** component focuses on the resilience of biological and physical systems.

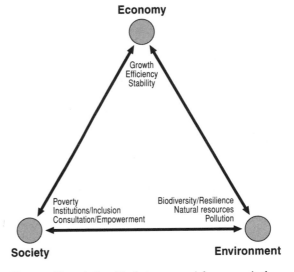

Fig. 1.1 The relationship between social, economical and environmental sustainability (after Jepma & Munasinghe 1998). Reproduced by permission of Cambridge University Press.

However, despite sustainability being well-developed as a theory and a political goal in many countries, a practical implementation is still far from a reality (McMichael *et al.* 2003). As social and economic injustice prevails, many of the fundamental functions and features of the ecosystems of the Earth are either at or beyond breaking point. The practice of ecological restoration and the science of restoration ecology are going to be major tools available to humankind for mitigating, arresting and reversing the adverse effects human activity has had on the Earth system, particularly since the Industrial Revolution. The challenges that we face include, among others (i) food, water and energy security, (ii) loss of biodiversity, (iii) global climate change and (iv) sea-level rise. The first two are a result of direct human pressures, with increasing population sizes demanding more resources; the last two are a result of amplification of the positive loops of natural feedback mechanisms, with the production of greenhouse gases from fossil stocks as the forcing function (Lovelock 1991). More discussion of this global picture will be found in the last part of this volume.

So, what do we mean by terms such as ecological restoration and restoration ecology? How are they related? How do they fit into society?

1.1.2 What is restoration ecology?

Many definitions exist of ecological restoration (e.g. NRC 1992, Jackson et al. 1995, SWS 2000). All incorporate the idea of reversing ecosystem developments that are considered negative through active human intervention. In this volume we adopt the definition presented by the Society for Ecological Restoration International (SER International; SER 2002; www.ser.org), and adopted by the International Union for the Conservation of Nature (IUCN), which is as follows:

> Ecological restoration is the process of assisting the recovery of an ecosystem that has been degraded, damaged or destroyed.

This definition is very broad but it shows clearly that restoration is not something theoretical without any practical obligations, but has to do with active engagement and intervention in current social and environmental affairs. In the present volume we will, like most other authors, use the words ecological restoration for the actual practice, and restoration ecology for the fundamental science upon which these actions *should* be based. Ecological restoration differs from rehabilitation, ecological reclamation, ecological engineering and landscaping in that all aspects of ecosystem structure and function are considered and addressed (see section 1.3.1).

1.2 Societal aspects of ecological restoration

Robertson and Hull (2001) have argued that the ultimate purpose of conservation science is to inform and affect conservation policy, and therefore those engaged in the production, review and application of conservation science should gauge their success in terms of their work's influence and impact on conservation decision making. They name this philosophy 'public ecology'. Whether you regard ecological restoration as an extension of conservation or a complement thereto, Robertson and Hull's appeal is equally applicable. We could also refer to the notion of human ecology, or sustainability science (see also Chapter 16 in this volume).

What is crucial, however, is that the arena in which restoration ecology and ecological restoration meet is societal. Societal involvement may imply volunteer work or a project carried out in a private back garden, or a multinational effort to sequester carbon by reforesting tropical rain forest, all beyond society's expert circles. Higgs (1997) has argued, most compellingly, that good restoration must encompass technical, historical, political, social, cultural and aesthetic components to offer any prospect of *sustainable* success. The issue of sustainability is central to restoration efforts. To simply execute a technical programme involving earth moving, tree planting and species re-introduction, to mention only a few, will not succeed in the long term if not embedded in a social context. It all demands inputs of time, effort and resources. There are several examples of how local communities may be involved, where non-expert information is crucial to success (Geist & Galatowitsch 1999). Light and Higgs (1996) and Swart et al. (2001) have illustrated that conflicts may arise when restoration programmes impact heavily on local populations. The level of community influence at each level may be summarized as in Fig. 1.2. Here, as the scale (maybe also the complexity) of a project increases, the direct influence of the communities involved decreases and the influence of scientific opinion increases. The community of 'non-experts' is incapable of forming accurate, or useful, judgements at such levels. This approach is becoming increasingly hard to sustain, as ultimately it is quite undemocratic, and is under

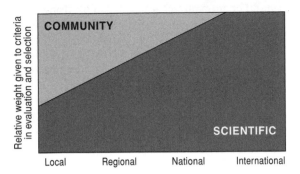

Fig. 1.2 Degree of weight given to community versus scientific arguments in relation to scale (after Harris *et al.* 1996).

increasing political pressure in the industrialized world (Kleinman 2000). It is likely that there will be an increasing move to 'citizens' juries' to come to decisions on topics of a quite complex technical nature (Kleinman 2000, Marris *et al.* 2001). Whether the old regime persists, or this new democratized approach is adopted, the onus for providing clear, objective advice to decision makers based on well-founded ethical experimentation and observation falls squarely on the shoulders of scientists.

So, are there formulations of how society can, and does, act on the findings of restoration ecology research programmes? In the European context there is increasing advocacy for going beyond simple agri-environment schemes, and combining them with carefully targeted large-scale habitat restoration (Sutherland 2002). This addresses several problems simultaneously, including sea-level rise, water catchment protection, flood defence and biodiversity issues. Before going into more detail we will address the choice of targets which themselves depend on the concept of nature that people have.

1.2.1 Concepts of nature

The debate over the meaning of what is natural is neither casual nor unimportant with respect to ecological restoration. Katz (1992) and Elliot (1997) have put forward the argument that restored landscapes are no longer natural, and therefore of little or no value. They base this argument on the definition of natural

as human intervention being absent. This absolutist approach would require humanity to be removed from the history of the Earth. The principal difficulty lies in the conflation of 'unnatural' with 'intentional'. If all intentional acts are committed by humans alone, this permanently bars all other animal life from claiming sentience, which is in itself indefensible, and there are few suggestions that a precise facsimile of a previous system is either feasible or desirable. Vogel (2003) and Light (2003) have argued against this absolutist stance, and in favour of ecological restoration as a culturally worthwhile project, while firmly confirming humanity and artefacts as part of nature.

The concepts of nature and naturalness are difficult to define easily. The principal difficulty crystallizes around humanity's role in nature – separate from or part of it? For example, Callicott *et al.* (1999) have detected a dichotomy in current schools of thought regarding nature conservation; compositionalism versus functionalism. They suggest that the latter sees humankind as part of nature, and the former separates humankind from nature; interestingly they suggest that ecological restoration belongs to the former school, i.e. humankind is separate from nature. We believe that this is a misapprehension, resulting from the strong focus of current restoration ecology on technical aspects: indeed, the scope of journals dealing with ecological restoration must be widened. A more comprehensive overview was presented by Swart *et al.* (2001). They identified three archetypes of nature, as follows.

1 The **wilderness** concept ('natural landscapes', see Westhoff 1952; 'natural ecosystems', see Christensen *et al.* 1996; 'primitive attitude', see Schama 1995). Central to this concept is the idea that nature regulates itself (Foreman *et al.* 1995). Humans do not play a significant role except for practical issues like legislation to protect areas, safeguard against poachers and the like. Typical discussions associated with this view are whether or not to introduce top predators into a restored area (e.g. Fritts *et al.* 1997) or the number of large herbivores that should be introduced to keep an area in a desired (i.e. open) state. Not surprisingly, this view is especially supported in less-densely populated areas like large parts of the USA, Russia and Africa. However, examples of this approach can also be

found in the densely populated areas in central and western Europe, including the present attempts to restore natural floodplains along all large rivers of western Europe.

2 The **arcadian** concept ('semi-natural landscapes', see Westhoff 1952; 'semi-natural ecosystems', see Christensen *et al.* 1996; 'arcadian view', see Worster 1977; 'pastoral attitude', see Schama 1995) is based on the long tradition of human interference with nature and is especially dominant in Europe. From the 18th century onwards the idea of 'nature' to European citizens was highly influenced by philosophers from the Age of Reason who depicted the simple, happy life in the countryside close to God. A second impulse came during the 20th century when ecologists realized that the highest (plant) species richness in history was found in these semi-natural landscapes. Present-day European nature conservation organizations often pursue this goal by mimicking ancient agricultural management techniques (Bakker 1989). Humans play a very active role: arcadian nature would not exist without massive human impact in the form of mowing, cutting back of hedges, slight drainage of species-rich meadows, etc.

3 The **functional** concept ('rural landscapes', see Westhoff 1952; 'intensively managed systems', see Christensen *et al.* 1996; 'imperialistic view', see Worster 1977) is heavily anthropocentric and considers nature as something to be used by humans. In fact this is the dominant view of nature. In the past, agriculture, forestry and fishery were the main domains of resource use, whereas modern use also includes other ecosystem functions such as clean air and flooding prevention, which can be associated with the notion of ecological engineering.

The above-mentioned example of floodplain regeneration is a typical case that was born from two completely different views of nature. Adepts of the functional approach emphasize the reduction of flooding risks, whereas adepts of the wilderness approach like this idea because a self-sustaining ecosystem is restored. The concept of nature determines the restoration goal in a given situation and this determines the necessary actions and spatial layout. Harms *et al.* (1993) compared the spatial consequences of alternative restoration targets, associated with different nature concepts (Fig. 1.3). Using these different scenarios as targets raises the additional problem of timescale – those scenarios nearer to the present state of the landscape will be achieved more readily and in less time than those that require extensive development of woodland, for example.

We must therefore ask, what are the components of the system that we are restoring, and for what purpose are we restoring them?

1.2.2 Biodiversity

One activity where restoration ecology has a clear mandate for action from society is in the field of conserving and enhancing biodiversity. The Rio Protocols (UN 1992a) and the European Union (EU) demand action under Biodiversity Action Plan orders. Here the requirement to audit and act on biodiversity targets with respect to particular species is now a well-established process. Further to this there has been legislation passed and likely to be adopted by the EU with respect to environmental liability. There will be a legal requirement for damaged ecosystems to be restored, or mitigated by restoration elsewhere, at the expense of the party responsible for the damage. Despite the widespread acceptance of biodiversity as a base for conservation and restoration, there is a lot of confusion associated with this term. In its original form it meant 'the whole variety of life on Earth' (Gaston 1996a) but this concept is not immediately useful for basing concrete actions upon. Therefore, the term is mostly used in association with a certain area. Biodiversity can be considered at three levels:

1 **genetic diversity** is the amount of variation within or between species;
2 **species diversity** means mostly just species richness;
3 **ecosystem diversity** is the variety of ecosystems on the planet (or part of it).

Most often the second meaning is used (species richness), but an implicit assumption in this use is that all species (of a given area) are known. This is evidently not true: at present there are about 1.75 million species described whereas the estimates of the total number of species vary between 7 and 20 million (Groombridge & Jenkins 2000). Independently of

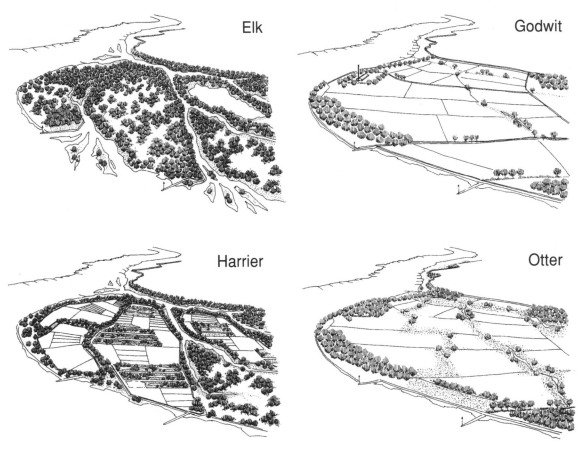

Fig. 1.3 Spatial consequences of four alternative restoration targets. Scenario Elk refers to restoring a wilderness, scenario Harrier to an arcadian landscape, scenario Godwit to a functional landscape with multiple users, and scenario Otter is a mix of the above with emphasis on improving the connectivity between landscape elements (after Harms *et al.* 1993). With kind permission of Springer Science and Business Media.

whether these estimates are correct or not, this means that only one-eighth of all living species has actually been described. It implies that our knowledge of bio-diversity is pretty limited and therefore difficult to use as a yardstick.

In practice this problem is generally solved by *choosing* certain target or indicator organisms that are used to measure (changes in) biodiversity in a given area. This can lead, however, to unwanted (and un-foreseen) conflicts. Restored areas are often colonized very quickly by mobile animals like birds whereas less-mobile organisms like plants or certain insect groups may (re-)appear much more slowly. If bird species rich-

ness is taken as a criterion such a project is regarded as very successful, whereas the project might be much less successful in terms of vegetation development.

A second point of attention is to use the concept of biodiversity in a sound way in restoration projects and always to compare the actual situation to a well-chosen reference situation. Species-rich but otherwise disturbed communities will be valued much more highly than certain species-poor targets such as bogs, heathlands or tundras. Even then, intermediate stages might be more species-rich than the actual restora-tion goal and this implies that elements of valuation come into play. If we examine the species-density/

productivity curve, i.e. the 'hump-backed distribution' curve, we note that different productivities are linked with characteristic species densities (Al-Mufti *et al.* 1977; see also Fig. 5.1 in this volume). Therefore, maximum species density *does not* equate to preservation of the most species, as those habitats requiring conditions found at either end of the curve (woodlands, tall herbs) will be excluded. Although often considered otherwise, biodiversity (in whatever sense) cannot be used as an objective yardstick and has everything to do with valuation and thus with the societal goals of the actors involved.

1.2.3 Ecosystem valuation

Economic valuation

One area in which the societal justification for ecological restoration may be tested is that of ecosystem valuation. The concept of ecosystem valuation is one undergoing much examination and development following seminal works by Helliwell (1969), and was given added impetus as a result of work by Pearce (1993), produced in response to the need for providing policy makers with a means of understanding carbon-emission control. A significant attempt was made for a global assessment of ecosystem services reported by Costanza *et al.* (1997). For a substantial review of the concepts and procedures involved we refer the reader to Daly and Farley (2004).

The basic premise upon which economic valuation systems rest is that the functions, goods and services that an ecosystem provides can be valued in monetary terms so that a cost-benefit analysis can be carried out. Dabbert *et al.* (1998) distinguished four types of **value** that ecosystems provide to society:

1 direct-use values, e.g. groundwater for drinking water and fish stocks;
2 indirect-use values, e.g. filtration and/or chemical alteration of pollution, and fixation of CO_2;
3 optional values: the values of using and experiencing areas, e.g. for recreational purposes;
4 non-use values: the intrinsic values of the mere existence of ecosystem, e.g. for species survival.

From top to bottom it is increasingly difficult to use classical economical valuation techniques. Whereas direct-use values can be expressed rather easily in monetary terms this is virtually impossible for non-use values. In the latter type of case economists often rely on so-called willingness-to-pay enquiries, e.g. house pricing in attractive areas. This approach can also be used in a negative way (avoidance costs), for example in an enquiry into how much money people are willing to spend on travel to avoid living close to the factory where they work. Useful as this willingness-to-pay approach may be, it also has certain disadvantages; for example, it is highly context-sensitive (people tend to regard natural values much more highly in a polluted environment that in a natural setting), and dependent on socio-economic circumstances. De Groot *et al.* (2002) followed a different approach (see below p. 9) and distinguished four types of **function** that ecosystems provide. Once this approach is made operational it would provide a basis for integrated cost-benefit analysis and balanced decision making not only for sustainable use and conservation of natural capital as the authors suggest, but also for resource-effective ecological restoration. It could also aid in the development of a 'restorability index' for sites in varying degrees of degradation.

One estimate (Boumans *et al.* 2002) suggests that the value of global ecosystem services is about 4.5 times the value of the Gross World Product. This approach appears attractive, as it talks to policymakers at an international level in units that they can readily understand. The danger lies in the fact that, accurate estimations of equivalency aside, it can lead to the assumption that given enough money anything can be replaced. This is not the case for species that are nearing extinction.

However, if a rational and sustainable case for this type of approach can be developed, it lends itself very readily to assessing the worth of ecological restoration as a societal endeavour.

Ecological valuation

Several novel approaches for considering ecosystem properties in a societal context have appeared recently, and we outline some of these below.

The **ecological footprint** concept has been developed in the last decade. The area of land required to support the consumption of anything may be calculated, from the level of an individual to the scale of the whole planet (Wackernagel & Rees 1996, Wackernagel *et al.* 2002). Once the area of land required is larger than that directly available, then the consumption is viewed as unsustainable. This provides a useful shorthand for the ecosystem valuation approach outlined above and is comprehensible to the non-expert. This approach has been criticized as being an over-simplification and being too dependent on scale (van den Bergh & Verbruggen 1999). These technical criticisms are resolvable, however, and we suggest that ecological restoration may be viewed as a tool for reducing our ecological footprint. By restoring ecological capacity to the degraded system in one region of a nation, we could reduce the footprint of that nation as a whole. This leads to the question, what contributes to the size of the footprint on the positive side? This is explored in the following paragraphs. This issue will be referred to once again in the last part of this volume.

A concept which has emerged in the last decade is that of **ecosystem health**. Ecosystem health can be defined as comprising a set of indicative factors (Rapport *et al.* 1998):

- vigour: activity, metabolism or primary productivity;
- organization: diversity and the number of interactions between system components;
- resilience (or counteractive capacity): a system's capacity to maintain structure and function in the face of stress.

The concept of ecosystem health is the focus of quite heated debate. Some commentators suggest that it is a ridiculous notion (Lancaster 2000), and others offer a critical analysis (Calow 1995). Harris and Hobbs (2001), however, suggest that ecological restoration could be regarded as one of the clinical tools for restoring ecosystem health, and provide numerous examples to that effect. In this sense ecosystem health provides the tools for assessment and ecological restoration the treatment (see also Chapter 2 in this volume).

De Groot *et al.* (2002) suggest that there are four principal functions supplied by ecosystems, without which human society could not function:

1 regulation functions: providing maintenance of essential ecological processes and life-support systems;

2 habitat functions: providing suitable living space for wild plant and animal species;
3 production functions: providing natural resources from which to make goods (consumable and structural);
4 information functions: providing opportunities for cognitive development.

This analysis has a considerable advantage over other approaches in that the economics of restoration are well characterized, and therefore have increased the chances of funding for, for example, restoration of floodplains by taking down levees or berms to re-instate flood storage.

1.3 A call for scientific analysis

1.3.1 Level of ambition

Restoration scientists usually stress the necessity to define and agree upon common targets in restoration projects (Hobbs & Norton 1996, Pfadenhauer & Grootjans 1999, Bakker *et al.* 2000). We agree with their plea for clear and measurable targets but would like to take the argument somewhat further. A definition of targets depends to a large degree on the level of ambition for which regeneration plans are being developed. Hobbs and Norton (1996) suggested a number of reasons why restoration might be carried out: (i) to enhance conservation values in protected landscapes, (ii) to enhance conservation values in productive landscapes, (iii) to improve productive capability in degraded productive lands and (iv) to restore highly degraded but localized sites such as mine sites.

These reasons may be consistent with the schema devised by van Diggelen *et al.* (2001), who distinguished three levels of ambition of programmes aimed at reversing degradation of ecosystems. The first and most ambitious level could be called **restoration** and consists of a reconstruction of a previous situation or self-sustaining target. This includes not only the re-establishment of former functions but also of the characteristic species and communities. The principal problem with taking this absolutist stance is that, if based on past environments and not current or future circumstances, it is likely to be unsustainable. If we accept, however, that we are intending to re-instate the maximum prior biodiversity, within maximized ecosystem services, then this term is again useful. The second level is often called **rehabilitation** and consists of the restoration of certain ecosystem functions (Mitsch & Jørgenson 1989, Wali 1992), such as the reduction of flood risks by creating the development of water-retention systems or restarting peat growth to fix CO_2 in peat layers. This option would make parts of the landscape as a whole more natural, but it would not necessarily result in a significant increase in biodiversity in the whole landscape. The third level is sometimes called **reclamation** and consists of attempts to increase biodiversity *per se*. The landscape as a whole would benefit from an implementation of such measures on a large scale but it usually does not contribute much to the protection of endangered red-list species. The definition given by Bradshaw (2002) of 'making the land fit for cultivation' is probably the easiest to comprehend, and most widely applicable.

The above-mentioned goals are generally associated with different *scales*, especially in densely populated areas where most restoration activities take place. Although it is technically sometimes possible to really recreate some former communities (true restoration) on a local small scale and at high cost, this is generally impossible at the landscape scale because of land-use conflicts, long-distance effects of other activities and lack of public support. Reclamation is often the only realistic option at this scale. Rehabilitation seems to be practical at an intermediate scale, often as a network within a certain landscape, for example riparian restoration (Kentula 1997). We suggest that simple recreation of past species lists is unlikely to succeed: **process** and **connectivity** must be taken fully into account, along with **biodiversity**.

A useful way of separating the reclamation and restoration definitions is to base them on the types of barrier that need to be crossed. It is possible to relate these terms to the conceptual arrangement of degraded, recovering and restored ecosystems in the light of a schema defined initially by Whisenant (1999) and further developed by Hobbs and Harris (2001), as illustrated in Fig. 1.4. Here, a number of putative stable ecosystem states, from degraded to intact, are related to ecosystem function. There are two principal barriers between degraded and restored (intact) systems. The first are abiotic barriers, which

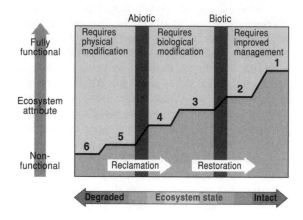

Fig. 1.4 Relationship between measured ecosystem attributes, biotic and abiotic barriers, and the processes of reclamation and restoration (modified from Hobbs & Harris 2001).

could be a lack of appropriate topology, contaminated substrate, too-high or -low a groundwater table, little or no organic matter, etc. These barriers all require physical modification to bring the systems to a new level of stability associated with a new 'higher' level of function. The second barrier is biotic; this may be as a result of a lack of appropriate species or interaction between them and abiotic components. Again, active modification allows these barriers to be overcome. We suggest that the first transition is the reclamation phase and the second is the restoration phase of a programme designed to restore ecosystem function and structure. Rehabilitation may be regarded as the transition from level 4 to level 3 on Fig. 1.4; that is, without crossing a significant threshold leading to a new, self-sustaining, ecosystem trajectory.

This gives us some clear goals to aim at, such as the re-instatement of a hydrological regime, or re-introduction of a keystone species. We must go further than simply measuring one feature or attribute. There are often conflicting interests from groups favouring plants, birds or animals, for example. The techniques designed to restore a species-rich meadow, with late cutting once a year, may be totally inappropriate for invertebrate populations, dependent on bare substrates for nesting and nectar plants for feeding: the late cut turns a feast into a desert overnight.

1.3.2 Ecosystem description

Before we can decide a target, we need to know, and be explicit about, what it is we are aiming for. We need a clear description of essential attributes of our target ecosystem. Hobbs and Norton (1996) provided a set of such attributes to be considered for measurement and manipulation in restoration programmes.

1 **composition**: species present and their relative abundances;
2 **structure**: vertical arrangement of vegetation and soil components (living and dead);
3 **pattern**: horizontal arrangement of system components;
4 **heterogeneity**: a complex variable made of components 1–3;
5 **function**: performance of basic ecosystem processes (energy, water, nutrient transfers);
6 **dynamics and resilience**: successional processes, rate and amplitude of recovery from disturbance.

These categories are conceptually clear, but not all of them are amenable to direct manipulation: essentially attributes 1–4 may be, to a greater or lesser degree, directly manipulated by such processes as civil engineering of soil contours, tree planting, seed sowing, altering hydrological regime and species re-introduction. Attributes 5 and 6 are a consequence of successful restoration, and therefore may be considered as prime indicators of progress towards a target. We must also note, however, that some recovering ecosystems will not be able to attract or sustain key indicator species (attribute 1 above) without some recovery in all six attributes: for example, there is no point in re-introducing predator species without there being prey present and abundant in the first place, which is dependent upon sufficient, robust, primary productivity.

In a similar vein, Vital Ecosystem Attributes (VEAs) and Vital Landscape Attributes (VLAs) have been suggested by Aronson *et al.* (1993a, b) and Aronson and Le Floc'h (1996a) respectively. These are intended as an aid to quantitative evaluation of whole ecosystem structure, composition and functional complexity over time. This entire coherent set of parameters is readily comprehensible, but not all may be amenable to routine inspection. This is not necessarily problematical, as some attributes (e.g. landscape structure) may change slowly, unless catastrophically altered, in which case they become the primary focus of attention in a restoration programme: that is, putting the original contours back.

1.3.3 Targets in restoration ecology

Ehrenfeld (2000) suggests three simple types of goal that we might consider for designing restoration prescriptions: species, ecosystem functions and ecosystem services. In many ways these are a re-phrasing of Hobbs and Norton's schema. These all have advantages and disadvantages, many outlined in Ehrenfeld (2000) and earlier by Ehrenfeld and Toth (1997).

Species

When we take a species-level approach to designing a restoration scheme we may use a number of targets:

• keystone species: related to particular functions;
• endangered species: often obscure as to wider function, but may be charismatic;
• assemblages: most likely to be summative as to total ecosystem status.

This approach may rescue particular species from extinction, locally or globally, and could definitely be said to increase biodiversity, but there are some potentially serious pitfalls. Landscape and ecosystem-level interactions may be ignored, with potentially adverse consequences in the long term, inadvertent pressure may be put on other species, and schemes overly focused on one species or group (e.g. birds) may lead to loss of habitat for other species, resulting in their decline.

Ecosystem functions

This level of analysis offers the potential to recognize explicitly the need for different components of ecosystems to be connected and working effectively. The types of processes and pools which may be examined are:

- material and energy flows, e.g. rates of transmission within, into and from the ecosystem, between biotic and abiotic components, change in pool sizes;
- biotic components, e.g. standing crop, species inventories, trophic connection, nutrient pools;
- abiotic components, e.g. nutrient pools, active components in nutrient cycling;
- ecosystem architecture, e.g. physical development and arrangement of the system.

This encourages due weight to be given to the interests of a diverse group of stakeholders, but relationships between, for example, biodiversity and functional stability are not yet clear. The overall problem of what an ecosystem is remains, leading to boundary problems, but this can be resolved by taking, for example, the catchment (watershed) to be the geographical context in which this activity should take place.

Ecosystem services

We have already outlined the four types of ecosystem service upon which human society depends – these are in many ways a combination of species and function classifications outlined above, but they may be easier to comprehend for a wider audience. Consequently, if we can devise suitable targets incorporating all three latter categories, while giving due weight to the public perception of past ecosystems being in some sense unsullied, there will be a greater chance of ecological restoration being adopted as a land–ocean management process.

1.3.4 Which type of target to choose?

Which restoration target to choose depends on the level of ambition of the project. If true restoration is chosen as a goal, a reference ecosystem is essential as a template for the restoration scheme to be based upon. The study of reference systems provides several types of relevant information. They enable the study and understanding of key processes that are relevant for the restoration trajectory. They provide some kind of yardstick that can be used both to estimate the degree of degradation of a disturbed ecosystem and the distance between the actual situation and the end point of the restoration. Finally they enable measure-

ment of tolerances of key organisms with respect to environmental conditions. If the ambition level is rehabilitation or reclamation, the demands are less strict and one might proceed without reference systems and just choose the function one likes to enhance. In that case, however, the line between landscape design and restoration becomes very thin and we do not advocate this approach here. Instead we argue that one should also use references in these cases and aim to restore certain species and/or functions of an ecosystem that have been lost through human-induced disturbance.

Historical reference ecosystems

One of the most superficially attractive targets for restoration projects is to re-instate the ecosystem 'prior to degradation' (Egan & Howell 2001). A number of sources of information may be available to varying degrees to allow the development of a restoration prescription, including historic photography, less-disturbed remnants on-site, similar sites, historical accounts and museum collections, and palaeoecological evidence.

A way of organizing this into types of information is shown in Table 1.1. Those techniques which rely on purely human records (historical photography and records) may not be available for many areas, and the palaeoecological evidence often begs the question, how far would you like to go back? All of the methods which reference the past suffer from what we might call moving-target syndrome, illustrated in Fig. 1.5. The original target system would have changed with time responses to environmental pressures, if left undisturbed. The question then arises, should our target prescription be set for where the system was, or where it would be now? What if key species have become extinct? We must also recognize the variability in natural systems in time and space (White & Walker 1997). The landscape, if unconstrained, would be a shifting mosaic of habitats at different successional stages, a newly exposed clearing here, a stand of ancient trees there, carved and renewed by the interactions of water, fire and sometimes even greater perturbations such as earthquakes and volcanic eruption. This does not, however, detract from the aspiration to restore some desirable characteristics of historical ecosystems, if they are sustainable.

Table 1.1 Information available for different types of reference system.

	Availability of information			
Type of system	Present status	Memory	Records	Forensic
Current	+	+	+	+
Recent	−	+	+	+
Past	−	−	+	+
Prehistoric	−	−	−	+

+, Available; −, not available.

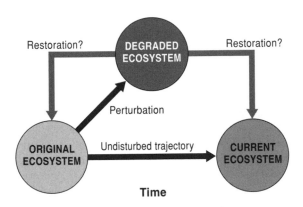

Fig. 1.5 Time changes an undisturbed ecosystem, making targets from the past hard to determine.

Modern reference ecosystems

Of more immediate practical use is the deployment of a reference system based on modern likely equivalents: what would be here in this defined topographic unit if degradation had not taken place? This immediately addresses the problem of moving-target syndrome, described above, as by definition all biotic components will at least be available to some degree, and we may take measurements aimed at capturing the key attributes of the modern reference system which we aim to achieve in the restored system. A problem with modern reference ecosystems is that we do not know whether they occur around their optimum or are in fact also found under sub-optimal conditions. For example, Grootjans and van Diggelen (1995) suggested that many sites with *Caricion davallianae* vegetation (EU priority habitat, heavily protected in most Euro-

pean countries, many red-list species, etc.) are in fact short-lived degradation phases of peat-forming communities. Another problem with modern reference ecosystems is that the reference ecosystem may itself be under threat due to the changes in the Earth system wrought by anthropogenic pressure, as outlined in the first section of this chapter; for example, global climate change might make lowland wet grasslands unsustainable due to changes in rainfall and hydrological regime.

Functional targets

When using functional targets we have some idea of the gross parameters we have to measure (primary productivity, energy flows, hydrological regime), and success in restoring these functions indicates a certain level of integration. Unfortunately for the restorationist they are nearly all emergent properties, and not directly capable of manipulation. As we have indicated above superficial success may be garnered by a massive species re-introduction, only to be followed by a catastrophic failure. But they do give us a coherent and cogent set of targets, and we should adopt their use because the success of these functions is self-consistent. The outcome does not rely on the judgement of human agency.

Ecosystem service to society targets

Setting our targets from the various categories above in an ecosystem-service context is the only way in which we are going to produce truly self-sustaining restored ecosystems. As Cairns (2000) reminds us: 'Major ecological restoration will not be undertaken

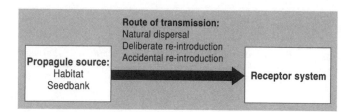

Fig. 1.6 Source, pathway and receptor system for re-introduction of biotic components to degraded ecosystems.

unless human society approves the goals and objectives of restoration. In addition, restoration will not persist unless human society has sufficient esteem for the restored ecosystem to protect its integrity.'

So we have to address all types of target, some directly by direct physical manipulations, others by incorporation of stakeholders and the full socio-economic dimension, in order to achieve truly sustainable restorations, and be cognisant of the need to reconnect processes and functions with the biological components of a system, after abiotic thresholds have been overcome (Hobbs & Harris 2001). This is a hard task, as resources often only allow, or are granted for, one of these goals. This is because of the multiplicity of funding streams aimed at each of these types of target, a confusing melange of government agencies, non-governmental organizations, charities aimed at single species (hedgehogs appear to be a particular favourite), and individual activists and volunteers. Funding may in fact be completely absent for particular groups, and legislation may make certain restoration goals actually illegal, as with the re-introduction of some native species into agricultural areas being prohibited as they are regarded as noxious weeds. In certain parts of Europe there is a cultural and legislative resistance to the active replanting and re-introduction of species, rather relying on the dictum 'build the field and they shall come'. This completely ignores the importance of migration rates of sources of biological diversity: how far away the sources of biological potential are, and what the routes of transmission are. We might suggest an inverse of the risk-assessment approach (Fig. 1.6).

Here the risk of failure of transmission is the object of interest – we must know this in considerations of re-establishment of complete species inventories. The route of transmission is often limited for technical reasons, but in some societies the greater obstacles are cultural.

1.3.5 How to measure the degree of restoration success?

In the SER Primer on ecological restoration (SER 2002) a set of aspects of the structure and functioning of ecosystems was pointed out that can be used to assess the effectiveness of the success of any given restoration project. In our opinion not all aspects are applicable at all ambition levels. In Table 1.2 we indicate what attributes we consider applicable for the several ambition levels. Ideally the effect of ecological restoration is as depicted in Fig. 1.7. The restored object changes under the restoration management towards the target in the same way as it would without human interferences, only much faster. In practice, however, failure occurs very often. It is, therefore, necessary to measure and monitor the degree to which the restoration has been successful and appropriate; timely monitoring programmes are an indispensable part of the restoration process.

Dale and Beyeler (2001) mentioned several criteria that ecological indicators should fulfil in order to be suitable as measurements of ecosystem quality. They should be easily measured, be sensitive to anthropogenic pressures, be anticipatory – giving early signals of a treatable larger problem, allow for adaptive management intervention, be integrative, have known responses to stress, disturbances and time, and have low variability in response. Without these, it is difficult to see how a sustainable measurement and monitoring programme would be sustained by society, so we must ensure that the full range of spatial and temporal variability is captured by the critical component states measured.

Clearly, using Hobbs and Norton's (1996) schema (section 1.3.2) we can choose indicators from all six of their attributes and apply them to the type of time-series analysis suggested in Fig. 1.7. This will only be of use, however, if we are able to distinguish change

Table 1.2 Suitability of ecosystem attributes to measuring the degree of restoration success at different ambition levels.

Attribute	Ambition level		
	Reclamation	Rehabilitation	Restoration
Characteristic assemblage of species	−	−	+
Indigenous species	−	−	+
All functional trophic groups	−	−	+
Normal functioning of cycles	−	+	+
Appropriate physical environment	+	+	+
Integrated into a larger landscape	+	+	−
External threats eliminated or reduced	+	+	−
Resilience to perturbation	+	+	−
Self-sustaining	+	−	−

+, Suitable; −, not suitable.

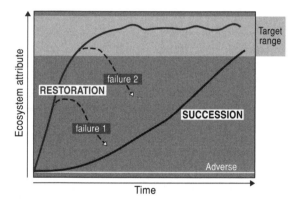

Fig. 1.7 Restoration success in relation to time: failures go undetected without appropriate monitoring.

in ecosystem trajectory from natural fluctuation and, crucially, are able to act upon the result of this analysis, swiftly and effectively. In order to do this we need to have a deep understanding of the ecology of our systems of interest.

The proper placing of technical aspects of ecological restoration within a societal setting is clearly a crucial task for the restoration ecologist. To be successful in this, however, we must be clear as to what we mean by those technical parameters, their limits and utility, in order to aid the larger project of integration into the societal context. Clarifying and identifying those technical and scientific issues in restoration ecology is the primary intention of this volume.

2

Concepts in restoration ecology

Jelte van Andel and Ab P. Grootjans

2.1 The scope of restoration ecology

Restoration ecology is the science of ecological restoration, recently re-defined as 'the process of assisting the recovery of an ecosystem that has been degraded, damaged or destroyed' (SER 2002; www.ser.org). In an influential editorial in the key scientific journal in the field, *Restoration Ecology*, then editor William A. Niering wrote: 'Ecological restoration will continue to provide important insights into the way that ecological communities are assembled and ecosystems function. There is no question that restoration ecology is the wave of the future' (Niering 1997a). Conservation biologist Truman Young (2000), wrote 'restoration ecology is the long-term future of conservation biology', and the eminent biologist and philosopher Edwin O. Wilson (2002) elaborated at length on the same theme in his latest book, *The Future of Life*. Nature conservation *sensu stricto* focuses on the maintenance of existing ecosystems, which often demands long-lasting application of the same management, such as grazing of mowing regimes. Ecological restoration, however, may demand a much more dynamic management approach, the more so when there is uncertainty about the feasibility of the planned trajectory and final goal of ecosystem development (see Fig. 2.1). In view of what has been explained in Chapter 1 in this volume, ecological restoration comes close to nature conservation in the case of the repair of a degenerated ecosystem back to its previous intact state (called true restoration), whereas rehabilitation and reclamation or ecological engineering can be considered a contribution to nature conservation or even the establishment of new nature.

Ecological restoration can focus on restoring entire ecosystems or communities, or on the rescue or re-introduction of certain target species. Ecosystems, communities and populations function in a **landscape** context (Bell *et al.* 1997; see Chapter 3 in this volume). Restoration efforts have profited much from landscape concepts, such as spatial heterogeneity and connectivity, and associated dynamic models. Landscape hierarchies – based on geomorphology, soil and vegetation – have also been used to identify reference ecosystems and associated plant communities to prioritize and guide restoration of disturbed ecosystems (Palik *et al.* 2000). At the same time restoration studies can be used to advance the field of landscape ecology, which can utilize the information provided by restoration projects to improve and test basic questions, especially those linked to habitat function and fragmentation. Restoration ecology has also benefited much from experience in **ecosystems** research (Ehrenfeld & Toth 1997; see Chapter 4 in this volume). Methods have been developed to delineate ecosystem boundaries, to assess nutrient fluxes within ecosystems and to measure ecosystem attributes such as primary production and structure. Methods also exist to help choose appropriate restoration goals, for example using vegetation types as references, and to select target species, indicator species and ecosystem attributes for evaluation. Going further, in order to restore ecosystem functioning, it is necessary to know how natural disturbances affect ecosystems. Such information can only become available if various scientific disciplines work together and provide methods and information concerning how restoration measures will affect damaged ecosystems. Restoration projects may benefit from available knowledge on

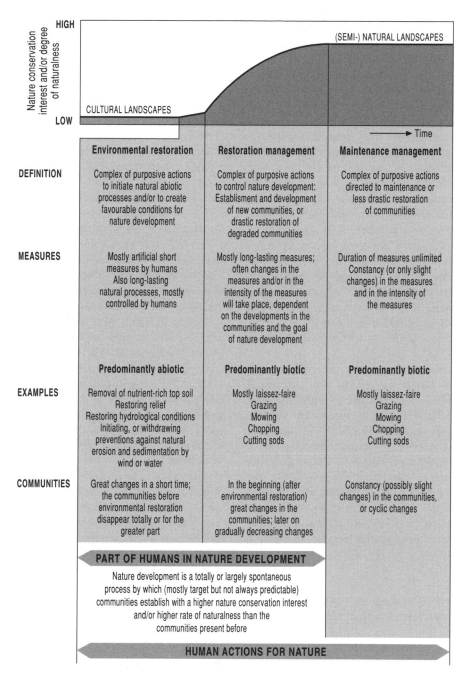

Fig. 2.1 Schematic overview of management aspects of ecological restoration and nature conservation. After Bakker and Londo (1998). Reproduced by permission of the authors.

community ecology (Palmer *et al.* 1997; see Chapter 5 in this volume), in particular on the vast amount of information of species interactions to understand the processes regulating foodwebs and keystone species, species richness and community organization, sometimes expressed in terms of assembly rules. From a restoration-ecological perspective, we may want to know whether and how to identify reference communities, a subject we will discuss in more detail. Finally, **population ecology** and genetics are of interest when populations of rare species are threatened with extinction and measures have to be taken to increase population sizes (Montalvo *et al.* 1997; see Chapter 6 in this volume). To re-introduce a population to a restoration area where target species have become extinct is one of the most obvious examples of potential problems in restoration projects (Falk *et al.* 1996; see Chapter 7 in this volume). From a restoration-ecological perspective, new questions come into focus, such as, how many plants and animals should be re-introduced and is the genetic variation sufficient to prevent inbreeding and genetic drift?

Several of the attributes discussed above were listed by SER (2002) as criteria for determining whether and when restoration has been accomplished, and they will be discussed in the present chapter. Recently, Walker and Del Moral (2003) illustrated how current knowledge of primary succession can be applied for rehabilitation purposes. Challenging new scientific questions arise from the search for and testing of new applications. In the following chapters several aspects will be dealt with in a certain order, one after the other, but the reader will recall that ecological restoration demands an integrative approach (see Chapter 1).

2.2 Objectives of ecological restoration: targets, references and trajectories

The goal of a particular restoration project may involve the return of an ecosystem to an approximation of its structural and functional condition before damage occurred, but it can also include the creation of a new ecosystem that had never existed before on the site selected for restoration. For instance, abandoned agricultural fields that have been fertilized

intensively for decades cannot easily, if at all, be transformed to the nutrient-poor heathlands that once covered the landscape. The increased nutrient stocks in the soil, combined with increased nitrogen deposition from the air, will cause the development of a much more productive ecosystem, even when most of the organic top soil has been removed at great cost (Verhagen *et al.* 2003).

Restoration ecology, therefore, often requires the evaluation of ecological concepts and approaches in a modern societal context. Higgs (1997) argued that good restoration requires an expanded view that includes historical, social, cultural, political, aesthetic and moral aspects. Indeed, targets for restoration can be identified by using historical ecological knowledge (White & Walker 1997, Egan & Howell 2001), but such targets should be considered to be feasible and cost-effective as well. Allen and Hoekstra (1987) held the opinion that the goal of a restoration is not to 'recreate' some ideal pristine ecosystem, but to establish an ecosystem in which processes can indeed fit into the available area of land, which is often limited by the claims of other land users. We cannot just formulate restoration goals that are unacceptable for other land users, or that have no backing by politicians and public opinion in general. A restoration scientist should gather as much information as possible on the historical development and anthropogenic transformation of the ecosystems and landscapes to be restored. This knowledge is necessary to formulate acceptable goals in restoration projects (Aronson & Le Floc'h 1996b). Recently, Ehrenfeld (2000) reviewed the relative merits and pitfalls associated with specifying restoration goals based on (i) species, (ii) ecosystem functions and (iii) ecosystem services, showing that ecological restoration implies the development of a desired ecosystem in a certain area in a societal context.

As in health care, the restoration of damaged ecosystems should start with a good diagnosis of the problem, followed by a suitable trajectory to repair the damage or otherwise eliminate the complaints by rebuilding new structures. The way towards such a goal may demand the recognition that there is no one paradigm or context for setting restoration goals, and that ecologists need to develop probabilistic laws, recognizing that developing ecosystems may undergo rapid transitions between different meta-stable states (Fig. 2.2). Very often removing a stress or disturbance

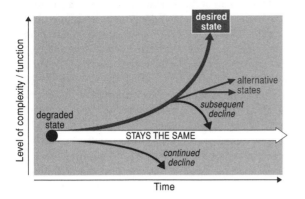

Fig. 2.2 Potential paths of ecosystem development after disturbance to a degraded state. Restoration objectives may or may not be met. After Hobbs and Norton (1996). Reproduced by permission of Blackwell Publishing.

factor will not result in recovery of components of the original ecosystem that have been lost, because it has caused irreversible damage to the ecosystem involved, which may cause switches that are sometimes described in terms of 'thresholds of irreversibility' (Aronson *et al.* 1993b). Drainage of wetlands, for instance, leads to irreversible changes in the soil structure (decomposition of peat, iron depletion, acidification, etc.). Just rewetting the drained wetland will not lead to the return of the original species of that ecosystem.

Restoration efforts may not result in a goal that is wanted, because the developing system may be locked in an alternative stable state. For example, reduction of nutrient loading in eutrophied turbid shallow lakes rarely leads to a satisfactory recovery of the clear state, even if the nutrient level is considerably reduced (Scheffer *et al.* 1993), which can be explained by their 'bi-stability theory'. The clear state is dominated by aquatic vegetation, and a turbid state by a high algal biomass. Nutrient reduction has to be accompanied by reduction of the fish stock in the lake to enforce the switch back to clear water. This may be due to a trophic cascade effect (reduction of the predation pressure by phytoplanktivorous fish allows populations of large-bodied zooplankton to peak and graze down the algal biomass, causing clear water in spring), and to the effect of reduced sediment resuspen-

sion (resulting from removal of fish that feed at the bottom of shallow lakes).

A reference ecosystem, therefore, serves as a model for planning a restoration project, and later for its evaluation (Egan & Howell 2001, SER 2002). The identification of a reference system is related to the level of ambition of a reconstruction project (see Chapter 1). For individual species, knowledge of a reference system requires the determination of its habitat (see for example Rickers *et al.* 1995); for communities it is the determination of an association or assemblage of species and its synecology (see for example Ellenberg 1986). However, historical data sets of plant species and the associated plant community members can become outdated within half a century (Bakker *et al.* 2000, Strykstra 2000). This potential discrepancy is a general problem, as it also limits the success of identifying functional groups from species traits. Evolutionary mechanisms do only partly result in future success. This holds for species, but more so for communities, even for so-called pristine tropical rainforests (van Gemerden *et al.* 2003). Considerations of natural dynamics imply that reference systems, which we know from the past, can therefore only serve as a point of orientation, not so much as a goal in terms of returning to the past. Restoration ecology demands insight into potential perspectives of nature development from the past towards the future, which can be modelled in terms of scenarios or trajectories. These ideas will be dealt with *in concreto* in Chapters 8–15 in this volume, where case studies will also be described to illustrate the practice of ecological restoration.

2.3 A systems approach: the concepts of disturbance and stability

In view of the central task of ecological restoration to restore or rehabilitate disturbed ecosystems, we will reflect upon the notion of disturbance. In general terms, a disturbance is a long-term disordering of a constant or steady state, due to an external event to which the system is not capable of responding through resistance or resilience (Fig. 2.3; see also Holling 1973, DeAngelis 1992, Mitchell *et al.* 2000, SER 2002). The latter terms refer to the notion of stability, a concept also to be dealt with below.

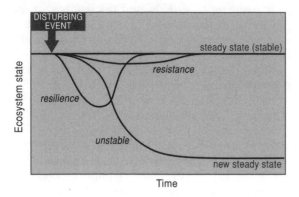

Fig. 2.3 Ecosystem responses to a disturbing event: resistance and resilience are modes of recovery, the unstable response implies a long-term disturbance which may or may not be reversible through restoration practices. Ecosystem state may be represented by productivity, species richness or other characteristics. Modified after Aber and Melillo (1991).

2.3.1 Disturbance

An external change of environmental conditions may or may not result in a disturbance of the system under consideration. In the case of disturbance, there is no way of returning to the initial state of the system, at least not in the short run, but the system may arrive at an alternative state, which also may appear to be a steady state (see Fig. 2.3). It is a task for restoration ecologists to judge the differences in quality between the two or more steady states, for example between a forest and a clear-cut forest (which then may be transformed to a semi-natural meadow).

In addition to the aforementioned general description of disturbance, various definitions of disturbance exist in the literature. Two definitions of disturbance are worth mentioning here.

1 **Disturbance** is a change in conditions, which interferes with the normal functioning of a biological system (van Andel *et al.* 1987).
2 **Disturbance** is a change in the minimal structure of an ecological unit caused by a factor external to the level of interest (Pickett *et al.* 1989).

Both definitions have their weaknesses. In the first definition the normal functioning of a biological

system is not defined, and in the second definition the same is true for the minimal structure of an ecological unit. Yet, disturbance has no real meaning without some notion of the normal functioning or minimal structure of a reference system. Both these definitions link causes and effects in one way or another, and indeed they are both relative definitions. White and Jentsch (2001) suggested measuring disturbance in absolute terms, for instance by the reduction of biomass (see also Grime 1979). In that way problems related to the relative definitions of disturbance could be avoided. For the time being we prefer a relative definition of disturbance, thus challenging the need to relate it to a well-defined reference system. Indeed, the terms resilience and resistance, applied to indicate the return to a steady state (stability, see below) are even defined in a relative sense.

For purposes of ecological restoration, reference systems have to be defined to determine whether a disturbed ecosystem is being restored or remains in a state of degradation. If a reference system is being defined in terms of a 'minimum structure' (Pickett *et al.* 1989), it may mean either that the system can be confined to a few keystone species, the other species being more or less redundant, or that all the species in the community are to be included in the minimum structure in order to restore ecosystem functioning, with no species being redundant. These two contrasting points of view are still being discussed. Some authors favour the redundant-species hypothesis (only a few keystone species contribute to the functioning of the ecosystem), while others advocate the rivet hypothesis (all or almost all species essentially contributing to some ecosystem function); see for example Ehrlich and Ehrlich (1981, 1992), Lawton (1997), Prins and Olff (1998), van Andel (1998a), Waide *et al.* (1999) and Chapter 3. It has long been assumed that an increase in the number of species in a community or ecosystem would imply an increase in connectance and interaction strength; such an increase in complexity would increase stability. But this seems to be the exception rather than the rule. It is still far too early to draw any general conclusions.

To avoid confusion, it is useful to distinguish between kind, frequency, intensity and scale (or extent) of disturbance (e.g. Connell & Slatyer 1977, Grubb 1985). The **kind** of disturbance depends on the environmental factor concerned, whether biotic or abiotic. The **degree** of disturbance is determined by the difference between the new conditions and the previous steady state (or reference) conditions. **Frequency** is important because of different effects from isolated, recurrent and continuous disturbing events; they can be irregular or regular and of differing durations. The **scale** refers to different spatial and temporal patterns, and to different levels of ecological organization: ecosystem, community, population or individual. In line with this, Turner *et al.* (1993) distinguished four major factors characterizing the dynamics of landscape, *viz.* (i) the disturbance frequency or its inverse, (ii) the rate of recovery or its inverse, (iii) the size or spatial extent of the disturbance events and (iv) the size or spatial extent of the landscape. The framework they proposed permits the prediction of disturbance conditions that lead to qualitatively different landscape dynamics and demonstrates the scale-dependent nature of concepts of landscape equilibrium.

The term stress is used frequently in a similarly fuzzy way to the term disturbance. The notion of stress originates from animal ecophysiology to indicate a specific negative response of individuals to a so-called stress factor (below or beyond optimal conditions; shortage or surplus of resources). This is, again, a relative approach, challenging one to define non-stress behaviour. Grubb (1998) reviewed and classified stress responses of plants, which is very useful for this purpose. In analogy to stress physiology of individual organisms, ecosystems may be considered as being under stress (i.e. out of optimal conditions). Grime (1979, 2001) suggested evolutionary pathways of plant species to result from their stress physiology, in addition to their capabilities of coping with disturbance, and classified plant communities according to this principle.

2.3.2 Stability

The notion of stability suggests a long-term steady state. The choice of parameters to measure ecosystem stability is of utmost importance. Much knowledge on long-term vegetation changes has been derived from permanent plots, which are patches of land that are recorded regularly. Measures of ecosystem stability are often confined to vegetation biomass or species composition. The longest regularly recorded permanent plots are situated in Rothamsted Park in England, better known as the Park Grass Experiment (PGE). Collins (1995) wrote: 'Although biomass has been shown to fluctuate from year to year, there has been no overall change in biomass production during the past 100 years. The overall conclusion from the PGE data is that biomass of functional groups has been relatively stable for the past 100 years despite dramatic changes in species composition over the time span of these experimental plots.'

What does the notion of stability imply? Although we appreciate Grimm and Wissel's (1997) analysis of the stability concept, derived from 163 definitions found in the literature, we have not noticed any follow up in the literature that made use of this approach. We refer to the more generally accepted definitions of stability and related concepts of resistance and resilience, which are still being developed for systems ecology (see Fig. 2.3):

- **Stability:** the capacity of a system to return – in spite of changes of environmental conditions – to a certain starting value, which is considered a steady state. This steady state is a dynamic equilibrium, not at stasis, and is kept within boundaries through resistance and/or resilience.
- **Resistance:** systems that show relatively little response to a sudden change in environmental conditions are said to be resistant; they maintain their structural and functional attributes. It is often the case that resistant systems will take a long time to return to their initial state after an external change that is strong enough to alter the state temporarily.
- **Resilience:** resilient systems can be altered relatively easily but will return to the initial state more rapidly. They regain structural and functional attributes that have been damaged due to changes in environmental conditions. The length of time taken to return to the steady state is inversely related to resilience; the faster the system returns, the more resilient it is.

Why are some ecosystems more resilient than others? Mitchell *et al.* (2000), in an attempt to implement the aforementioned definitions of resilience and resistance, proposed a multivariate modelling approach to cope with this problem. They combined measurements of species and environmental variables, which can represent attributes of either ecosystem structure or ecosystem function. They provided an example for the conservation management of lowland heaths in Dorset, UK, which seems generally applicable.

In view of possibilities for ecological restoration, it is interesting to know whether a disturbance is definitely irreversible or can be counteracted by restoration measures. Systems can be disturbed in such a way that there is no way back to the initial steady state, unless particular measures are taken. For example, if artificial fertilizer is applied to transform a semi-natural hayfield into an intensively used agricultural grassland, the latter can be considered a new steady state (degenerated from a nature-conservationist point of view, intact from an agricultural point of view). The way back to the earlier steady state can be difficult to pave. Similarly, a semi-natural hayfield that is left unmanaged may turn into a woodland or forest. In nature, a system may be triggered – sometimes quite suddenly – to turn into one or another alternative stable state. The patchy spatial distribution of tree savanna, grass savanna and bare ground may result from different frequencies and intensities of herbivore grazing and browsing that may induce irreversible changes in vegetation structure (Noy-Meir 1975; Rietkerk & van de Koppel 1997; van de Koppel *et al.* 2002). In such semi-arid grazing systems the interactions between water infiltration or nutrient retention and plant density potentially give rise to the existence of alternating stable vegetation states and threshold effects, even without the effect of non-linear herbivore functional response or plant competition (Rietkerk & van de Koppel 1997). These interactions may trigger a positive feedback between reduced plant density and reduced resource availability, and lead to a collapse of the system. Similarly, Grootjans *et al.* (1998) have shown that the pioneer plants in wet coastal dune slacks prevent the accumulation of organic matter in the soil as long as they are capable of keeping up a positive-feedback mechanism, in this case by creating an oxic–anoxic gradient in the rhizosphere resulting in a subsequent combination of nitrification and denitrification (Adema *et al.* 2001). The pioneer stage of succession (with *Littorella uniflora* and *Schoenus nigricans*) can be sustained for an extended period of time (decades), and then suddenly move towards the next, more productive, stage of succession (with *Carex nigra* and *Calamagrostis epigejos*). This process is irreversible, unless measures such as sod-cutting are applied to reset the successional clock. Problems in the context of restoration ecology have been reviewed recently by Suding *et al.* (2004). Indeed, degraded systems can also be resilient through internal feedback that constrains restoration.

Although it seems only a small step to elaborate the notion of stability towards defining ecosystem health and ecosystem integrity, these terms may suggest that an ecosystem can be considered as some kind of a superorganism, which it is not. We have to cope with the term, recognizing that it is being used in a more societal context, where ecological and socio-economic valuation systems meet.

2.3.3 Ecosystem health and ecosystem integrity: useful metaphors

Ecosystem health can be described as the state or condition of an ecosystem in which its dynamic attributes are expressed within normal ranges of activity relative to its ecological state of development (SER 2002). As with the notions of disturbance and stability, we make a choice for using a relative definition of ecosystem health. From an ecological point of view, ecosystem health can be evaluated in terms of the ecosystem's overall dynamic state at a given time based on ecosystem functioning (Winterhalder *et al.* 2004). However, the notion of ecosystem health implies more than an ecological valuation; socio-economic criteria are generally also taken into account. Rapport (1995) stated that 'evaluating ecosystem health in relation to the ecological, economic and human health spheres requires integrating human values with biophysical processes, an integration that has been explicitly avoided by conventional science'. Boulton (1999), dealing with river health assessment, similarly emphasized that, in contrast to definitions of healthy ecosystems based on solely ecological criteria, judgements of river health must include human values, uses and amenities derived from the river system. Indeed, rivers are not just ecosystems, but can also be considered in terms of functions for mankind, for example, as sources of clean water for drinking and washing, for industrial and agricultural purposes, as conduits for pollutants, and as places for recreation and aesthetic pleasure. The notion of ecosystem health, therefore, adds a socio-economic valuation to the strictly ecological qualifications in terms of stability and disturbance (see Chapter 1).

Ulanowicz (1997) argued that ecosystem health relates to how well an ecosystem is functioning at a given point in time, while ecosystem integrity can only be evaluated over a longer time period, including the ability of a system to deal with unforeseen circumstances in the future. Indeed, whereas in the use of the term **ecosystem health** the focus is on ecosystem functioning, the term **ecosystem integrity** includes a strong biodiversity component. It is the state or condition of an ecosystem that displays the biodiversity characteristics of a reference system, expressed in terms of species composition and community structure (SER 2002). We recognize that ecosystem functioning and species richness are not independent ecosystem characteristics; there is an increasing body of literature on functions of species richness (see Waide *et al.* 1999 for a review, and Loreau *et al.* 2002). A stable ecosystem may be resistant to invasions of alien species that ultimately can disturb the system. Although this is already quite a difficult matter to deal with in a strictly ecological sense, it becomes even more complicated in confrontation with other disciplines. Nevertheless, 'increasing our understanding of these interactions will involve more active collaboration between the ecological, social and health sciences' (Rapport *et al.* 1998).

2.4 The biodiversity approach

The term biodiversity (biological diversity) emerged in the 1980s as a general catchword for the whole variety of life on Earth (Gaston 1996a), but from a scientific point of view biodiversity has existed since the emergence of life on Earth over 3500 million years ago. Since then, through the process of speciation and (mass) extinction, new species have appeared and others become extinct (Hsü 1986, Wilson 1994). This resulted in the biodiversity encountered by modern humans since their appearance on Earth some 200,000 years before the present. Concern about biodiversity during recent centuries is mainly based on human impact and its decline. It has become a scientific issue since the worldwide adoption of the Convention on Biological Diversity, which arose from the Rio Declaration on Environment and Development (UN 1992a) and Agenda 21 (UN 1992b; see also WCMC 1992, Groombridge & Jenkins 2000). Sala *et al.* (2000)

distinguished five major drivers of biodiversity change, with different relative importances in each of the biomes: (i) changes in land use, (ii) atmospheric CO_2 concentration, (iii) nitrogen deposition and acid rain, (iv) climate and (v) biotic exchanges (introduction of aliens).

The issue of restoring biodiversity is central in the field of restoration ecology (Falk *et al.* 1996). Diversity or variability is characteristic of life on Earth and can be observed at different biological levels. Here we focus on (i) taxonomic diversity at the level of ecological communities: among species of plants, animals and/or micro-organisms, and (ii) genetic diversity at the species level: among individuals and populations within a species.

2.4.1 Biodiversity at the community level

Bakker *et al.* (2000) recommended the expression of biodiversity not in general terms, but rather in terms of specific diversities of plants, birds, insects, etc. to enable an adequate monitoring and evaluation. This is in line with the definition of communities as adopted in Chapter 5. How can restoration ecology cope with the community as an ecological entity? The species pool concept has been revisited for the purposes of ecological restoration, and we have seen a renewed interest in topics such as functional groups, keystone species and indicator species.

The species pool

A regional species pool is defined within a biogeographic region, extending over spatial scales many orders of magnitude larger than those of the local community. Local communities assemble from this pool through a series of filters, or stages. Pärtel *et al.* (1996) and Zobel *et al.* (1998) related the concept of species pools to plant community types and applied it to restoration ecology to identify the 'target community'. They considered a species pool as an ecological species group selected from a flora and distinguished three levels according to the following spatial scale (Fig. 2.4).

1 **Regional species pool**: the set of species occurring in a certain biogeographic or climatic region which are potential members of the target community.

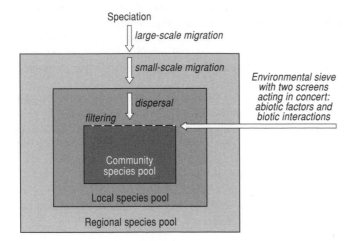

Fig. 2.4 The role of large- and small-scale processes in determining species richness. After Zobel (1997). Reproduced by permission of Elsevier.

2 **Local species pool:** the set of species occurring in a type of ecosystem or landscape (e.g. a river valley).
3 **Community species pool:** the set of species present in a site within the target community, including the soil seed bank.

Species pools can, according to Zobel *et al.* (1998), be determined on the basis of environmental similarity, functional similarity and/or phytosociological similarity between the species, or by applying an experimental approach (evaluating the results from sown seed mixtures). Weiher and Keddy (1999) provided several examples of how communities are assembled from species pools. They mentioned two developing paradigms for the assembly of communities: (i) the island paradigm, dealing with mainlands, islands, immigration and coexistence, and (ii) the trait–environment paradigm that begins with pools, habitats and filters, and convergence. The term 'assembly rules', used to describe the problem of assembly communities from pools, had been derived from Diamond's (1975) original usage of the word. Wilson and Gitay (1995) defined it as a set of ecological restrictions on the observed patterns of species presence or abundance, based on the presence or abundance of one or more other species or groups of species (not simply the response of individual species to the environment). These approaches are reflected in the scheme of Fig. 2.4, where the environmental filters (resources, abiotic conditions and biotic interactions) could be considered assembly rules.

Functional groups and keystone species

The terms functional groups and functional types (Walker 1992) have been used in studies on global change and also in restoration ecology to cope with the complex response of very many species to pronounced changes in the environmental conditions. Unfortunately, the terminology is inconsistently employed. While the term **group** seems to indicate a grouping of individually known species in one particular class, the term **type** suggests a (concrete or abstract) representative of a class of species. In the present volume we stick, therefore, to using the term **functional groups**. Several ecological classifications are being used to escape from dealing with individual species and to focus on ecological species groupings. Use can be made of earlier ecological classifications that aimed at identifying species groupings to describe the structure and functioning of communities or ecosystems. Examples are, in order of appearance, life forms (Raunkiaer 1934), guilds (Root 1967), *r-K* strategies (MacArthur & Wilson 1967), C-S-R strategies (Grime 1979, 2001), adaptive syndromes (Angevine & Chabot 1979, Swaine & Whitmore 1988) and groups that respond similarly to disturbances (Friedel *et al.* 1988, Lavorel *et al.* 1997) or resource shortfalls (Grubb, 1998).

We have not noticed any reflection in the literature on the notion of function or functioning of species groupings in communities. Instead, several approaches have been developed to identify functional groups, often even without an indication of which type of function is associated with the species group (see Woodward & Cramer 1996, Gitay & Noble 1997). This holds strongly for inductive approaches, in which it is taken for granted that functional groups exist and multivariate techniques are used to seek clusters of species; the search for syndromes fits within this approach (see several examples in Lavorel & Cramer 1999). Deductive methods are based on the *a priori* statement of the importance of particular processes or properties in the functioning of an ecosystem; a feasible set of functional categories (e.g. annual or perennial species, C_3 or C_4 grasses, nitrogen-fixing Leguminosae) is then deduced from these premises. If a shared class of resources is used as the criterion, then the functional group is a guild (Root 1967). The species assigned to a guild need not be taxonomically related, and a species can be member of more than one guild (Calow 1998).

Removal experiments carried out by Wardle *et al.* (1999) over a 3-year period have shown that removal of plant functional groups can have important effects on the composition of the remainder of the flora, and that this can influence vegetation dynamics, biomass, productivity and diversity. They provided evidence indicating that above-ground responses have the potential to induce corresponding below-ground responses, affecting soil food webs, community composition of soil organisms and their diversity, and ultimately ecosystem properties and processes. This type of analysis is derived from the efforts to identify keystone species. The term was coined by Paine (1966) and has been used rather loosely by very many authors thereafter. **Keystone species** are defined by having a key function in controlling the structure of a food web or the functioning of an ecosystem, which can be identified by removal of the species from the system. The main information on keystone species has been derived from removal experiments (Paine 1980, Wardle *et al.* 1999), from exclosure experiments (Drent & Prins 1987, Drent & van der Wal 1999) and by estimating interaction strengths (Menge *et al.* 1994). Keystone species can be identified among plants, herbivores, predators, parasites and all other components of a biocoenosis. Tanner and Hughes (1994), working on coral reefs, showed that the importance of a species to the dynamics of an assemblage may be unrelated to its abundance at equilibrium. Note,

therefore, that rare species groups can have a greater impact than more common ones.

Apart from intended and scientifically designed field experiments, incidental natural disturbances have also provided insight into the role of particular species in their respective ecosystems. For example, some authors suggested that a sudden bush encroachment in African savannas was probably caused by an outbreak of anthrax and rinderpest among impalas, leading to a very high survival of saplings of acacia trees (Prins & van der Jeugd 1993, Sinclair 1995). Similarly, outbreaks of myxomatosis among rabbit populations resulted in succession of grassland vegetation due to the establishment of seedling trees; for example the recruitment of oaks in England following rabbit myxomatosis (Crawley 1983). Lesser snow geese, ever increasing in population size, overexploited the tundra vegetation in the Hudson Bay salt marshes, which led to irreversible degeneration of tundra vegetation (Jefferies 1999). These examples illustrate that the fate of keystone species (be it a population collapse or a dramatic increase in population size) may have far-reaching consequences for other components of a food web. This phenomenon is often labelled 'trophic cascade'. This term has its origin in Paine's work in marine intertidal systems (Paine 1980) and has been applied to several other types of ecosystem where primary producers and herbivores interact (e.g. Carpenter *et al.* 1985, Jefferies 1999). The literature on this topic was reviewed by Persson (1999), who defined a trophic cascade as the propagation of indirect mutualism between non-adjacent levels in a food chain, but the term has also been used to include horizontal interactions (see also Chapters 4 and 6).

Indicator species

In restoration projects, indicators can be very usefully applied to characterizing the status of success. While keystone species play a relatively central role in the functioning of a food web or an ecosystem, this is in general not the case for indicator species, which can even be a single individual of a species. Indicators are supposed to be or have been proven to be indicative of certain abiotic ecosystem conditions. In cases of a single environmental factor (e.g. wetness characteristics of a mire vegetation; Tüxen 1954), the relationship is not very difficult to prove. In most cases, however, indicator species are used to point at a complex set of environmental conditions, either at the ecosystem scale (soil pH, eutrophic conditions) or extrapolated to indicate regional-scale processes (seepage water, atmospheric deposition). As indicator values are always based on field experience and often difficult to test quantitatively, it should be recognized that the values are of a qualitative nature; they can be used in a comparative way, and only within a certain well-defined region (e.g., Ellenberg *et al.* 1991, Wheeler & Shaw 1995a, Kotowski *et al.* 1998). Indicator species may indicate different sets of environmental conditions even at different sites within a region; several examples were elaborated by Everts and de Vries (1991) for plant species along the courses of brook valleys.

For aquatic systems, Cairns and McCormick (1992) and Cairns *et al.* (1993) recognized different types of indicator for applied purposes, once the goals have been established: (i) early-warning indicators, signalling impending deleterious changes in environmental conditions before unacceptable conditions actually occur, (ii) compliance indicators, assessing the degree to which previously stated environmental conditions are maintained and deviations from acceptable limits, (iii) diagnostic indicators, determining causes of deviations outside the limits of unacceptable conditions, and (iv) biogeochemical and socio-economic indicators, which are mutually affected by environmental degradation and restoration as well.

With regard to restoration of damaged soil ecosystems, Bentham *et al.* (1992) have pointed out the possibility of using microbiological indices to measure damage and restoration, far superior to conventional measurements. They used the following characteristics: (i) size, or the amount of biomass in the system, (ii) activity, or the rate of turnover of materials within the system and export/import of nutrients, and (iii) the degree of biodiversity within the system. These parameters are thought to indicate successional changes in the quality and quantity of organic material available to the soil microbial community (see also Harris *et al.* 1996).

Whereas keystone species are considered to be essential for proper ecosystem functioning, indicator species have a role in human quality control of the environmental conditions. They can be used in the process of piloting a system along a target trajectory that often is defined in terms of restoring species richness.

2.4.2 Biodiversity at the species level

In Europe and elsewhere many plant and animal species of concern to conservation and restoration ecologists have to bridge larger and larger distances between local populations in an increasingly fragmented landscape (see also Chapters 2 and 6). Dispersability properties that have evolved and been selected for in earlier times may not be sufficiently adapted to present requirements in the human-dominated landscapes. Strykstra *et al.* (2002) suggested that species may have adapted their dispersal capabilities to formerly existing environmental conditions and safe-site availability in communities, incompatible with today's situation of generating new habitats for restoration purposes. Re-introductions are meant, and sometimes executed, to repair an ecosystem by restoring species that – according to ecologists – were lost but could still play a role in the ecosystem involved. The aim is to restore a degenerated population and to improve ecosystem quality (see Chapter 7).

Re-introduction as compared to invasion

Intended re-introduction should be distinguished clearly from invasions of alien species. Biotic invaders are species that establish a new range in which they proliferate, spread and persist (Elton 1958), in many cases to the detriment of the environment (Mack *et al.* 2000), whereas a re-introduced species is supposed to have disappeared from a community that it still belongs to. So, we are not at all dealing with the problem of alien species or even unwanted invaders, but we may have to cope with alien populations. A problem that should also be discussed here is whether re-introduction of native species in their natural environment (from which they had been lost) includes the risk of introducing species with characteristics of invasive (alien) species. Should risky effects, such as those observed from the introduction, for economic purposes, of Nile perch into Lake Victoria (Oguto-Ohwayo 1999), be taken into account when ecological re-introduction of species is considered? Should population differentiation (intraspecific variability, ecotypes) be recognized here? To tackle such problems, we need answers to three questions.

1 Can we successfully distinguish invading aliens from non-successful invaders and from native species?
2 Can the abiotic environmental conditions have changed to such an extent that the statement that the species to be re-introduced belongs to the community cannot stand any longer (in other words, is the reference community obsolete)?
3 Can populations of the species to be re-introduced ecologically and genetically have changed in such a way that the intended re-introduction can turn out to be an unintended invasion?

Is there a risk of re-introduced species behaving like alien invaders? Sakai *et al.* (2001) reviewed aspects of the population biology of invasive species, such as their life-history characteristics, their genetics and evolutionary potential, the susceptibility of communities to invasion and the invaders' impact on the invaded communities. Though a lot is known about invasive species and their impact, it remains difficult to predict the behaviour of populations of invaders (see also Mack *et al.* 2000). There is, we think we are justified in concluding, no *a priori* risk of re-introduced species behaving as alien invaders, but how then can criteria for re-introduction be identified so as to exclude the risk? Here we refer to Strykstra (2000), who proposed to address the following topics in any plan for re-introduction: (i) the significance of the re-introduction, both for the target area and for the species involved, (ii) the match between the area and the species involved in (bio)historical terms, (iii) the suitability of the environment in the area for the species involved and (iv) the material and the method used. These issues will be elaborated in Chapters 5 and 7. Here, we will pay particular attention to the latter aspect, the question to what extent intraspecific variation should be taken into account in re-introduction projects.

Does intraspecific differentiation matter? van Andel (1998b) reviewed the results from reciprocal transplant studies that had been performed to illustrate population differentiation and local adaptation, rather than to test potential risks of transplantations. In essence, these data are apt to be applied to the new problem. The author concluded that, although local adaptation does occur, a small reduction in the fitness of transplanted populations does not imply any risk, apart from the risk of failure of the transplanting activity in cases of strong ecotypic differentiation. It seems better, he argued, to restore a population with its fitness reduced by a few per cent, than to restore no population at all. Keller *et al.* (2000), however, warned that these suggestions should not be applied uncritically within declining populations. In cases of re-introductions of animal species, similar discussions and viewpoints can be found in the literature. Abrams (1996) argued that current theory is insufficiently developed to provide guidance in predicting what might happen to either trait values or population densities of other species after the addition or deletion of a species. And, indeed, both positive and negative effects have been found (see Chapter 7).

In summary, failures of re-introduction may be considered a significant risk, rather than unintended aspects of success, provided that the habitat is of adequate quality. Species characteristics such as competitive aggressiveness in an ecological sense are related to the introduction or invasion of aliens, rather than to re-introduction of species *sensu stricto*.

2.5 Concluding remarks

In this chapter, we have provided a brief overview of the field of restoration ecology. We started by discussing the concepts of stability and disturbance representing a systems approach, applicable to ecosystems, and communities or populations, because steady states and disturbances may occur at any organizational level. The biodiversity concept is confined to tangible organisms and has only during the last decade been linked to a systems approach (e.g. Tilman 1996, Loreau *et al.* 2002; see also Chapter 4). Biodiversity is fundamentally a societal concept (see Chapter 1), adopted within restoration ecology with the challenge to get the issue operationalized (this chapter). Evolutionary theories, though in the core of thinking about the origin of biodiversity, have so far not been included in the field of restoration ecology. These fields of interest are, apparently, too far away to be made applicable, but it is our hope that this may change quite soon.

In the five theoretical chapters that follow (interactions at the level of landscape, ecosystem, community and population) we will follow the classic hierarchical approach, to explore which of the fundamental ecological theories, established long before the emergence of restoration ecology, have gained new importance in this era when ecological restoration is rapidly gaining prominence. In the examples of ecological restoration, presented in Chapters 8–15, restoration ecology will appear as an integrated science, transcending borders between and among classical academic disciplines.

Part 2

Ecological foundations

3

Landscape: spatial interactions

Rudy van Diggelen

3.1 Relevance of landscape processes for restoration ecology

Not so long ago ecological restoration was mainly seen as a technical activity where measures were taken at a certain spot in case of an unwanted situation or unwanted developments. However, it is becoming increasingly clear that ecosystems do not function independently from their surroundings and that spatial relations matter a great deal. It is here that restoration ecology is starting to adopt ideas from the field of landscape ecology. When MacArthur and Wilson (1967) published the theory of island biogeography, they stated that extinction increased with decreasing size of the island and immigration decreased with increasing distance from the mainland. This theory has been applied in the case of isolated nature reserves within an intensively used landscape and has had a large impact on theories of landscape planning. By the beginning of the 1980s the field of **ecohydrology** was developing in north-west Europe (Succow 1982, Grootjans *et al.* 1996) and it became increasingly clear that water flows connect wetland sites. Human activities in one particular area may cause disturbance in another, sometimes distant, area and can indeed affect restoration perspectives significantly (van Diggelen *et al.* 1995). The same is true for airborne pollution (Bobbink *et al.* 1998). Insights in landscape relations are therefore essential when evaluating the restoration perspectives of degraded sites.

3.2 Concepts

3.2.1 Landscape

The scientific concept of landscape has its roots in central Europe and was mainly developed by German scientists (Troll 1939, Schmitthüsen 1963). At that time the concept seemed rather clear and was exclusively viewed from a human perspective. A typical *Landschaft* was many square kilometres in size and included several villages, farms, etc. For those researchers there was no doubt that cultural elements should be included in the concept. Later work, especially in less densely populated areas, did not automatically assume that a landscape should be viewed at the human level and much less emphasis was put on the cultural aspect. As a consequence the concept became increasingly vague and two subsequent analyses of papers in *Landscape Ecology* by Wiens (1992) and Golley (1995) showed no clear trend in parameters such as scale of research, level of organization or subjects of study. The only elements the papers had in common were that they were all concerned with spatial relations and they were all published in *Landscape Ecology*! In the present chapter we will adopt a practical approach and define landscape as a spatial matrix at the human scale in which interactions of biotic and non-biotic elements take place. In comparison to other ecological concepts such as ecosystem or population, a landscape is a very concrete part of the surface of the Earth, with boundaries and a history. Typically it has a size of at least a few square kilometres and can be photographed or put on a map.

3.2.2 Scale

There is a clear correlation between the scale at which a process operates and its dynamics (see Chapter 2 in this volume). Small-scale processes tend to be more dynamic and more erratic than large-scale processes. Traditionally landscapes are seen as the highest practical scale that is influenced by human activities and which humans readily perceive. In this view they constitute the third level in the conventional ecological hierarchy after the biosphere and biomes and before ecosystems. Recently there is a tendency to expand the term landscape and use it for any spatial arrangement, depending on the view of the organisms considered. Experimental studies (Wiens 1976) showed that fine-scale spatial structures were highly relevant for beetle movement and suggested that a landscape from their perspective looked entirely different than that from the perspective of a human being (see also Haskell *et al.* 2002). In this view a landscape is a more abstract concept and can be seen as a context in which (meta)populations operate (Wiens 1976, Allen & Hoekstra 1992). Intuitively this may seem very obvious but in the context of restoration it implies that scales at which the target organisms operate and at which disturbing processes occur should be identified explicitly. For the restoration of marsh birds for example, it might be sufficient to flood 5 ha of degraded wetland whereas it might be necessary to restructure a whole catchment when the aim is to restore specific groundwater-fed plant communities in the same site.

3.2.3 Patches and patterns

All landscapes are heterogeneous. Parts of the landscape that are considered to be uniform are called **patches**. Again, the scale of a patch depends entirely on the point of view of the organism under study. A part of a landscape that is a patch for an elephant can be a highly diverse region from the point of view of a butterfly. A group of patches is called a **pattern**. Topography (e.g. elevation, slope), natural disturbances (e.g. fire, flooding, storms) and increasingly also human activities (e.g. farming, building) are the principal causes of pattern.

3.2.4 Connectivity and corridors

Connectivity has become one of the central concepts in landscape ecology. It was realized that patches are not isolated entities but rather are connected to other patches, thus enabling an exchange of matter and organisms. The relation between such flows and the landscape pattern is often very complex. While the present-day structure determines actual flows, this structure itself depends to a large degree on past flows. The size of the flows depends on the connectivity between patches, which in turn depends on the amount of suitable habitat. Model simulations (Gardner *et al.* 1989) show that in purely 'random' landscapes c.60% of the space must be covered by suitable habitat in order to ensure that organisms can travel from one side of the area to the other without having to leave the habitat; that is, be part of a functioning metapopulation. In real landscapes this percentage can be much lower, due to the presence of **corridors**, relatively narrow strips of suitable habitat that connect patches. The above-mentioned simulations showed also that presence or absence of corridors makes a dramatic difference for the percolation of organisms within a landscape. One could say that corridors function as communication channels within the landscape. However elegant this concept may be, the recognition of corridors in reality is not easy and may differ from one organism to another. Hedgerows may be corridors for beetles, but are unlikely to be so for ospreys. In the latter case the river that flows through a hedgerow landscape may be the real corridor.

3.3 Flows between landscape elements

3.3.1 Transport of matter

Transport of matter by wind

Large quantities of material are transported by airflows, especially along the edges of continents where winds are generally stronger than in more interior areas. Under natural conditions this is a major mechanism for the exchange of chloride between sea and inland areas. Rainwater analyses from different sites showed a decrease in Cl⁻ content from over 100 to less than

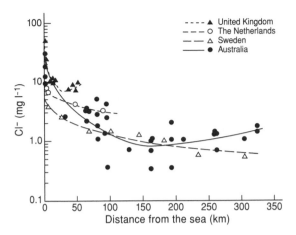

Fig. 3.1 Chloride concentration in precipitation in relation to distance from the coast (after Voigt 1980).

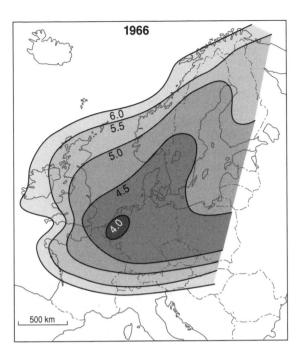

Fig. 3.2 pH of precipitation over Europe in 1966. (after Likens *et al.* 1972). Reproduced by permission of Helen Dwight Reidd Educational Foundation.

1 mg l^{-1} between coast and inland (Fig. 3.1). In more human-affected landscapes air transport of pollutants has become an important issue as well. Examples include the transport of sulphur dioxide and mineral nitrogen from industrial and agricultural sources, but also the transport of radioactive gases after the accident at Chernobyl in the former USSR. Distances covered by wind transport can be large. The acidification of Scandinavian lakes through the deposition of sulphuric acids is mainly caused by industrial emissions in western Europe, a distance of 500–1000 km! (Fig. 3.2). Not only can the distance covered be significant, but also the quantity of material transported by wind. Nitrogen deposition in the Netherlands has increased from *c.*5 kg ha^{-1} yr^{-1} under natural conditions to over 30 kg ha^{-1} yr^{-1} at present (see the Dutch Rijksinstituut voor Volksgezondheid en Milieu (RIVM) environmental data compendium at http:// arch.rivm.nl/environmentaldata). In nutrient-poor soils this source accounts for more than 25% of the nitrogen in the system (Verhagen *et al.* 2003) and goes beyond critical loads (Achermann & Bobbink 2003).

Water flow

Only a minimal part of all the water on the planet is found in the atmosphere and on the continents (Table 3.1) but water is at the same time indispensable for all terrestrial life forms and has huge effects on the spatial structure of landscapes. For the Earth as a whole the exchange between atmosphere and continents amounts to 106×10^{12} kg of water yr^{-1} while about two-thirds (69×10^{12} kg) of this precipitation returns directly to the atmosphere by evaporation. The remaining precipitation surplus of 37×10^9 m^3 yr^{-1} is divided very unevenly over the Earth but in most areas there is – at least temporarily – water flow towards seas and oceans. This flow can take place either through rivers and lakes as surface water or through the soil as groundwater. Eventually, the large majority of groundwater flows do also discharge in surface-water systems but this route may take a long time, often centuries and sometimes even millennia. In the meantime these systems may sustain wetlands outside the limits of areas that are flooded regularly with surface water. The size of groundwater systems differs greatly from catchment to catchment and depends both on the topography and the permeability of geological strata, but most systems are at least several square kilometres in size. Groundwater flow, especially in cases

Reservoir	Mass (10^{15} kg)	Proportion (%)
Ice sheets	26,200	80
Groundwater	6,200	19
Lakes, fresh water	126	0.4
Lake Baikal	22	
Great Lakes, North America	32	
East African lakes	36	
Lakes, brackish water	105	0.4
Caspian Sea	80	
Soil, unsaturated zone	150	0.5
Rivers	2	

Table 3.1 Freshwater reservoirs on the land.

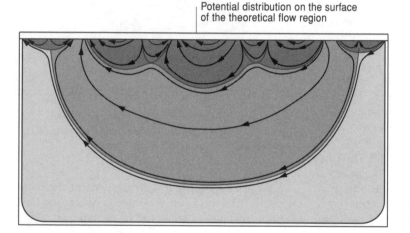

Potential distribution on the surface of the theoretical flow region

Fig. 3.3 Groundwater flow systems in a homogeneous subsoil after Toth (1963). Higher order systems lie nested within lower order systems. Reproduced by permission of American Geophysical Union.

where deep permeable layers exist in almost flat landscapes, can be very complex. Toth (1963) developed the concept of nested systems (Fig. 3.3) where small flow systems are embraced by larger systems. These systems may differ considerably, not only in flow rates but – depending on the chemical composition of the soil strata – also in water chemistry.

Transport of water by water flows

Inhabitants from temperate latitudes often forget that the most relevant substance transported by water flow is water! Of course this is highly relevant in arid regions but it can also be of significant importance for the survival of wetlands during dry periods in more humid areas. This is clearly the case along the fringes of rivers and lakes where wetlands can occur under any climate but it is also true in groundwater-fed systems in climates where there is a precipitation surplus on a yearly basis. During wet periods the groundwater stock is replenished and this water can sustain wetlands in large areas during dry periods. Such groundwater-fed mires once covered substantial areas (Succow & Joosten 2001; see Chapter 9 in this volume). Today the hydrological balance is changed in many wetlands by human activities but, interestingly, this sometimes leads to an increase in wet areas instead of the more commonly observed decrease. Loucks (1990) presents the case study of the Columbia wetland where the amount of upwelling groundwater was more than tripled due to the creation of a cooling lake near a power plant.

Transport of matter by water flows

Apart from water, rivers do also transport large amounts of solutes, sediments and sometimes biomass. The latter is deposited on floodplains during flooding periods and this causes natural fertilization in otherwise nutrient-poor landscapes. As long ago as several millennia BC the agriculture in Egypt was based on regular flooding by the river Nile and this could sustain a large human population in an environment which was otherwise very hostile. Other ancient cultures also knew the value of water. Nabateans, Persians and Carthaginians harnessed ephemeral rivers and used these for water harvesting.

Both surface and groundwater contain also dissolved substances such as nutrients (soluble compounds of N, P and K) and other ions. The most common ones are called major ions (Hem 1959) and they constitute up to 99% of all solutes in water. Which ions belong to this category differs somewhat from region to region but on a worldwide scale the most common cations are Na^+, K^+, Mg^{2+} and Ca^{2+} whereas HCO_3^-, Cl^- and SO_4^{2-} are the most important anions. The exact ion content depends on the source of the water and is sometimes used as a fingerprint to trace its origins (Stuyfzand 1989, Komor 1994). In the case of groundwater it depends on the chemical composition of the soil layers that the water has flown through and on the residence time of the water because some reactions are very slow. Groundwater in areas with thick layers of mineral-poor sand often shows a very typical evolution in water composition (Fig. 3.4). Both rainwater and groundwater of upper layers are generally very poor in dissolved solutes. When going deeper this water reaches layers where soluble minerals have not been washed out and these go into solution and increase the mineral load. Quantitatively the most important minerals are several Ca^{2+} compounds, especially $CaCO_3$ and $CaSO_4$, but other minerals can also be found. If groundwater flows from different depths to the surface alongside each other this can result in a steep hydrochemical gradient at the surface.

Transport of matter by human activities

Apart from sometimes quite significant human impact on natural flows, huge amounts of matter are also transported on purpose by humans. Certainly the most important objective for doing so is agriculture. At present, in quantitative terms the most important flows are transport of water for irrigation and of fertilizer to increase crop production in less fertile areas. Water transport is practised mainly in warmer climates and consists of a redistribution of water from areas with sufficient precipitation, for example mountains, to sites where there is not sufficient water for crop

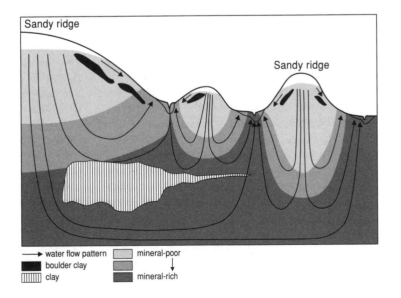

Fig. 3.4 Changes in groundwater composition along its way through the subsoil. Mineral-poor rainwater infiltrates the subsoil and takes up solutes along the way. Arrows indicate direction of water flow, darkness increases with mineral content of the groundwater.

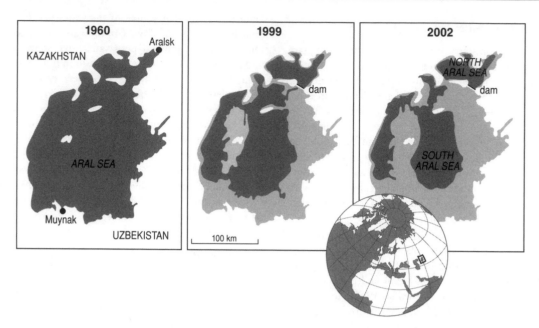

Fig. 3.5 The changed shape of the Aral Sea since 1960 (from Jones 2003). Reproduced by permission of *New Scientist*.

growth. Examples can be found in the south-western USA where water from the Colorado River is used for irrigation so intensively that the river dries up completely before it reaches the Salten Sea. Another, well-known example is that of the shrinking Aral Sea (Jones 2003; Fig. 3.5). In recent times there is also a development towards transporting water to areas where there are only temporal water shortages, i.e. during the dry season. For example, in the Netherlands huge canals enable the transport of water from the River Rhine to all corners of the country in order to optimize conditions for crop growth throughout the year. The impact this may have on the re-arrangement of aquatic plant and animal communities is still largely unexplored; see also Chapter 11 in this volume.

Not only a redistribution of water, also the use of fertilizers has enabled an enormous increase in crop production in previously marginal areas. The average nitrogen gift in the Netherlands increased from 15–30 to 150 kg ha^{-1} yr^{-1} between 1900 and 1960 for meadows (Bakker & Olff 1995) and from 200 to 450 kg between 1950 and 1985 for intensively used arable fields (Keuning 1994). The same trend can be witnessed all over western Europe and resulted on

average for the whole European Union in a productivity increase in commercial crops of 1.9% ha^{-1} yr^{-1} between 1950 and 1991 (Strijker 2000). The effects of these developments are not restricted to agricultural areas; the yearly nitrogen load from agricultural sources into the River Rhine at the German–Dutch border has increased from c. 100 × 10^6 kg of N in the period 1954–7 to more than 150 × 10^6 kg of N in the period 1993–5 (Behrendt 1997). Similar developments also occurred in other parts of the western world. Despite its dry climate and nutrient-poor soils, Israel has become one of the world's major exporters of citrus fruits due to an improved water redistribution and nutrient application. At the same time this has led to a severe pollution of groundwater and created large problems for the production of drinking water (Ronen & Magaritz 1991).

In densely populated regions deliberate transport of matter is probably quantitatively more important than natural flows. As a result, natural differences between landscapes fade away and the boundaries between different areas of countryside become increasingly blurred. Broadly stated, organisms that are adapted to moist and eutrophic conditions are stimulated, whereas other types of organism are hampered. This trend is especially pronounced in easily cultivable – that is, generally flat – areas. Mountainous regions are much more difficult to cultivate and geomorphology-based differences between landscapes are much more preserved.

3.3.2 Movement of organisms

Dispersal by airflows

In flat, homogeneous landscapes wind transport of organisms is diffuse and little affected by existing corridors. The study of this phenomenon has been intensified during the last few decades both in the field and theoretically, the latter especially with the help of computer models. Despite a general assumption that wind is an important long-range dispersal vector for plants this has only been measured under exceptional conditions. Whelan (1986) showed that diaspores could be found at heights of 3 km or more in the thermal updrafts generated by large fires. Consequently the distances covered can be quite large, depending on the characteristic speed of fall (**terminal velocity**)

of the species. Under more normal conditions the height at which seeds are found and the distances covered are much more restricted. Studies in wind tunnels (van Dorp *et al.* 1996, Hammill *et al.* 1998, Strykstra *et al.* 1998) and in the field (Verkaar 1990, Coulson *et al.* 2001, Jongejans & Telenius 2001) found that hardly any seeds reached distances of more than 10 m from the parent plants. This general pattern was confirmed by modelling studies. Jongejans and Schippers (1999) developed a model for seed dispersal by wind based on the parameters terminal velocity, seed-release height, number of seeds per surface unit and wind speed. Simulations showed that the majority of species covered small distances of up to a few metres. Only species that produced large numbers of very light seeds high above the surface of the ground, for example large marsh plants such as *Phragmites australis* and *Typha angustifolia*, were capable of reaching greater distances from the parent plant (Table 3.2). This is even more true for species with extremely small seeds such as orchids or spores, as in the case of mosses and ferns. Van Zanten (1993) showed that moss spores could survive in the troposphere and suggested that in this case wind transport could take place over

Table 3.2 Effectiveness of wind dispersal of some wetland species (van Diggelen, unpublished observations). Results were obtained with the simulation model of Jongejans and Schippers (1999) in a spatial context. Seed production was measured in the field, terminal velocity was measured in the laboratory under standard conditions (Askew *et al.* 1997) and height of release was taken from standard flora. The last column shows the size of the receptor area as a fraction of the source area of 576 m^2. A value of 1.00 means that the receiving area is as large as the source area.

Species	Seed production (seed m^{-2})	Terminal velocity (m s^{-1})	Height of release (cm)	Source area/ receptor area
Agrostis stolonifera	25	1.64	57	0.2921
Anthoxanthum odoratum	2,258	1.52	32	0.3357
Holcus lanatus	7,383	1.68	60	2.2452
Lemna minor	–	–	–	0
Mentha aquatica	16,296	2.18	55	0.5099
Phalaris arundinacea	2,072	1.39	150	5.6288
Phragmites australis	18,000	0.21	200	44.4788
Plantago lanceolata	1,222	3.8	32	0.2032
Ranunculus acris	262	3.14	65	0.3862
Ranunculus repens	224	2.7	37	0.2599
Typha angustifolia	2,600,000	0.14	150	1,443

thousands of kilometres and was highly relevant for gene flow between continents.

The importance of wind dispersal changes dramatically in more heterogeneous landscapes. Steep slopes and irregular warming of diverse surface materials lead to an increasing importance of thermal updrafts for airflow. Unlike the situation in homogeneous landscapes, air transport in heterogeneous landscapes does follow more or less distinct corridors, namely those parallel to steep slopes or along edges between different surface types. This can occur on a scale of several hectares but also on much larger scales. Tackenberg *et al.* (2003) modelled seed dispersal of rare plant species that grew on small porphyric outcrops in an otherwise intensively used agricultural landscape, illustrating that heat-induced air turbulence enables seed exchange between sites with distances of up to some hundreds of metres for many species despite a small size and a low number of seeds. Pennycuick (1972) studied bird movement by thermal updrafts and showed that both daily movements of vultures over distances of several hundreds of kilometres and seasonal eagle migration over thousands of kilometres were energetically only possible because of this type of airflow.

Dispersal by water

Several field studies (Johansson *et al.* 1996, Bill *et al.* 1999, Middleton 2000, Andersson & Nilsson 2002) showed that running waters contain large numbers of seeds and this suggests that they are most likely to be a major dispersal agent in wetlands. In comparison to wind, water has some typical characteristics. Water dispersal takes place only via corridors and it is highly unidirectional, i.e. downstream. It is only omnidirectional in the case of standing water, but seed dispersal in these systems is quantitatively probably of minor interest, as transport rates are only a fraction of those of running waters. Laboratory measurements on seed buoyancy of characteristic plant species of bogs (Poschlod 1990) and riparian habitats (Danvid and Nilsson 1997, van den Broek *et al.*, personal communication) showed that seeds of most species were capable of floating for many weeks, including seeds without clear morphological adaptations. This picture was confirmed by palaeobotanical studies when diaspores of a large number of species were found in buried drift-line material (Cappers 1994, Knörzer 1996). Moreover, even seeds with a restricted floating capacity can probably travel large distances in running waters. Nilsson *et al.* (1991) calculated that only 2.5 days should be sufficient to cover a 230-km long river stretch in northern Sweden during the spring floods. Studies of riparian vegetation have indeed shown a large floristic similarity between upstream and downstream areas, which decreased slowly with distance (Tabacchi *et al.* 1990, Johansson *et al.* 1996). Tabacchi *et al.* (1990) found a sharp decrease in similarity only when the river merged with major side-branches, suggesting that at such points 'propagule flows' of two (or more) areas with different species pools (see Zobel *et al.* 1998; see also Chapter 2 in this volume) are combined.

Dispersal by animals and humans

With the exception of waterfowl and fruit-eating birds (Vivian-Smith & Stiles 1994, Galatowitsch & van der Valk 1996, Handel 1997) animals are considered poor dispersal agents in the modern landscape. In contrast to the past when livestock was transported from winter to summer areas and vice versa over distances of up to hundreds of kilometres, seed exchange between areas by moving animals has probably almost disappeared (Poschlod *et al.* 1996, Poschlod & Bonn 1998). This has been considered the reason why gene exchange between isolated nature reserves has become minimal (Bokdam & WallisDeVries 1992, Grashof-Bokdam 1997, Mouissie 2004).

Humans have always played a role as a 'moving corridor' (Poschlod *et al.* 1996). Plants lack all morphological adaptations to dispersal by human beings, yet people have probably been the most effective long-distance dispersers since the Middle Ages (Sukopp 1972, Sykora *et al.* 1990, Poschlod *et al.* 1996). For example, modern traffic facilitates rapid dispersal of halophytes along motorways and a rapid invasion of alien weeds along railway lines (Ernst 1998). At the same time the decreasing degree of agricultural activities in most nature reserves has probably led to less exchange with other areas (Poschlod & Bonn 1998). The significance of humans for the dispersal of rare plants to such reserves is, therefore, likely to be very low except when a deliberate management scheme is set up (Strykstra *et al.* 1996a).

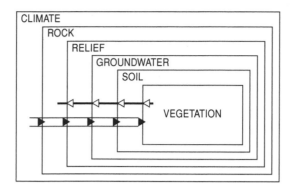

Fig. 3.6 Dependence of vegetation on major factors. Note that the relationships are not symmetrical. Higher-order factors determine lower-order factors to a much higher degree than vice versa.

3.4 Patterns

Actual landscape patterns are often very complicated and are caused or governed by major factors and complex interactions between them. At the same time the relations between these factors are not symmetrical (Fig. 3.6). Factors of a higher order such as climate and geology have a much larger impact on lower-order ones like hydrology and soil type than vice versa. This leads to a repetition of patterns of lower-order elements given a certain similar constellation of higher-order ones. One can observe for instance a large degree of similarity between little-disturbed landscapes in the north-eastern part of the USA and their equivalents in southern Scandinavia. The climate is quite similar in these two areas and the geomorphology of both regions is determined by glacial activity during the last ice age. At the same time this natural template is increasingly influenced by another, independent dominant factor, namely human impact. Climatic conditions and geomorphology in New Jersey may not differ much from those in the area around Helsinki, but this does not mean that the actual landscapes are very similar. Differences in population density, types of land use and history result in completely different landscape patterns in Manhattan and villages near Helsinki.

Having said this, there is often a remarkable similarity in vegetation patterns, both over larger areas and over larger timescales. If only one major environmental factor determines a spatial gradient, this leads to the development of **zonation**, perpendicular to this factor. Typical examples include altitudinal gradients in mountainous areas and hydrological gradients in river valleys (Fig. 3.7), but also gradients in land-use intensity in the past. Even in cases where more major factors determine species patterns one can often witness a relatively simple zonation, because the factors often coincide to a high degree in the gradient. Until recently, land use was highly correlated with natural productivity and ease of cultivation (Fig. 3.8). In mountainous areas the villages were situated in the valleys where at the same time both climate and soil fertility are most favourable for agriculture. Human activities led to a further increase of already existing differences in productivity between higher and lower areas. During the last century these gradients vanished, especially in western countries. Technological developments in drainage techniques, a steep increase in fertilizer application and the use of agricultural machinery resulted in a huge increase in agricultural productivity, and it became possible to produce nutrient-demanding crops almost everywhere. Only small relics of the former gradients have remained but they are often no longer connected (**fragmentation** of the landscape).

3.5 Landscape development

3.5.1 Natural landscapes

The actual geomorphologic structures in North America and the northern half of Europe are almost entirely the result of erosion and sedimentation processes during the last glacial period. Glaciers and polar deserts were the dominant landforms by then and the climatic conditions were so harsh that hardly any organisms could persist. The great majority of organisms had to (re)colonize the released area during the following interglacial period and trees in particular have been quite successful. Delcourt and Delcourt (1983) and Davis *et al.* (1986) have presented a detailed reconstruction of this process. They showed that colonization rates differed significantly between species and this might explain actual differences in tree distribution to a certain degree. Several studies on dispersal and establishment (e.g. D'Antonio *et al.*

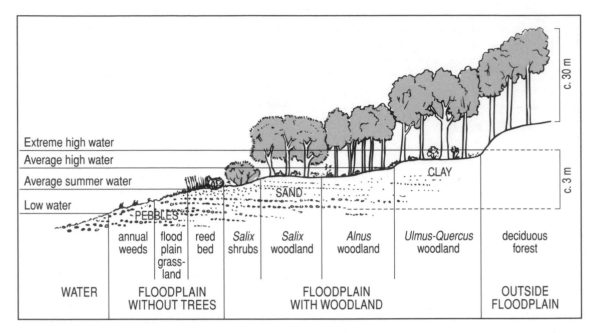

Fig. 3.7 Vegetation zonation along rivers in relation to flooding frequency (after Ellenberg 1986). Reproduced by permission of Verlag Eugen Ulmer.

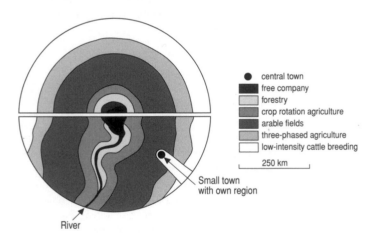

Fig. 3.8 Land use patterns around a town according to the schema of the German agricultural economist Von Thünen (1826, cited in Bieleman 1992). The zonation in this model is mainly caused by the factor *transport costs*. Reproduced by permission of the publisher.

2001, de Meester *et al.* 2002) showed that sites with very similar abiotic conditions could differ considerably in species composition. These results fit very well with the inhibition model of Connell and Slatyer (1977), which states that in the case of organisms with low colonization rates such as plants or low-mobility insects chance alone would lead to different species assemblages in otherwise similar sites during the first phases of the colonization processes. Indeed, existing theories on invasibility (Elton 1958) state that the resistance of a given area to species enrichment increases with the number of species already present and this implies that the first stages can have a significant impact on the final community composition. However, this theory is also criticized (e.g. Levine & D'Antonio 1999).

The natural situation lasted in Europe until the first centuries AD and in North America until the arrival of European settlers.

3.5.2 Human-affected landscapes

Central and western Europe

Human impact on landscapes in north-west Europe dates back to the Neolithic era (7000–5000 yr before present) but at first this influence was only locally relevant. Human densities were very low (<0.5 inhabitants per km^2) and the landscape consisted mainly of woodland and mires. Open spots with low-intensity agriculture were found only in the direct vicinity of human settlements. At first these fields were created by burning a certain area of woodland and used until the nutrients were depleted, typically after a few years, and then the settlements were broken up and a new area was sought. There the whole cycle started again (**slash and burn culture**). This **near-natural landscape** existed until the beginning of the Middle Ages (*c*.500 AD).

The next, **semi-natural** stage is associated with the development of a more sedentary form of agriculture. The settlements became fixed and human impact on the landscape increased. Uniform landscapes were split up, at least partly, and large-scale gradients were replaced by smaller gradients of human impact. Close to the villages human impact was sometimes considerable whereas this influence became almost negligible further away. This process continued with time and human impact became ever more dominant. In certain areas humans had altered the structure of the landscape completely by the end of the Middle Ages (*c*.1500 AD), for example in England where most woodlands had been cut for fuel and people had to switch to peat for heating. Another example is the western part of the Netherlands where people started to protect themselves against flooding by building dykes and creating polders from the 15th century onwards (Bos *et al.* 1988). However, in general landscape use in north-west Europe was more or less determined by natural restrictions and remained semi-natural until the end of the 19th century.

Developments in natural sciences stood at the basis of the next phase with **industrial agriculture** and led to landscape changes in rural areas at an ever-increasing speed and scale. Chemistry enabled the development of artificial fertilizer, and engineering sciences made large-scale mechanization possible. In Europe this process became especially apparent after the Second World War when politicians feared famine and stimulated large-scale reallotment programmes and further rationalization of agriculture. The process was further accelerated with the formulation of the Common Agricultural Policy in the European Community during the 1960s. This policy has been so successful that Europe has become a major exporter of food.

North America

To a certain degree the afore-mentioned developments also occurred in North America. There are, however, two major differences with Europe. First, the time-scales are different. Whereas the natural situation ended at the onset of the Middle Ages in Europe this stage continued until the arrival of European settlers in North America. Even then intensification occurred on a comparatively small scale during the first two centuries and was mainly restricted to the East Coast. Large-scale developments in North America started especially after colonization of the whole area by white settlers in the second half of the 19th century, but then developments took place at a much faster pace than in Europe. One of the consequences was that the semi-natural situation, which is a major reference for ecological restoration in Europe, hardly occurred in North America. In most regions the natural stage was followed immediately by the industrial agricultural phase. A clear example is the situation in the Mid-West where prairies were converted directly into intensively used arable fields, which almost led to the extinction of large herbivores such as bison.

A second important difference between North America and north-west Europe is the large difference in human population size. Whereas the high population density in Europe made it necessary also to produce food in less suitable areas with infertile soils, insufficient moisture supply, etc., this was much less the case in North America. Here land-use pressure was much lower and food production could be restricted to more favourable areas. An important consequence is that larger areas with relatively untouched wilderness still exist in North America, whereas such areas are no longer present in western Europe. It is therefore not surprising that wilderness is the major reference for restoration in North America.

3.5.3 Fragmentation

Gradients in a natural landscape are closely linked to geomorphologic structures and soil types. As long as these features are relatively uniform – for example, in flat areas – the landscape is also rather monotonous and gradients are large. Human activities often split such larger gradients partly into smaller ones and the landscape becomes fragmented to some degree. This is probably not a big problem for most organisms as long as the patches are connected, though large mammals might experience problems even at a low degree of fragmentation (Soulé & Terborgh 1999). Under little-to-moderate human impact this process often leads to a local increase in biodiversity, creating additional suitable patches for species that had a very limited distribution in pristine landscapes (Sukopp & Trepl 1987; Fig. 3.9).

The situation changes dramatically when land use becomes so intense that corridors between patches disappear. Island theory (MacArthur & Wilson 1967) predicts that the probability of local extinction increases with a decreasing patch size and that the

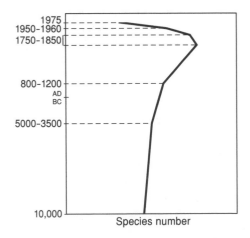

Fig. 3.9 Species richness of higher plants in Europe in time (after Sukopp & Trepl 1987).

probability of recolonization decreases with increasing distance. Present ideas in conservation theory are that this is also true for small patches of suitable habitat in otherwise unsuitable surroundings such as nature reserves surrounded by agricultural fields (Soulé & Simberloff 1986). Though this hypothesis has never been proven unambiguously, it is supported by the fact that small habitat fragments are generally poorer in species than larger ones (Rosenzweig 1995, Cook *et al.* 2002). What is certainly clear is that exchange of species between patches, at least in the case of sessile organisms like plants, has decreased significantly (Poschlod & Bonn 1998). Probably there is still some interchange between sub-populations but it is unknown whether this is sufficient to sustain a functioning metapopulation (Opdam 1991, Hanski 1999; see Chapter 6 in this volume). Nevertheless, it is feared that present-day distribution patterns of many organisms are in fact spatially isolated relics from a former continuous population that has become fragmented. The danger is very real that many of these small populations will die out and species will become locally extinct due to chance processes alone.

Changes in land use did not only change species-exchange patterns: water flows have also been modified by human activities. The building of dams and canals, but also the straightening of river tracts, affected water flow rates and fluctuation patterns to a high degree. This sometimes even resulted in a complete reversal of water flow direction (see Chapter 11). Such activities have a large impact on wetlands downstream. The straightening of the Upper Rhine along the French–German border resulted in the loss of 87% of characteristic floodplain woodland, on both the German and French sides (ICPR 1998).

Changes in groundwater flows are less obvious but can potentially have equally large effects. The most obvious effects occur when the water regime is changed. Lowered water tables during the dry season lead to increased mineralization of organic matter and hence increased nutrient availability. In low-productivity wetlands this results in a substantial increase of non-characteristic ruderal plants at the cost of typical species of low competitiveness (Grootjans *et al.* 1985). Changes in water chemistry are more insidious but can have similarly large impacts. A groundwater-abstraction facility does not necessarily lead to lowered water levels in a nearby groundwater-fed mire because surface water or rainwater may replace the groundwater. In the long run, however, the increased influence of nutrient-rich surface water will result in increased plant productivity (Koerselman *et al.* 1990) whereas a replacement with acidic rainwater will lead to acidification of the top soil (van Diggelen *et al.* 1994). Both processes result in completely altered vegetation composition but it may take up to several decades before the effects are clearly visible. The relative importance of these processes depends to a high degree on the local situation.

3.6 Landscape restoration

Human impact on the landscape has been enormous, not only on the spatial distribution of organisms but especially also on the conditions that enable their occurrence. If one wants to reverse this process and especially if this should be done in a sustainable way, one should pay attention to the spatial context in particular (Hobbs & Morton 1999). This implies that restoration of complete landscapes gives the best chances for long-term survival of self-sustaining ecosystems. In practice this occurs very little, mainly because of practical and financial constraints. The first attempts are, however, being carried out.

The spatial relation between wetlands and regional hydrology made it clear relatively early on that landscape restoration is unavoidable for rehabilitating degraded systems. Most cases concern riverine systems, especially in dry areas where the effects of human actions are most clearly visible, for example in the western USA (Kentula 1997). Typically such processes are now in the planning phase. Elaborate techniques to analyse causes of degradation and to evaluate alternative restoration scenarios are now being developed (van Diggelen *et al.* 1995, Kershner 1997) and sites are being prioritized.

An example of restoration of one of the largest wetlands is the plan for a *c.*290,000 ha lowland river landscape in the state of Mecklenburg-Vorpommern (north-east Germany). The goal of this plan is to re-establish functioning river valleys, including the most relevant parts of the infiltration areas. The concept consists of a combination of plans for true restoration (re-establishment of a former situation), rehabilitation (re-introduction of certain ecosystem functions) and

conservation; see Chapter 1 in this volume. Existing natural values, both of pristine mires (8100 ha) and species-rich semi-natural grasslands (3600 ha), will be strictly protected and managed. Reclaimed polder areas will be partly reflooded (37,000 ha) to develop into eutrophic swamps and partly managed in an environmentally friendly manner (139,000 ha). Also part of the infiltration area (111,000 ha) will be managed in an environmentally friendly manner. Inevitably the implementation of such a large plan takes considerable time. Whereas the conservational part of the plan will be implemented for the greater part before 2005 this is certainly not the case with the restoration part. In the largest river valley, that of the River Peene, 2800 ha of the planned 19,000 ha were restored by 2000, 8 years after the start. The main reasons for this slow progress are economic. The estimated costs of the project amount to €130.5 million for an implementation period of 20 years (Landesregierung Mecklenburg-Vorpommern 2000; see www.um.mv-regierung.de).

During the second half of the 1990s plans for the restoration of the Everglades, in Florida, USA, were beginning to emerge. Not only has the surface of the Everglades decreased dramatically in historic times but there are also large hydrological problems in the remaining part due to a reduction of the incoming water flow by 70%. About 7.5 million m^3 are diverted to the Atlantic Ocean instead of entering the reserve. Consequently, water shortage occurs, not only in the reserve but in the public water supply of Miami as well. A so-called comprehensive plan was presented to the US Congress in July 1999 and consisted of a combination of over 200 projects. Interestingly, these are less focused on increasing the size of the reserve than on restoration of the hydrology. Topics to be dealt with are quantity, timing, quality and distribution. The end point has not been decided upon yet but 50% of the restoration goals should be achieved by 2010. Total costs are estimated to amount to $7.8 billion, $1.4 billion of which have been made available by the signing of the Everglades Restoration Act in December 2000. State and federal sponsors must return to Congress every 2 years to get new projects authorized as the restoration moves forward (see www.evergladesplan.org).

A somewhat different approach is followed in the development of the *Ecologische HoofdStructuur* (EHS;

meaning ecological infrastructure) in the Netherlands, now also further elaborated for western Europe. Contrary to the previous plans, this project is not focused on restoring sites to increase ecological values on the spot, but instead to enhance flows between sites to counteract fragmentation. Nevertheless large areas and large amounts of money are included in this project as well. At present the Dutch nature reserves cover c.150,000 ha (4% of the national territory) which are all semi-natural eco-systems that must be managed regularly. Additionally the country has about 300,000 ha of woodland with nature protection as a secondary goal. The implementation of EHS will last from 1992 until 2018 and consists of a mix of conservation and restoration measures. Existing areas with high nature-conservation interest but without legal protection yet are being bought by the State and by private nature-conservation organizations and will be designated as strict nature reserves. An additional 50,000 ha will be taken out of agricultural use, restored ecologically and function as corridors. A final 100,000 ha will act as a buffer zone that can be used for agriculture under certain restrictions. The costs of the plan are estimated at €3.3 billion; for €300 million financial cover is still lacking (Price Index 2000; see www.minlnv.nl/international). The progress of the plan was evaluated in 2000. Both the acquisition of areas with high nature-conservation interest and buffer zones were well on schedule but restoration had only reached 50% of what was planned to be finished by that time. The main cause was difficulties with the acquisition of grounds (see RIVM natuurbalans at www.rivm.nl/milieu).

4

Ecosystems: trophic interactions

Jelte van Andel

4.1 Introduction: ecosystems and biotic communities

Shugart (1998) provided a nice overview of the development of the ecosystem concept, from which we take some information. The concept has its roots in theories regarding the organization and dynamics of natural systems, which were often seen as a super-organism (Clements 1916): a highly organized and co-evolved assemblage of plants and animals interacting in a dynamic system analogous to the manner in which cells in an embryo interact to produce an organism. Tansley (1935) defined the term ecosystem as 'the whole system (in the sense of physics), including not only the organism-complex, but also the whole complex of physical factors forming what we call the environment of the biome – the habitat factors in the widest sense'. In brief, the biotic community plus its abiotic environment act within a space-time unit of any magnitude. Lindeman (1942), who explicitly considered a lake as an ecosystem, placed an emphasis on understanding the material flows into and out of components of the system. He asserted that a lake could best be considered as an ecological unit in its own right. Lindeman's theory nicely integrated the biotic approach with the functional approach in his so-called 'trophic dynamic viewpoint', including community change, successional patterns and energy flows. Odum's (1953, 1971) textbook, *Fundamentals of Ecology*, was organized around the concepts of ecosystems and their structure and function, with emphasis on the flows of energy and matter. The processing of material and forms of energy through terrestrial systems and the residence time of materials in different ecosystems links terrestrial ecosystems directly to other global systems. As to the scales in space and time, Shugart (1998) referred to Delcourt *et al.* (1983), who considered the time and space scale of different disturbance factors, the ecological mechanisms that are excited by these phenomena and the patterns produced by the interactions between the disturbances and the ecological mechanisms (Fig. 4.1).

The Society for Ecological Restoration (SER International; SER 2002; www.ser.org) listed nine attributes to provide a basis for determining when restoration has been accomplished. In brief:

1 the restored ecosystem contains a characteristic assemblage of the species that occur in the reference ecosystem;
2 the restored ecosystem consists of indigenous species to the greatest practicable extent;
3 all functional groups necessary for the continued development and/or stability of the restored ecosystem are represented or have the potential to colonize;
4 the physical environment of the restored ecosystem is capable of reproducing populations of the species necessary for its continued stability or development;
5 the restored ecosystem apparently functions normally for its ecological stage of development;
6 the restored ecosystem is suitably integrated into a larger ecological matrix or landscape;
7 potential threats to the health and integrity of the restored ecosystem from the surrounding landscape have been eliminated or reduced as much as possible;
8 the restored ecosystem is sufficiently resilient to endure the normal periodic stress events in the local

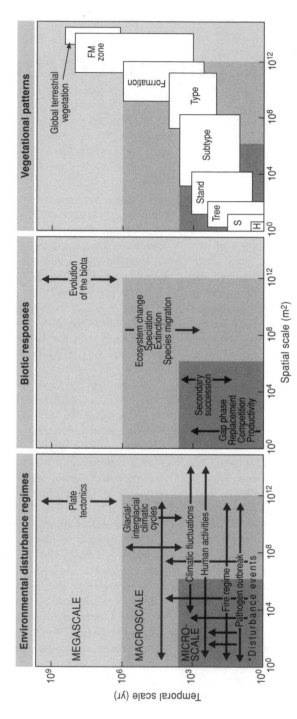

Fig. 4.1 Environmental disturbance regimes, biotic responses and vegetational patterns viewed in the context of space–time domains in which the scale for each process or pattern reflects the sampling intervals required to observe it. FM, formation zone; H, herb; S, shrub. From Shugart (1998); after Delcourt et al. (1983). Reproduced by permission of Elsevier.

environment that serve to maintain the integrity of the ecosystem;

9 the restored ecosystem is self-sustaining to the same degree as its reference ecosystem, and has the potential to persist indefinitely under existing environmental conditions.

This list demonstrates clearly that the focus of ecological restoration is on ecosystems, while recognizing that the biotic community – all the biota in an ecosystem – is the most important component. Assemblages of organisms can be recognized by their species composition as pattern or structure, but also by their functional roles in the ecosystem as for example primary producers, herbivores, carnivores, decomposers, nitrogen fixers or pollinators, in which case they are known as functional groups.

Research on the biotic community in an ecosystem is often limited to food-web studies. The trophic structure has been considered in terms similar to those used in ecosystem studies: in terms of equilibria and stability. For example, Noy-Meir (1975) explained the stability of grazing systems in the Serengeti as a special case of predator–prey systems, with two steady states: a high-biomass, high-productivity steady state which is stable to fluctuations within a certain range, and a low-biomass, low-productivity steady state. Either of the two states may be the more stable. Holling (1973) pointed out that an equilibrium-centred view is essentially static and provides little insight into the transient behaviour of systems that are not in equilibrium. He proposed that the behaviour of ecological systems could well be defined by two distinct properties: resilience and stability (see Chapter 2 in this volume). From Holling's point of view, the notion of an interplay between resilience and stability might also resolve the conflicting view of the role of diversity and stability in communities, because instability in numbers might result in more diversity of species and in spatial patchiness, and hence in increased resilience.

4.2 The diversity–stability debate

A large pile of literature has become available on the topic of the diversity–stability relationship. Lindeman (1942) suggested that dynamic processes within an ecosystem tend to produce certain obvious changes in its species composition. Processes were thus thought to govern structure. Other authors, for example Paine (1980), laid more emphasis on the biotic interactions, from which the processes would result. Removal of a weakly interacting – that is, a functionally insignificant – species would yield no or slight change, and removal of keystone species may have a cascade of effects on the community composition, transmitted by a chain of strongly interacting links. According to Pimm *et al.* (1991) food-web theory predicts that highly connected, complex communities should be most sensitive to the loss of species from the top of the web because secondary extinctions propagate more widely than in loosely connected, simple communities, whereas simple communities should be more sensitive to the loss of plant species than complex communities because in simple communities the consumers are dependent on only a few species and cannot survive their loss. Keystone species can occur at all trophic levels. According to classical results from mathematical food-web theory, omnivory (feeding on more than one trophic level) would destabilize ecological communities (Pimm & Lawton 1978), whereas more recent conceptual syntheses suggest that omnivory should be a strongly stabilizing factor in food webs (Polis & Strong 1996, Fagan 1997). For aquatic systems in particular, it should be recognized that various components of a food web may demand different timescale-dependent dynamics to be included in food-web modelling (Kerfoot & DeAngelis 1989).

McCann (2000) reviewed the diversity–stability debate. Recognizing that much of ecological theory is based on the underlying assumption of equilibrium population dynamics, he proposed to distinguish between two categories of stability in ecology: stability definitions that are based on (i) a system's dynamic stability and (ii) a system's ability to defy change (resilience, resistance). Differential responses of populations may sum, through time, to result in stable community dynamics. Field tests at the scale of the food web are few in number. McNaughton (1985), studying herbivores in the grazing ecosystem of the Serengeti under naturally variable conditions, has shown that greater diversity reduced the relative magnitudes of fluctuations in productivity induced by seasonal change. Similarly, Prins and Douglas-Hamilton (1989) illustrated that the consumption pressure (grazing and browsing)

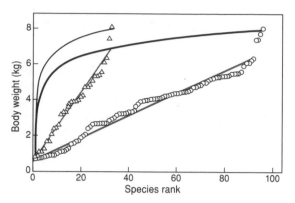

Fig. 4.2 Ranked body mass (body weight, in kg) of grazing species, heavier than 2 kg, plotted against species rank number (Ri) for the whole of Africa (○, thick lines; 96 species) and the Serengeti ecosystem (Δ, thin lines; 33 species). For each assemblage the observed patterns (○, Δ) are given with the linear regressions (straight lines), as well as the patterns predicted assuming an equal probability of each body mass being found (curves). After Prins and Olff (1998). Reproduced by permission of Blackwell Publishing.

in a savanna ecosystem remained quite constant over a number of years, in spite of relatively large fluctuations in the population densities of the herbivore guild. Apparently, these fluctuations may compensate for one another as measured by overall consumption, in spite of the fact that each herbivore species may play a specific role (Prins & van der Jeugd 1993). Later on, Prins and Olff (1998) explained the species richness of assemblages of African grazers in different ecosystems by a small number of parameters (Fig. 4.2). The average body-mass ratio between subsequent grazing species, if ordered from light to heavy, is always constant within each assemblage. This weight-ratio parameter reflects niche differentiation between grazers, and determines whether competitive displacement or facilitation occurs. The second most important parameter explaining species richness is the maximum mass difference between the smallest and the largest grazer in the system. In areas with high primary productivity this parameter can have a higher value than in less-productive areas since large grazers need more vegetation biomass to cover their intake requirements. These two parameters together determine species richness among the herbivores at the present.

Advances in ecological theory have pointed to the interaction strength as a key to the understanding of realized food webs (e.g. de Ruiter *et al.* 1995). Several model investigations have revealed that natural food-web structures can indeed enhance ecosystem stability. Increasing diversity can increase food-web stability under one condition: the distribution of consumer–resource interaction strengths must be skewed towards weak interaction strengths. McCann (2000) suggested that MacArthur's (1955) hypothesis, that greater connectance drives community and ecosystem stability, seems a strong possibility provided that most pathways are constructed from weak interactions. Theoretical and empirical findings suggest that despite the presence of some strong interactions, weak interactions indeed prevail in communities.

4.3 Trophic interactions

As to the question of how species diversity affects ecosystem productivity, three postulates have been brought forward (Fig. 4.3):

1 a positive, non-linear relationship (the redundancy hypothesis);
2 a positive, linear relationship (the rivet hypothesis);
3 no obvious relationship (the idiosyncratic hypothesis).

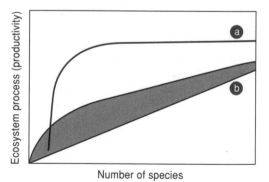

Fig. 4.3 Three potential hypotheses between species richness and ecosystem productivity as an ecosystem process: (a) redundant-species hypothesis, (b) rivet hypothesis, (c) idiosyncracy (not shown). After Lawton (1997); from van Andel (1998a).

The mutual relationship between ecosystem productivity and species richness has been reviewed by Johnson *et al.* (1996) and Waide *et al.* (1999). In general, empirical results have shown a positive, asymptotic relationship between ecosystem processes, measured as productivity and species richness. Only in a few cases did species richness enhance productivity and stability. These results suggest that once all functional groups are present, the addition of species with redundant functions has little effect on ecosystem properties. Reviews of the literature concerning deserts, boreal forests, tropical forests, lakes and wetlands lead to the conclusion that extant data are insufficient to conclusively resolve the relationship between species richness and primary productivity, or that patterns are variable with mechanisms equally varied and complex. For nutrient-related ecosystem attributes, such as primary productivity, functional groups will reflect resource-acquisition strategies of member species. Functional redundancy between species is likely to be highest where plasticity or other mechanisms facilitate coexistence of species that otherwise compete for consumable resources. However, careful attention should be paid to Grime's (1998) observations for plant communities: 'Attribution of immediate control to dominants does not exclude subordinates and transients from involvement in the determination of ecosystem function and sustainability. Both are suspected to play a crucial, if intermittent, role by influencing the recruitment of dominants. Some subordinates may act as a filter influencing regeneration by dominants following major perturbations.'

Predator–prey interactions are among the most intensively studied and modelled relationships, probably because in practice predators and preys are, in general, easily tangible as individuals, whereas plants are generally expressed in terms of biomass. Rosenzweig and McArthur (1963) proposed a hump-shaped prey isocline in the phase plane. The exact shape of the prey curve depends on the demographic characteristics of the prey and the carrying capacity of the environment (Fig. 4.4). Predator numbers should increase when prey numbers are high, but at high prey density predators stop increasing because of other limitations, such as territorial behaviour. Though the Rosenzweig–McArthur model of predator–prey interactions reveals a wide variety of dynamic behaviours,

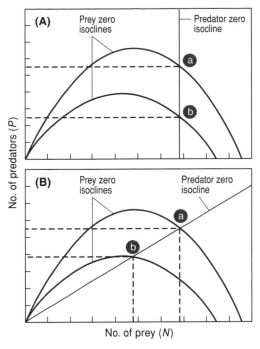

Fig. 4.4 Predator–prey isoclines in (A) the classical Rosenzweig–MacArthur model and (B) the ratio-dependent model. Two prey isoclines are shown for more-productive (a) and less-productive (b) habitat. The equilibrium intersection points are shown by dotted lines. After Krebs (2001), p. 211. Reproduced by permission of Pearson Education, Inc.

from stability to strong oscillations, Krebs (2001) emphasized that all these predator–prey models make a series of simplifying assumptions about the world, including a homogeneous world in which there are no refuges for the prey or different habitats, and the investigated system is one predator eating one prey. I mention just two examples showing the complexity of even a one-prey/one-predator system. Firstly, Scheiner and Berrigan (1998) measured production and maintenance costs of plasticity in the freshwater crustacean *Daphnia pulex* in response to the presence of chemical signals from a predator, the insect *Chaoborus americanus*. They found scant evidence for either production or maintenance costs of plasticity, probably due to a decrease in metabolic rates in the presence of *Chaoborus* extract, which may have compensated for any cost increases. The second example refs to asking the question: what determines the abundance of herbivores? Krebs *et al.* (1999) came to the following set of hypotheses:

- vertebrate herbivores are normally limited in abundance by predators, unless they can achieve escape in space or time;
- large mammals can assume active defence or simply become too large for predators to be effective, and these species normally become food-limited;
- migratory species can escape predator limitation in space when predators are territorial or have limited movements;
- smaller species can evade predator limitation by escape in time (through hibernation, for example) or occasionally by escape in space by behavioural adaptations;
- predator limitation may also be frustrated by excessive intraguild predation, or by predators killing other predators;
- disease and parasitism rarely limit herbivore abundance but may act synergistically with food limitation;
- small rodents may breed at a sufficiently high rate that predators cannot catch them, unless social limitation through spacing behaviour becomes limiting first.

For the sake of clarity I will present a number of trophic relations on their own, as parts of a food chain rather than a food web, selected in view of their importance in restoration ecology, soon thereafter recognizing that there are a number of both direct and indirect interactions in food webs.

4.3.1 Plant–herbivore interactions

Several theories have been developed to explain plant–herbivore interactions along productivity gradients, often associated with views upon the role of plant competition.

According to the M-S model (Menge & Sutherland 1976, 1987) environmental stress – at the lower end of the productivity gradient – affects higher trophic levels more severely, which implies that grazers are absent or not important, and that plants are regulated by environmental stress. Plant competition should be most important in habitats exposed to intermediate stress and a moderate intensity of grazing, whereas at high plant productivity herbivores should become dominating, which decreases plant competition.

The exploitation ecosystem hypothesis (EEH) follows seemingly opposite reasoning (e.g. Hairston *et al.* 1960, Oksanen *et al.* 1981, Fretwell 1987; see also DeAngelis 1992), but it actually only includes the role of carnivores, which come into play at the highest level of primary productivity, and the associated herbivore density. At low productivity, vegetation is too sparse to support herbivores. An increase of primary productivity should result in increased herbivore grazing pressure, maintaining a low standing crop until herbivore biomass is high enough for a population of predators to be sustained. Thus, with increasing grazing intensities along a successional gradient, plant competition should decrease until the grazers are regulated by their predators. During the last decade, the role of positive interactions in plant communities has been given more attention.

The abiotic stress hypothesis (ASH) proposes that the importance of facilitation in plant communities increases with increasing abiotic stress or increasing consumer pressure. This is because neighbours buffer one another from extremes of the abiotic environment and herbivory. The importance of plant competition would increase when abiotic stress and consumer pressure are relatively low (Bertness & Callaway 1994, Bertness & Leonard 1997).

Krebs *et al.* (1999) questioned simplifications such as the suggestion that there are 'generic' herbivores and predators that behave as units within the limits set by ecosystem productivity. They argued that trophic interactions in vertebrates critically depend on species quirks, so that, for example, one cannot classify all the Serengeti ungulates as 'herbivores' without knowing more about their individual life-history strategies. The EEH suffers from two serious deficiencies: it is a food-chain model rather than a food-web model, and hence fails to emphasize many kinds of indirect effect, and it fails to incorporate resource-quality responses, for example inedibility of certain primary producers. Indeed, Ritchie and Olff (1999a) found evidence for both compensatory and additive effects of multiple herbivore species on plant species composition, diversity and spatial heterogeneity. By feeding preferentially on weaker competitors, a single herbivore can accelerate species replacement in favour of stronger competitors. Alternatively, a herbivore may feed preferentially on stronger competitors, thereby producing a positive indirect effect on weaker plant species, allowing them to coexist or even exclude a stronger competitor.

Following Noy-Meir (1975), a number of studies have stressed the importance of herbivore saturation for the existence of stable states and threshold effects. However, multiple stable states can also occur in the absence of herbivore saturation; that is, at a plant standing crop where herbivore saturation does not yet play a role. Van de Koppel *et al.* (1996) have shown that several herbivore species can co-occur at intermediate standing crop, whereas at high standing crop plants may become inaccessible for herbivores, which together results in a hump-shaped pattern of herbivore density (Fig. 4.5). At low standing crop, various mechanisms, most notably soil degradation, may depress plant growth (van de Koppel *et al.* 1997) due to plant–soil feedbacks; a herbivore-induced decrease in plant standing crop has led to soil degradation and reduced plant growth. Positive feedback between reduced plant standing crop and deteriorated soil conditions has thereby contributed to irreversible vegetation destruction. It is conceivable that several mechanisms act simultaneously in many natural systems. As a result, the growth and consumption curves intersect twice in a phase plane, even in the absence of herbivore saturation.

4.3.2 Plant–parasite interactions

Host–parasite interactions can be represented by resource–consumer models, just as in the case of predator–prey and herbivore–plant interactions.

A few classic studies, referred to by Zadoks (1987), showed that pathogenic fungi, selectively parasitizing plant species in a plant community, may accelerate vegetation succession. For example, the willow rust *Melampsora bigelowii* killed many seedlings of the willows *Salix pulchra* and *Salix alexensis*, pioneer species which formed nearly pure stands on gravel banks of the River Yukon in Alaska, once the ice had receded. This might have accelerated succession to birch and spruce. Another example he referred to is the massive wane of submarine *Zostera* beds in Dutch

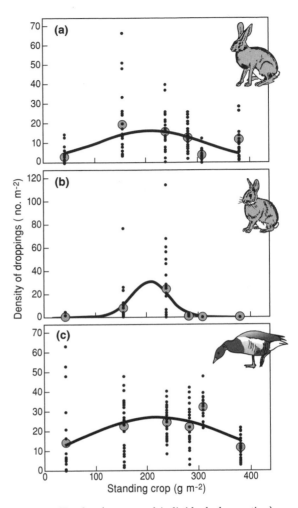

Fig. 4.5 Number (means and individual observation) of annual droppings of (a) hares, (b) rabbits and (c) barnacle and brent geese in relation to vegetation standing crop on the coastal salt marsh of the Waddensea Island of Schiermonnikoog, The Netherlands. After van de Koppel *et al.* (1996). Reproduced by permission of the Ecological Society of America.

estuaries in the early 1930s, partly due to the pathogen *Labyrinthula macrocystis*. There is increasing interest in the direct and indirect effects of pathogens and parasites on the structure of plant communities. Some recent examples of the roles of pathogens in determining plant community structure and the consequences of pathogen impact on the functioning of ecosystems have been reviewed by Dobson and Crawley (1994). These authors mentioned the phenomenon that fungal blights removed *Castanea dentata* from the eastern deciduous forests of the United States, *Tsuga mertensiana* from the Pacific northwest of Canada and the USA, and *Ulmus* species from much of western Europe, and a whole range of species from *Eucalyptus* forests of western Australia. In each of these cases, the removal of a dominant species led to the development of forests dominated by less competitive species from earlier successional stages, or opened up the canopy for colonization by earlier successional species.

Brown and Gange (1989) were among the first to pay attention to the effects of both above- and below-ground plant consumers (herbivores and pathogens) on plant succession. In an early stage of grassland succession, plant species richness and diversity increased as a result of the application of soil insecticide and, by the second season, were depressed by foliar insecticide. Three major life-history groupings – annual and perennial herbs, and perennial grasses – responded differently to herbivory, with a considerable effect on the pattern of early succession. Soil-borne diseases appeared to be also involved in the degeneration of *Ammophila arenaria* and *Hippophaë rhamnoides*, two plant species that dominate the coastal foredunes of Europe and which are widely planted for sand stabilization (van der Putten *et al.* 1993). Endoparasitic nematodes appeared to be responsible for reduced vitality of *A. arenaria*, thus favouring *Festuca rubra* ssp. *arenaria*. Nematodes of the genus *Longidorus* are capable of damaging the root system of *H. rhamnoides*, including nitrogen-fixing nodules, and the related mycorrhizal system, thus reducing the uptake of phosphate and other nutrients. This damage may result in acceleration of succession to, for example, *Sambucus nigra*, *Ligustrum vulgare* and *Rosa rubiginosa* on calcareous soils, or to *Empetrum nigrum* on acid soils. In general, the spatial and temporal dynamics of above- and below-ground herbivores, plant pathogens and their antagonists can differ in space and time. This affects the temporal interaction strengths and impacts of above- and below-ground higher-trophic-level organisms on plants (van der Putten *et al.* 2001).

A world-famous example of the indirect effect on vegetation caused by an animal disease is provided by the infection of rabbits (*Oryctolagus cuniculus*) by the *Myxoma* virus in southern England (see Dobson & Crawley 1994). Myxomatosis was introduced into Australia in 1950 and into France in 1952, from where it spread throughout western Europe, reaching Britain in 1953. The initial *Myxoma* virus in 1953 was a highly virulent strain and the 1950s rabbit population was reduced by about 99% within a few years. The rabbits remained extremely scarce for the following 15 years. Once rabbits had almost disappeared, acorns buried in grassland by jays had a vastly greater chance of producing seedlings and becoming established. The reduced rabbit grazing was responsible for the transformation of Silwood Park from an open grass parkland in 1955 into an oak woodland (*Quercus robur*) with occasional clearings within 15–20 years. This change was irreversible, even after the recovery of the rabbit population in the 1970s.

4.3.3 Tritrophic interactions

In the plant species *Silene alba* and *Silene dioica*, flowers are the site of infection by the host-sterilizing anther smut fungus *Ustilago violacea*, as well as the site of oviposition by the noctuid *Hadena bicruris*. In diseased populations of the two host species, the noctuid can act as a pollinator, fruit predator and vector of fungal spores. Hence, host traits that enhance or reduce visitation rates by the noctuid could affect fruit set, fruit predation and infection probability. Timing of the production of susceptible host stages can have a large impact on a plant's probability of infection by pathogens and on the extent of damage caused by herbivores. Biere and Honders (1996) investigated the impact of intraspecific differences in the timing of anthesis in these two closely related host species on susceptibility to the herbivore and the pathogen. The fungal pathogen and the seed predator both had a large impact on female fecundity in the two *Silene* species, although the fungal infection had little impact on host reproduction in the first year of infection. Effects of flowering phenology on the probability of infection differed between the host species, but the intraspecific differences in the timing of anthesis can significantly affect fruit set, fruit predation and infection risk in natural plant populations. The phenological rank required for high fruit set, low fruit predation and low infection risk differed between host species, but generally not within host species. In contrast to the expectation for pollinators that are also seed predators and vectors of disease, the authors found that the effects of timing of anthesis via fruit set on the one hand and via predation and infection on the other were generally not in opposing fitness directions within host species; they were either uncorrelated or affecting reproductive success in the same direction, allowing for simultaneous maximization of fruit set and minimization of damage. This example illustrates the complexity of the biotic interactions that modulate the effects of a simple host trait, timing of anthesis, on a host's reproductive success. Thrall *et al.* (1995) were able to model the disease transmission in the *Silene–Ustilago* host–pathogen system as an epidemiological process.

Plants have evolved an overwhelming diversity of mechanisms that can act as direct defences against herbivores (e.g. van der Meijden & Klinkhamer 2000). Because of much genetic variation in both concentrations and composition of plant chemical defences, specialist and generalist herbivores and their natural enemies could be differentially affected, which may change their effect on plant fitness. In addition, many studies have demonstrated indirect defence mechanisms, and here tritrophic interactions come into play. A wealth of rather recent literature is available on plants benefiting from the active chemical attraction of natural enemies of their herbivores (e.g. Dicke & Vet 1999, Vos *et al.* 2001). An example is the parasitoid wasp *Cotesia glomerata* which infests caterpillars of *Pieris brassicae*, both herbivore and parasitoid being chemically attracted by volatiles released from the leaves of *Brassica oleracea*. Such an indirect defence by the plants can be considered a combination of bottom-up and top-down effects. It can involve the provision of shelter or of alternative food, as floral and extrafloral nectar, or of information that can be used by carnivores during foraging. Chemical information on herbivore presence and identity may indeed be essential for successful location of herbivores by carnivores. Both the production of volatiles and the responses of arthropods to these volatiles are subject to variation, and natural selection acts on both sides of the interaction. Superimposed on this part of the food web, there is apparently an information web, which may even be more complex than the

food web itself, because information conveyance occurs irrespective of trophic relationships. Vos *et al.* (2001) examined the effects of herbivore diversity on parasitoid community persistence and stability, mediated by non-specific information from herbivore-infested plants in a *Brassica–Pieris–Cotesia* system. Parasitoids were attracted by infochemicals from leaves containing non-host herbivores. Thus, when information from the plant is indistinct, herbivore diversity is likely to weaken interaction strengths between parasitoids and hosts. In general, more than one herbivore will feed on a plant, and a single herbivore species is generally attacked by several carnivore species. The authors therefore modelled parasitoid–herbivore systems increasing in complexity, referring to experiments and field data. A simulated increase in herbivore diversity promoted the persistence of parasitoid communities. However, at a higher threshold of herbivore diversity, parasitoids became extinct due to insufficient parasitism rates. Thus, they concluded, diversity can potentially drive both persistence and extinctions.

4.3.4 Trophic cascades

Paine (1980) was the first to use the term 'trophic cascade'. Pace *et al.* (1999) suggested that empirical studies from a variety of systems indicate that trophic cascades are widespread, although many factors regulate their occurrence. They described trophic cascades as 'strong interactions within food webs that influence the properties of the system', thereby including a much wider spectrum of interactions, and suggested that cascades are no longer the sole province of lake and intertidal ecologists but clearly occur in a diversity of ecosystems on land and in the ocean. Polis *et al.* (2000) questioned this generalization and, referring to the review of 41 studies by Smith *et al.* (2000), suggested distinguishing between species-level and community-level cascades. Starting from the plant as a primary producer, species-level cascades occur within a subset of the community or compartment of a food web, such that changes in predator numbers affect the success of a subset (one or a few) of the plant species, whereas community-level cascades substantially alter the distribution of plant biomass throughout an entire system, in a manner consistent with the EEH. Although these definitions refer explicitly to plant–herbivore–predator systems, they would also apply to any multi-link food-web interaction. The central question would be: why are community cascades apparently absent or rare in terrestrial habitats? According to Polis *et al.* (2000), all the cascades that Smith *et al.* (2000) referred to in terrestrial systems measured interactions within subsets of a community. Many of the impediments to community-level cascades would arise from the complexity of natural systems, and support even for species-level cascades in terrestrial systems would be limited. However, a study of butterflies on calcareous grasslands (Steffan-Dewenter & Tscharntke 2002) indicated that the specialists of higher trophic levels (monophagous and strongly oligophagous butterflies) are more sensitive to habitat fragmentation than are species of lower trophic levels, in the plant community. In Chapter 5 in this volume I will, once again refer to the problem of trophic cascades, then in the context of metapopulation dynamics, which also cannot be considered without keeping an eye on the complexity of the communities they are part of.

4.4 Mutualism

We define mutualism as a bidirectional facilitation, as an interaction between individuals of different species that lead to an increase of fitness of both parties, often based on mutual assistance in resource supply. The benefits usually exceed the costs for both partners (Bronstein 1994). Mutualistic relationships can be facultative; leguminous plants for instance can live with or without *Rhizobium*. They can also be obligate, a condition for survival: lichens are a symbiosis between a fungal and an algal component. For vascular plants in general, mutualistic relationships with mycorrhizal fungi are of utmost importance. Many experimental investigations have shown that both plant and fungal symbionts benefit from the reciprocal exchange of mineral and organic resources, but sometimes the relationship may turn into parasitism. The case of plant–pollinator relationships is similarly complicated, because these interactions may ideally be mutualistic, but in many cases there is a bias in the balance in favour of one or the other component. Mutualism may thus be accompanied with mechanisms indicating a balanced antagonism. The current situation results from selection and evolution, a process that is continuous.

4.4.1 Plant–mycorrhiza interactions

The majority, say 80%, of species of temperate, subtropical and tropical plant communities are infected by vesicular-arbuscular mycorrhizal fungi (VAM fungi, nowadays called AM fungi). There is every indication that in such communities a vigorous semipermanent population of fungal symbionts with low 'host' specificity is involved in an infection process which effectively integrates compatible species into extensive mycelial networks (Francis & Read 1994); the number of these fungal species is only about 200.

Ectomycorrhizal fungi (ECM) occur mainly on woody plants and only occasionally on herbaceous and graminaceous plants and involve more than 5400 fungal species. Ericoid mycorrhizae occur mainly in the Ericales and are physiologically comparable with ECM. Non-mycorrhizal plants occur mainly in very wet or saline ecosystems and in ecosystems with a high nutrient availability and/or with recently disturbed soil (see references in Ozinga *et al.* 1997). Orchid–mycorrhiza relations are a special case (see below; Dijk *et al.* 1997).

AM fungi are presumably especially efficient in the uptake of inorganic P (and other relatively immobile ions such as Cu^{2+}, Zn^{2+} and NH_4^+) and are capable of increasing the P uptake more in nutrient-rich patches than in soils with a uniform P distribution (Cui & Caldwell 1996a, 1996b), whereas ECM and Ericoid mycorrhizae are more efficient in N-limited ecosystems. In contrast to AM fungi, some ECM and Ericoid mycorrhizae fungi have abilities to take up N from organic matter and translocate these nutrients to their host plant (Read 1991). Enzymatic degradation by ECM and Ericoid mycorrhizae fungi has been shown for proteins, cellulose, chitin and lignin. Changes in the proportion of nutrients in inorganic or organic form may create changes in the competitive abilities provided by different mycorrhiza types.

The presence of mycorrhizae has been shown to change the outcome of plant competition in many cases, both for AM plants and for ECM plants, and is thus a determinant of plant community structure (Ozinga *et al.* 1997, van der Heijden *et al.* 1998). Moora and Zobel (1996) tried to answer the question of whether a symbiotic interaction with AM fungi can help young plants resist the competition of older ones with which they naturally coexist. They chose two common subordinates in the lowest sublayer of species-rich calcareous grasslands; *Prunella vulgaris* as a target species and *Fragaria vesca* as a neighbouring species. In interspecific competition with old *F. vesca*, the shoots of target plants were 22% greater when inoculated with AM fungi than when non-mycorrhizal. Thus, if a young *P. vulgaris* plant has established itself somewhere in a natural gap, the presence of AM fungi might make intraspecific competition more severe, but may decrease the strength of interspecific competition. Similarly, Kiers *et al.* (2000) suggested that tropical mycorrhizal fungal (AM fungi) communities have the potential to differentially influence seedling recruitment among host species and thereby affect community composition. Earlier, Grime *et al.* (1987) demonstrated in a microcosm experiment that ^{14}C could be transported through a mycorrhizal network from dominant to subordinate species, which led to an increase in biomass of the inferior competitors. Mycorrhizal linkages were shown, indeed, to transport ^{15}N and ^{32}P within and between plant species (Chiarello *et al.* 1982, Finlay *et al.* 1988).

Orchids demand particular attention in the context of restoration ecology. Dijk *et al.* (1997) reviewed the nutritional relationships of orchid species. In the first heterotrophic and subterranean phase of orchid development, the growth of the protocorm is entirely dependent on mycorrhizal fungi. The nutrient metabolism of developing orchid individuals is adapted to this symbiosis: reductions in orchid nitrogen metabolism are permitted which can be considered adaptations to the parasitic habit during at least this phase. Mycorrhizal infection is restricted to subterranean tissues only, to the subepidermal zone of the protocorm and root parenchyma. After the initial infection the development of mycorrhizae can easily derail, and in symbiotic cultures a range of interactions can be met from a loss of mycorrhiza via normal mycorrhizal infections to pathogenic effects. The primary function of mycorrhizal infection in the juvenile phase lies in the transport of C compounds to the developing seedlings. Translocation of sugars towards protocorms has been demonstrated by radioactive labelling in classic studies. Apart from interfering with the carbon metabolism, mycorrhizal infection has a pronounced influence on the uptake of mineral macronutrients (P and N). These subtle processes in the juvenile phase should have a chance before we can ever expect adult and flowering orchids as a result of restoration efforts.

Mycorrhizal fungi not only play an important role in plant nutrition; they can also fulfil other functions which may be important for survival in the long term. Some species with a compact structure increase resistance against pathogens, heavy metals and polyphenolic substances (Ozinga *et al.* 1997). There may be a trade-off between protecting functions as a compact structure and efficiency in nutrient uptake. Moreover, mycorrhizal fungi and their positive effects on plant performace may benefit from the effects of mycophagous animals such as the order Collembola if the latter feed preferentially on non-mycorrhizal fungi in the rhizosphere (Gange 2000). It has also been shown that the presence of Collembola in the soil can lead to a decrease in reproduction of the aphid *Myzus persicae* when feeding on *Trifolium repens* (Scheu *et al.* 1999). Still another mechanism by which Collembola might positively affect AM fungi is through spore dispersal.

4.4.2 Plant–pollinator interactions

Plant–pollinator interactions can be considered a non-symbiotic mutualism. The most basic evolutionary outcome that is common across both plants and pollinators is the efficiency of both in exploiting what is for each a valuable or critical resource. But the mutualism is neither symmetrical nor cooperative. The mutual exploitation interest may be skewed towards a consumer–resource relationship between the two parties. One common manifestation is opportunism and flexibility on the part of pollinators toward plants, and vice versa.

In their review on 'endangered mutualisms', Kearns *et al.* (1998) pointed out the phenomenon that over 90% of 250,000 modern angiosperm species are pollinated by animals. Among the nearly 300,000 flower-visiting animal species are insects, birds, bats and small marsupials. Habitat fragmentation and other effects of land use, like agriculture, grazing, herbicide and pesticide use, and the introduction of non-native species, have a crisis-like impact on plant–pollinator systems. Specialist relationships are, in themselves, much more vulnerable than generalist relationships, but plant–pollinator interactions are only seldom specific to the species level; relatively few plant–pollinator interactions are absolutely obligate in a strict sense, in particular in much of Europe and the eastern and northern parts of North America (Johnson & Steiner 2000). Many flowers show specialization in floral traits, yet they are often visited by diverse assemblages of animals. Mutualism is mainly an interest of the plant community and the pollinator community, rather than that of two specific plant–pollinator populations. In their review, Kwak *et al.* (1998) illustrated that pollen and gene flow in fragmented habitats not only depend on the investigated plant populations as such, but also on the neighbouring species of the plant communities and the flowering phenologies of the component species. In small habitat fragments, less-attractive plant species may receive fewer pollen visits and a higher proportion of heterospecific pollen grains, thereby reducing pollination efficiency and gene flow.

4.4.3 Mutualism among animal species

Associations between tick birds, or oxpeckers, and their mammalian hosts are among the best known of mutualisms between vertebrates. Another example is fossorial blind snakes (*Leptotyphlops dulcis*) eating insect larvae in the nests of screech owls (*Otus asio*), and hence potentially reducing larval parasitism on nesting owls. Interactions between insect species are numerous, for instance between ants and aphids. Among plant–insect mutualisms, I refer to the mutualistic system of anti-herbivore defence in acacias brought about by their ant inhabitants. According to Dickman (1992), asymmetric commensal and symmetric mutualistic interactions occur frequently between species of terrestrial vertebrates. I do not refer to commensal interactions in terms of mutualism, but call it facilitation (see Chapter 5). Potential advantages for individuals in mixed-species associations are very diverse, and include reduction in parasite load, reduced risk of predation and increased access to food and other resources. In contrast to obligate mutualisms, associations between species of terrestrial vertebrates

are not permanent but may last for periods of a few minutes to several months, and the benefits often differ in kind and amount. Individual birds and mammals often achieve a higher rate of energy intake in mixed-species associations than when foraging alone or with conspecifics, and mixed-species flocks of bird may confuse predators.

4.5 Concluding remarks

The dynamics of seemingly distinct systems are intimately related by spatial flow of matter and organisms, which has led Polis *et al.* (1997) to try and integrate landscape ecology and food-web ecology. Indeed, ecosystems are closely bound to one another. The relationship between species diversity and ecosystem functioning is complex. For example, studies of interactions between below-ground herbivores have shown both competition and facilitation (Mortimer *et al.* 1999). Facilitation may occur through substrate modification or the increased attractiveness of damaged root to other herbivores. Specialist herbivores cause changes in plant community composition through their effects on specific host plants, primarily as a result of the alteration of competitive interactions between coexisting plant species. Generalists can also affect plant community composition, as a result of differences in their preference for plant species or host plant susceptibility.

Within food webs, the large array of indirect interactions may be at least as important as the direct interactions (Wootton 1994): for example exploitation competition, trophic cascades, apparent competition, indirect mutualism or interaction modifications. Indirect effects occur when the impact of one species on another requires the presence of a third species. They can arise in two general ways: through linked chains of direct interactions and when a species changes the interactions among species.

Simulation models indicate that some indirect effects may stabilize multi-species assemblages. On the one hand, such complexity may stress the notions of chaos and unpredictability. On the other hand, this has not given rise to hopelessness, but to a better recognition of choosing adequate temporary and spatial scales for the analysis of mechanisms. Grover (1994), recognizing that the assembly history of nearly all communities is unknown and that it thus may be feared that community ecology will be plagued by unpredictability and inexplicability, tried to counter this by developing an understanding of the 'assembly rules' specifying the possible pathways by which extant biotic communities are created. He started his exercise with analyses of the mechanisms of local population dynamics and interactions in biotic communities composed of simple food chains, each comprising a nutrient-limited plant and its specialist herbivore. Such communities are characterized by the presence of keystone species, whose deletion induces a cascade of extinctions and which can be identified, *a priori*,

from their role in local interactions. Food-web modelling remains a useful tool with which to evaluate the state of knowledge and identify perspectives for future research. On the other hand, it is promising to see how a general principle, such as the dependence of primary production on climate and soil fertility – translated into terms of quantity and quality of food for herbivores following Breman and de Wit (1983) – could be applied to estimating hotspots of diversity of mammalian herbivores on a global scale (Olff *et al.* 2002). Theoretical and empirical approaches increasingly suggest scale-dependence in the relationship between species richness and productivity.

5

Communities: interspecific interactions

Jelte van Andel

5.1 Introduction: the community concept

The term **ecological communities** is being used for different kinds of species assemblage. In the previous chapter I dealt with the **biotic community**, which includes all the biota in an ecosystem (Odum 1971, SER 2002; www.ser.org). We distinguish biotic communities from species assemblages classified on the basis of the species' taxonomic status (e.g. an insect community) or life form (e.g. a tree community), which we define as 'the set of individuals of two or more species that occur in the intersection of the local distribution areas of these species' (Looijen & van Andel 1999, 2002); in the present chapter I will call these assemblages **communities**. Population dynamics and genetics cannot be included in community ecology in a straightforward manner, because community boundaries do not coincide with population boundaries, and as a result different individuals of a population may be part of different communities. This is probably one of the reasons why vegetation science has traditionally been linked much more to ecophysiology than to population ecology.

Communities develop through a process called community assembly, in which individuals of species invade, persist or become extinct. The search for assembly rules makes ecological knowledge explicit, rather than implying a philosophy on the identity of a plant community. Though a community is not just the summing up of its individual components, it should not be considered as an organismal entity either.

Interactions between species are generally considered as mechanisms that affect community structure and provide the community with emergent properties as compared to the sum of the component individuals (see Looijen & van Andel 1999). Due to the complexity of these interactions, not much progress seems to have been made in understanding the assembly rules since Diamond (1975) coined the term to deterministically explain stable communities, despite a more recent attempt by Weiher and Keddy (1999). The analytical review by Belyea and Lancaster (1999) is of great help in understanding the literature dealing with assembly rules. It clearly distinguishes within trophic levels from across trophic levels, and between environmental and dispersal constraints. In the present chapter, I will pay attention to the dispersal-assembly perspective and the niche-assembly perspective of communities (see Hubbell 2001), leaving the discussion on assembly rules out of scope. (See also Chapter 16 in this volume.)

5.2 The dispersal-assembly perspective

5.2.1 The species pool

According to Hubbell (2001), the dispersal-assembly perspective asserts that 'communities are open, non-equilibrium assemblages of species largely thrown together by chance, history, and random dispersal'. In Chapter 2 in this volume I have explored the concept of the species pool and how to identify the adequate species pool from a more general species assemblage through environmental 'filters'. Species can colonize new or existing plant communities either from a seed bank or by spatial dispersal. Seed banks do not, by definition, occur in the early stage of a primary succession. Here, the species composition during the colonization phase depends on the arrival of propagules from elsewhere. This is in concert with the classic approach of 'relay floristics' as proposed by Egler (1954), which does not exclude that one species may facilitate the establishment of another species, and could give rise to 'obligate succession' (Horn 1976). The reappearance of plant species in restoration projects may depend on their persistence in the soil seed bank as a 'memory' of the original plant community (Bakker et al. 1996b). Many species of the north-west European flora have only a short-term persistent seed bank (Thompson et al. 1998), which implies that dispersal is the bottleneck in regeneration. If a species has been lost from the persistent soil seed bank, it has to be transported to the site of reappearance by some vector, for example wind, water, animals or humans, and incorporated into the fresh seed bank, waiting for safe sites for germination and establishment. Seed dispersal is an important key for the establishment of target communities or species (Poschlod & Bonn 1998). Also for aquatic systems, Palmer et al. (1996) argued that dispersal needs to be viewed as a regional process that may routinely influence local benthic dynamics, because fauna can move to and from water-column dispersal 'pools' and may do so at frequent intervals; benthic invertebrate assemblages may be influenced in an ongoing fashion by dispersal.

5.2.2 Alien invasive species

In Chapter 7, risks of re-introductions will be discussed, meant to restore or to complete a community species pool. Here, I will deal with the problem of the unintended invasion of alien species, not belonging to the community species pool (see also Chapter 2). According to McCann (2000), the current evidence indicates that, although most species invasions have a weak impact on ecosystems, the occasional invasive exotic species alters an ecosystem profoundly. It is, therefore, necessary to carefully distinguish between native and alien (exotic, non-indigenous) immigrants into communities. Invasions of single-species populations being a natural phenomenon in terms of island biogeography, they may remain ineffective in case of dispersal constraints and they may come to play a particularly negative role in affecting communities when the species are alien to the local community. The vigour and success of aliens in areas where they have been introduced has been attributed not only to more favourable environments, but also to the release from natural phytophagous enemies (Crawley 1987). For Lythrum salicaria, introduced to North America some 200 years ago, both increased competitive ability and release from herbivores have been suggested as an explanation of increased alien vigour (Blossey & Nötzold 1995). The invasion of Senecio inaequidens, a perennial and highly self-fertilizing pioneer plant transported in the early 20th century with sheep's wool from South Africa to Belgium and the Netherlands, coincided with its adaptation to the Atlantic climate by early flowering, with flowering time moving from August to May, and prolongation of the flowering period to the end of December (Ernst 1998).

A survey of plant traits in the Czech flora (Pyšek 1995, Pyšek et al. 2002) revealed that the alien species differed significantly from the native flora in the frequency distribution of life forms and life strategies. Therophytes and phanerophytes showed remarkably higher representation among aliens, whereas hemicryptophytes and hydrophytes were under-represented. Across 302 nature reserves, the mean proportion of aliens was 6.1% (range 0–25%). Vegetation type alone explained 14.2 and 55.5% of the variation in the proportion of aliens in regions of mesophilous and mountain flora, respectively. Taken together, the environmental variables used (altitude, climate and human impact) explained 44% of the variation in the representation of all alien plants. The positive relationship between the occurrences of neophytes and native species could indicate that the two groups do not directly compete. To the contrary, Levine and D'Antonio (1999) and Alpert et al. (2000) concluded from their survey of spatial pattern and invader addition studies that exotic and native species respond to the same factors in a similar way; it has proved difficult to identify particular traits that are consistently associated with the tendency of plant species to invade. The traits that seem to best explain invasiveness are probably a broad native range and rapid dispersal. Rapid dispersal has been associated with traits such as short generation time, long fruiting period, large seed number, small seed size, prolonged seed viability and transport by wind or animals. Rejmánek (1999) mentioned, in addition, phenotypic plasticity, small genome size and vegetative reproduction (in aquatic habitats).

None of the afore-mentioned characteristics imply, *a priori*, aggressiveness. Nevertheless, it should be kept in mind that invasiveness depends not only on the characteristics of the invading species, but also on the interactions with the abiotic and biotic factors in the new habitat. In this respect changing environmental conditions are of importance. Dukes and Mooney (1999) suggested that, although climate change might not directly favour alien plant species over natives, many invasive plant species share traits, such as rapid dispersal, short regeneration times and higher tolerances, that could increase their dominance in a climate undergoing transition. Sometimes, a species that has been a native member of plant communities can start behaving as an aggressive invader. For example, *Phragmites australis* has been a member of tidal plant communities in New England, USA, for thousands of years, and an extensive increase in its occurrence has been noted only during the last 100 years or so (Orson 1999). Currently, the species can form dense monocultures with less diverse plant associations, probably due to large-scale changes in the hydrology.

In spite of the reality that invasions cannot be predicted quantitatively because parameter estimation is technically close to impossible for conditions still alien to a species (Hengeveld 1999), progress is being made in predicting the fate of invaders. Studies by Kolar and Lodge (2001), using quantitative statistics to identify characteristics that distinguish groups of species more or less likely to become established or invasive, revealed in general clear relationships between the characteristics of releases and the species involved, and the successful establishment and spread of invaders. For example, the probability of bird establishment increased with the number of individuals released and the number of release events. Also, the probability of plant invasiveness increased if the species had a history of invasion and reproduced vegetatively. Davis *et al.* (2000) formulated a number of testable predictions for plant invasibility, in brief:

- environments subject to pronounced fluctuation in resource supply will be more susceptible to invasions than environments with more stable resource supply rates;
- environments will be more susceptible to invasion during the period immediately following an abrupt increase in the rate of supply or a decline in the rate of uptake of a limiting resource;
- invasibility will increase following disturbances, disease and pest outbreaks that increase resource availability;
- invasibility will increase when there is a long interval between an increase in the supply of resources and the eventual capture or recapture of the resources by the resident vegetation;
- the susceptibility of a community to invasion will increase following the introduction of grazers into a (nutrient-rich) community;
- there will not necessarily be a relationship between the species diversity of a plant community and its susceptibility to invasion;
- there will be no general relationship between the average productivity of a plant community and its susceptibility to invasion.

5.2.3 Species richness

According to MacArthur and Wilson's (1967) dispersal-assembly hypothesis, species richness on islands results from an equilibrium between immigration and extinction rates of the species concerned. In view of this, species richness is a neutral theoretical concept which implies that species number can be quantified for theoretical purposes without recognizing differentiation among them (see Hubbell 2001). Species-area curves represent the same phenomenon (Rosenzweig 1995). For nature conservation and restoration purposes, species richness has been valued, often with the short-cut notion that high species richness is a goal of the management regime. This was triggered by the observation that the application of artificial fertilizers since the early 20th century had contributed to a large extent to the decline in species. It was empirically shown that species richness in general exhibits a hump-shaped relationship with ecosystem productivity (see Fig. 5.1; Grime 1973, Connell 1978, Rosenzweig & Abramsky 1993, Tilman & Pacala 1993).

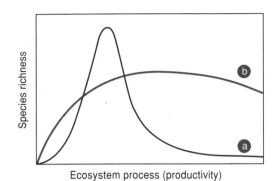

Fig. 5.1 Species richness as a function of ecosystem productivity: (a) spatial gradient model according to Grime (1973, 1979) and Tilman & Pacala (1993); (b) spatial and succession model after Connell (1978). After van Andel (1998a).

The productivity-diversity pattern is scale-dependent. In lakes, the relationship is hump-shaped at the local scale such as within ponds, for example, and linear at the regional scale, between ponds (see Chase & Laibold 2002). This dependence on scale results because dissimilarity in local species composition within regions increases with productivity. By using this knowledge, a relatively high species richness could be achieved in local communities through arranging an intermediate primary productivity, associated with intermediate environmental stress and intermediate disturbance, for example by removal of produced plant biomass, even if knowledge of the mechanisms behind the relationship is still largely speculative.

Species richness as a neutral concept is interesting from a scientific point of view, for example in search of an explanation for the relationship between species diversity and ecosystem stability (see Chapter 4), but it is insufficient as a criterion for nature-management purposes. Indeed, it is not just a high number of species that is highly valued, but particularly the values of each of the component species that count. Rare or subordinate species are often more highly valued as a tool for nature conservation than common or abundant species, and aggressive alien invaders are not wanted. This is why target communities are not only defined as a species list, but as a syntaxon, a community of species known to be co-occurring frequently, and numerically identified as belonging to a natural or semi-natural community, which provides them with a value of naturalness. This is how restoration ecologists are capable of distinguishing between native species, which are worthy of being reintroduced if they have become extinct, and alien invaders.

5.2.4 Abundance, commonness and rarity

Within a community, in general a few species are abundant or dominant and a larger number is rare or subordinate. This is reflected in so-called dominance-diversity curves (see Whittaker 1965, 1975), and in the distinction between dominants, subordinates and transients (Grime 2001). Similarly, at a regional scale some species are common and many species are rare. This is reflected by Preston's (1962) classic lognormal distribution. Hubbell (2001) provided a nice historical overview of these two aspects of commonness and rarity, at the community scale and at a geographic scale.

Some two decades ago, seven forms of rarity were identified by Rabinowitz (1981), according to combinations of the geographic range, the habitat specificity and the local population size of a species (Table 5.1). According to Gaston (1997) – who proposed to leave out habitat breadth as a parameter – broad consensus favours a definition of rarity based on abundance and/or range size, with species of low abundance or small range size being regarded as rare. Here range size is treated as a measure of the area of the spatial distribution of a species, with geographical range size being that measure for the full global breadth of the occurrence of species. Virtually all definitions of rarity explicitly mention at least one parameter based on abundance or range size. Gaston (1996b) argued that species-range-size distributions have received remarkably little attention, in contrast to species-abundance distributions. Assuming the lognormal species-range-size distribution to be real, it can be viewed in terms of evolutionary and ecological determinants of species occurrences, although their relative significance remains unclear. The general, positive, interspecific relationship between local abundance and range size has been explained in terms of similarity of the ecological characteristics that permit species to become locally more abundant and

Table 5.1 A classification of rare species based on three characteristics: geographic range, habitat specificity and local population size. From Krebs (2001); after Rabinowitz (1981).

| | Geographic range | | | |
| | Large | | Small | |
Habitat specificity . . .	Wide	Narrow	Wide	Narrow
Population size				
Large, dominant somewhere	Locally abundant over a large range. In several habitats geographically.	Locally abundant over a large range. In a specific habitat geographically.	Locally abundant in several habitats, but restricted.	Locally abundant in a specific habitat, but restricted.
Small, non-dominant	Constantly sparse over a large range and in several habitats.	Constantly sparse in a specific habitat but over a large range.	Constantly sparse and geographically restricted in several habitats.	Constantly sparse and geographically restricted in a specific habitat.

geographically more widespread. This suggests that similarities in species-abundance distributions and species-range-size distributions could have some common mechanistic basis. Whatever the causes, the strong right skew to species-range-size distributions has some important consequences, particularly for inventories of faunas and floras and for conservation.

5.3 The niche-assembly perspective: interspecific interactions in communities

5.3.1 Types of interaction

When individuals of two species meet, the interaction between these two can be positive, which results in advantage, negative, which is disadvantageous, or indifferent to one or both interacting species. The advantage or disadvantage can be measured in terms of an increase or decrease in fitness or fitness components, as compared to a control with no interaction (see Table 5.2). In the present chapter, I will confine myself to interspecific interactions within communit-

ies in a strict sense (taxonomically defined; see the introduction to this chapter), including guilds, and leaving interspecific interactions in biotic communities (representing all the biota in an ecosystem) out of scope.

Competitive interactions are mutually detrimental, or one species is inhibited and the other is not affected. The latter is sometimes called amensalism, but we avoid using this term, because this can be caused by either competition or allelopathy. It seems useful to distinguish between the latter two, because competition is mostly two-sided and refers to space or resources, while allelopathy is one-sided and results from the release of toxic organic compounds from one of the species into the environment.

Parasitism is a relationship where one of the species, the consumer, benefits, whereas the other, the resource, suffers. This can be a relationship between trophic levels in a biotic community, like predation and herbivory, but it can also take place within a community, for example among plants.

Facilitation and mutualism are interactions where one, two or more species benefit. We avoid the term commensalism for a one-sided advantage, because it

Table 5.2 Simplified presentation of different interactions between two species (A and B), when they meet (On) or do not meet (Off).

	On		Off	
	Species A	*Species B*	*Species A*	*Species B*
Competition	−	−	0	0
Allelopathy	0	−	0	0
Parasitism	+	−	−	0
Facilitation	0	+	0	0
Mutualism	+	+	−	−

− Disadvantage; + advantage; 0 indifference.

is difficult to know if the facilitator is unaffected or suffers in a way, for example by becoming outcompeted.

5.3.2 Competition

Competition between organisms can be direct, for space or territory, or indirect, for resources. Direct or interference competition reveals a winner and a loser, or the two parties remain equal, although in this case there is a cost of energy anyway which implies a relative loss as compared to having no interference. In indirect or exploitation competition there are usually two losers. If a large plant competes for nitrogen with a small plant, each of them has a negative impact on the other. In view of the absolute amount of nitrogen taken, the exploitation competition can be asymmetric, but in terms of the relative loss of fitness of each of the plants the competition can be symmetric. In animal ecology a distinction is made between contest competition and scramble competition. Contest competition is a form of asymmetric competition and may equal interference competition, whereas scramble competition is a form of symmetric competition and is associated with exploitation competition. Mostly, loss of fitness between two parties is measured in a relative way, as compared to the potential fitness of an organism, and not in terms of the total amount of resource captured.

Begon *et al.* (1996) provided a useful working definition of exploitation competition: 'An interaction between individuals, brought about by a shared requirement for a resource in limited supply, and leading to a reduction in the survivorship, growth and/or reproduction of the competing individuals concerned'. The latter part could be summarized by stating that the process leads to a reduction in one or more fitness components.

A large amount of literature is available on interspecific resource competition as a phenomenon, starting in the early 1930s with Gause's famous experiments with *Paramecium* species feeding on either the same food resource (bacteria) or on different food resources (bacteria and yeasts). From these experiments Gause's principle of competitive exclusion was derived, implying that the number of species that can coexist cannot exceed the number of limiting resources. Competition theory was at that time mathematically related to population growth, in terms of the Lotka–Volterra logistic growth curves, which suggest r and K being important parameters of success. MacArthur and Wilson (1967) elaborated on this approach by formulating the concept of r- and K-selection, resulting in a colonizing strategy and a competitive or maintenance strategy respectively. In this concept, competitive ability is assumed to have been evolved at the expense of colonizing ability of species; there would be a trade-off between r- and K-characteristics. Fitness of individuals is considered a compromise between these contrasting choices, expressed in terms of the relative importance of different fitness components, as for example generative versus vegetative reproduction of plants.

Gause noticed that the order of competitive abilities between two species of *Paramecium*, both feeding on one and the same resource, did not depend on the densities at the start of the experiment, which implies that population growth rate could not easily explain which species would win the competition. Whereas Grime's (1979) explanation of competitive abilities of plants is largely based on differences in plant characteristics such as relative growth rate and plant morphology in the context of vegetation processes, Tilman (1982) was able to predict competitive hierarchies, tested with unicellular organisms as in Gause's experiments, from their patterns of resource depletion (related to population growth). In theory, the winner is the species with the lowest minimum resource requirement, expressed as the resource concentration at which growth and mortality are equal. This point of view seems to contrast with Grime's view, saying that the species with the highest resource capture (related to growth rate) will win the competition. However, the two views are not mutually exclusive: species A can gain dominance in the early phase of an experiment, at non-equilibrium transient conditions, because it has the highest growth rate and captures a higher amount of the resource, whereas species B can become the ultimate winner at equilibrium conditions, because it utilizes the resource more efficiently; that is, it has a lower minimum requirement. Eventually, species B may competitively exclude species A; that is, if it survived as a subordinate in the transient period, and if an equilibrium can be achieved anyway (Fig. 5.2). Unicellular organisms of different species in continuous-flow cultures do have an opportunity to arrive at one equilibrium state. In plant communities, succession can become inhibited before any potentially final equilibrium or climax stage can develop, exemplified by the dominance of *Molinia caerulea* in wet heathlands (Berendse & Elberse 1990), of *Brachypodium pinnatum* in calcareous grasslands (Bobbink *et al.* 1989) and of *Elymus athericus* in coastal salt marshes (Bakker 1989).

Studies of plant interactions focus on only a few dominant species in plant communities under more or less homogeneous environmental conditions. The increasing interest in biodiversity issues motivated researchers to wonder how subordinate species can remain coexisting in competitive plant communities. This resulted in a renewed interest in environmental heterogeneity and unpredictability. Huisman and Weissing (1999) offered a solution to the so-called plankton paradox, based on the dynamics of the competition itself, by showing that (i) resource competition models can generate oscillations and chaos

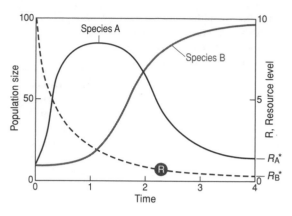

Fig. 5.2 Population responses of two species (A and B) competing for a single limiting resource (*R*), showing that species A can be dominant in the early phase of competition because it can utilize the resource rapidly, whereas species B can take over due to its lower minimum resource requirement (*R**). Population size, resource level and time are given in arbitrary units; these will vary depending on the organism. After Tilman (1988). Reproduced by permission of Princeton University Press.

when species compete for three or more resources and (ii) these oscillations and chaotic fluctuations in species abundances allow co-existence of many species on a handful of resources. So, whereas classical competition theory predicts competitive exclusion of species with similar requirements, recent ideas stress that species diversity may be explained by a multitude of processes acting at different scales, and that similarities in competitive abilities often may facilitate coexistence (Bengtsson *et al.* 1994). It is still to be seen to what extent these ideas fit in processes at the level of terrestrial animal and plant communities.

Competitive abilities in developing communities in ecological restoration are difficult to predict, because in such cases the initial conditions and the early-colonizing species may have a long-term rather than a transient impact. In recent competition theory it has been shown that competitive hierarchies or coexistence among a number of species may very well depend on the initial species composition and abundance, which may imply chaos and unpredictability (Huisman & Weissing 1999).

5.3.3 Allelopathy

Allelopathy can be considered a form of interference competition among plants. It is a unidirectional process, the phenomenon of plants of one species negatively affecting plants of other species, generally by releasing toxic organic compounds. Examples are tannins from *Pteridium aquilinum* and volatile oils from *Eucalyptus* species, substances that are supposed to have an anti-herbivore function as long as the plant organ is alive, and which in plant communities simultaneously may prevent changes in species composition and thus inhibit vegetation succession. We follow the definition given by Calow (1998), which is in concert with the afore-mentioned definition of competition: 'Allelopathy is a form of interference competition, brought about by chemical signals, i.e. compounds produced and released by one species of plants which reduce the germination, establishment, growth, survival or fecundity of other species'. In contrast with what have been named positive-feedback switches, which means that individuals of a species modify their abiotic environment in such a way as to promote their own persistence in the area, these organisms benefit from their own fitness promotion, thus also potentially reducing the competitive abilities of other component species. The classic work of Rice (1974) gives an overview of the state of affairs as resulting from early research on allelopathy among plants and microorganisms, in both terrestrial and aquatic systems. There are many thousands of such compounds, but only a limited number of them have been identified as toxins involved in allelopathy, and even if they potentially have detrimental effects they might not cause allelopathy in reality.

Kuiters (1990) reviewed the role of phenolic substances in forest soils. They include simple phenols, phenolic acids and polymeric phenols as condensed tannins or flavonoids. Once released in the soil environment, they influence plant growth directly by interfering with plant metabolic processes and by effects on root symbionts, and indirectly by affecting site quality through interference with decomposition, mineralization and humification. Effects of phenolics on plants include almost all metabolic processes, such as mitochondrial respiration, rate of photosynthesis, chlorophyll synthesis, water relations, protein synthesis and mineral nutrition. Interestingly, the author noticed that phenolic substances affect plant performance especially under acidic, nutrient-poor soil conditions. In calcareous soils, most phenolic compounds are rapidly metabolized by microbial activity and adsorption is high.

In the Swedish boreal forest, the ground-layer vegetation in late post-fire successions is frequently dominated by dense clones of the dwarf-shrub species *Empetrum hermaphroditum*, one of the most widespread plants in the European arctic and boreal biomes. The plants produce large quantities of phenolics, in particular batatasin-III (a phenolic in the compound class of stilbenes, C_6-C_2-C_6), which is held responsible for the strong negative effects exerted by *E. hermaphroditum* on tree-seedling establishment and growth, microbial activity and plant-litter decomposition rates, thus contributing to humus accumulation and reduced nitrogen availability (Nilsson *et al.* 1998). Further, this plant species is conspicuously avoided by herbivores. Nilsson (1994) tried to determine the relative impacts of chemical inhibition (allelopathy) and resource competition by *E. hermaphroditum* on seedling growth of *Pinus sylvestris*, by adding finely powdered, pro-analysis activated carbon as an adsorbent to the soil surface to remove the allelopathic effect, while exclusion tubes were used to subject seedlings to allelopathy in the absence of belowground competition by *E. hermaphroditum*. Both allelopathy and root competition had a strong, negative influence on seedling growth of the Scots pine.

The interaction between allelopathy and fire is an interesting issue. According to Zackrisson *et al.* (1996), with the prolonged absence of fire boreal forest can be dominated by *Picea abies* and *Empetrum nigrum*, whereas fire at intervals of 50–100 years is conducive to domination by *Pinus sylvestris* and the ground-layer species *Vaccinium vitis-idaea* and *Vaccinium myrtillus* in mesic and nutrient-poor sites. Indeed, charcoal was shown to absorb phytotoxic active phenolic metabolites from *E. hermaphroditum* solution, and wildfires were shown to play an important role in boreal forest. The authors proposed that charcoal particles can act as foci for both microbial activity (biodegradation) and chemical deactivation of phenolic compounds through adsorption.

The role of allelopathy in ecological restoration is largely unknown, but it cannot be ignored, because the composition of the organic matter may determine the chances of germination of early colonizers.

5.3.4 Parasitism

Calow (1998) defined parasitism as 'an intimate and usually obligate relationship between two organisms in which, essentially, one organism (the parasite) is exploiting resources from the other organism (the host) to the latter's disadvantage'. It is, therefore, an interaction between individuals of different species, brought about by a consumer–resource relationship, leading to a gain in fitness of the consumer species and a decrease in fitness of the resource species. Here I focus on parasitism at the level of plant communities; parasitism in biotic communities has been dealt with in Chapter 4.

Parasitism between plants is a widespread phenomenon (Kuijt 1969), with over 3000 species of parasitic plants occurring worldwide. Parasitism in the plant kingdom does occur among trees, shrubs, long-lived perennials and annuals, and all plant parasites are dicots in only a few lineages. Parasitic plant species that are thought of as weeds belong to the families Cuscutaceae, Loranthaceae, Viscaceae, Lauraceae, Orobanchaceae and Scrophulariaceae. *Santalum album* (sandalwood) is one of the well-known woody parasitic species. A parasite depends on a host for its fitness, whereas the host can live without the relationship; it only suffers from the parasite if it is present. It is not in the interest of the parasite to kill its host, but it can occur, for example in the case of *Cuscuta* species (Devil's guts). Among plants, holoparasites, such as species of *Orobanche* (broomrapes) and several orchid species exploit both root and shoot products from the host, while hemiparasites, such as species of *Rhinanthus* (hay rattles) and *Striga* (witchweed) exploit the root products only and are capable of photosynthesis themselves. Holoparasites do not contain chlorophyll and are heterotrophic; they not only depend on the host for water and minerals, but also require its organic compounds. All holo- and hemiparasites are connected with roots or shoots of host plants by means of a haustorium. Water, minerals and a wide variety of organic substances are transported through this organ. It is always a one-way

flow, but the degree of dependence varies; some species can be grown to flower and set seed without a host, whereas others do not even germinate without a host stimulus; root exudates of the host stimulate germination. The effect on the host is variable, too; it can be dramatic or hardly measurable and difficult to detect in other cases. Strict host specificity does not seem to exist.

The role of interplant parasitism in determining plant community structure is poorly understood. Pennings and Callaway (1996) investigated the impact of *Cuscuta salina*, a common and widespread obligate parasitic annual in saline locations on the west coast of North America. Their results suggested that the parasite is an important agent affecting the dynamics and diversity of vegetation. Because *C. salina* prefers to parasitize the marsh-dominant *Salicornia virginica*, it indirectly facilitates the rare species *Limonium californicum* and *Frankenia salina*, thus increasing plant diversity, and possibly initiating plant vegetation cycles. For hemiparasites such as species of *Rhinanthus*, *Odontites*, *Euphrasia* and *Melampyrum*, it is clear that the parasites depend on host vegetation to some extent, but in which way do they affect the vegetation? Is the vegetation open because of the presence of the parasite, or is the parasite present because the vegetation is rather low and open? The data on the role of vegetation structure show that the latter certainly has some effect (ter Borg 1985), which may be negative, neutral or positive (Pennings & Callaway 2002).

5.3.5 Facilitation

Facilitation can be defined as an interaction between individuals of different species, where one species changes the environment in such a way that it is beneficial to the other, either in space or in time. It does not seem useful to distinguish between direct and indirect facilitation, as in the case of competition, because the effects are always indirect, via an impact on the environment. Plants transform the physical or chemical soil conditions, for example by aerating a wet, anoxic soil, increasing the moisture content of a dry soil or by soil formation due to nitrogen-fixing plants, or act as a shelter against harsh above-ground conditions. Large herbivores can pave the way for smaller herbivores. Shelters, nests and burrows constructed by many species of terrestrial vertebrates are often co-utilized by heterospecifics.

Facilitation in plant communities can result in a change in plant species composition or even in vegetation succession in which case the beneficiary species outcompetes its facilitator (Connell & Slatyer 1977). Facilitation in primary successions is often mediated through plants with nitrogen-fixing microorganisms in their root systems, be it *Rhizobium* or *Frankia* species. This phenomenon is described in almost every textbook dealing with plant succession. But often it is part of a more complex set of interactions. For example, Pugnaire *et al.* (1996) have shown in south-eastern Spain that the leguminous shrub *Retama sphaerocarpa* strongly improved its own environment and facilitated the growth of *Marrubium vulgare* and other understorey species, and at the same time obtained benefits from sheltering herbs underneath. The interaction between these two species was indirect, associated with differences in soil properties and with improved nutrient availability under shrubs compared with plants grown on their own. Facilitation may, however, also result in the coexistence of different species.

Positive spatial associations between seedlings of one species and sheltering adults of another species are common, and have been widely referred to as the 'nurse plant syndrome'; see the review by Callaway and Walker (1997). According to the latter authors, the importance of facilitation of seedlings by adults of other species has been supported by studies in deserts, savannas and woodlands, tropical forests, Mediterranean-climate shrubland, salt marshes and grasslands. In many of these cases, seedlings of beneficiary species are found spatially associated with nurse plants, whereas adults are not, which suggests that the balance of competition and facilitation shifts among the various life stages of the beneficiary and the benefactor. The mechanisms that may act have in some cases been discovered by field manipulations, showing that there are effects on nutrients, light, temperature, humidity and other abiotic factors.

Facilitation among animals was illustrated in early work on mammalian herbivores in the semi-arid savanna of the Serengeti (Vesey-Fitzgerald 1960, Bell 1971, McNaughton 1976). Facilitative processes have since been demonstrated in marine, freshwater and terrestrial systems, and across a large range of body sizes. The phenomenon of facilitation between small vertebrate herbivores has been studied in north-west European temperate salt marshes, between brown hare and brent geese (van der Wal *et al.* 2000b), and between barnacle geese and brent geese (Stahl 2001). Brown hare (*Lepus europaeus*) facilitates grazing by dark-bellied brent geese (*Branta bernicla bernicla*) by retarding vegetation succession. Winter grazing by hares prevents *Atriplex portulacoides* from becoming dominant in younger parts of the salt marsh, since *Atriplex* can only poorly withstand removal of above-ground tissue. Goose grazing pressure decreases when bushes of *Atriplex* were planted, while removal of *Atriplex* swards leads to an increase in goose utilization. In the absence of hares, a large part of the core feeding area of geese would be unsuitable for grazing by the birds. Not only hares, but also other herbivores such as rabbits, domestic cattle, sheep and horses can prevent or delay *Atriplex* from becoming dominant in salt marshes along the north-west European coast. The authors claimed facilitation to be of utmost importance to brent geese, setting the carrying capacity along the north-western European coastline for this goose species. The mutual relations between barnacle geese (*Branta leucopsis*) and brent geese (*B. bernicla bernicla*) are more complex (Stahl 2001). It appeared that brent geese preferred to forage on vegetation previously grazed by barnacle geese; they probably reacted to the enhanced quality of the regrowth, thereby discounting the higher biomass in the ungrazed alternative, in concert with the concept of grazing lawns and grazing optimization as pursued by McNaughton (1984), but now shown to happen in a guild of two species of bird herbivores.

Overall, facilitation has been increasingly recognized as a significant interaction in the structuring of communities, and it can play a role next to competition and other interactions, thus revealing a large number of mostly weak and complex interaction strengths. The role of complex species interactions in ecological restoration is still largely ignored. The inclusion of facilitation in ecological theory 'will change many basic predictions and will challenge some of our cherished paradigms' (Bruno *et al.* 2003; see Fig. 5.3).

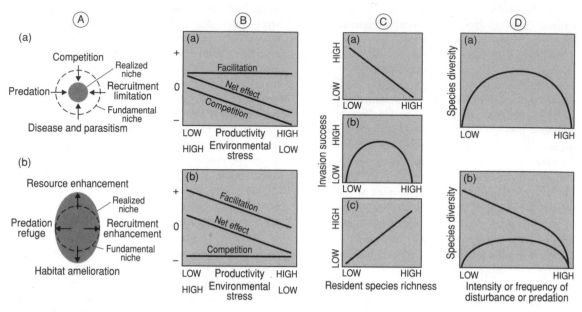

Fig. 5.3 Four fundamental models of ecology, with and without facilitation, illustrating paradigm shifts if facilitation is included. (A) The realized niche can be larger than the fundamental niche due to facilitation; (Aa) without facilitation, (Ab) with facilitation. (B) Facilitation may affect the competitive abilities of species along an environmental gradient; (Ba) facilitation weak, constant, (Bb) facilitation strong, variable. (C) Facilitation may have an impact on the success of invaders; (Ca) without facilitation, (Cb or Cc) with facilitation. (D) Facilitation may change the relationship between species richness and ecosystem productivity; (Da) without facilitation, (Db) with facilitation (lower line, primary space holders; upper line, secondary space holders). From Bruno *et al.* (2003). Reproduced by permission of Elsevier.

This may imply, for instance, that the realized niche of a species can be larger than its fundamental niche, affecting the net effect of competition. Positive interactions could also affect the relationship between diversity and invasibility.

5.4 Complex species interactions affect community structure

The intensity of interspecific interactions, be it competition, facilitation, parasitism or any other relationship, may change during community development, and several interactive mechanisms may be at work simultaneously, structuring communities in space and in time (e.g. van Andel *et al.* 1993). Callaway and Walker (1997) provided many examples to illustrate that species interactions may involve a complex balance of competition and facilitation. For example, *Quercus douglasii* trees have the potential to facilitate understorey herbs by adding considerable amounts of nutrients to the soil beneath their canopies. However, experimental tree-root exclusion increased understorey biomass under trees with high shallow-root biomass, but had no effect on understorey biomass beneath trees with low shallow-root biomass. Thus, the overall effect of an overstorey tree on its herbaceous understorey is determined by the balance of both facilitation and competition. In wetlands, shifts in facilitation and competition among aerenchymous plants were shown to occur in anaerobic substrates when temperatures change. *Myosotis laxa*, a small herb common in wetlands of the northern Rocky Mountains, benefits from soil oxygenation when grown with *Typha latifolia* at low soil temperatures in greenhouse experiments. At higher soil temperatures, the significant effects of *Typha* on soil oxygen disappear, presumably because of increased microbial and root respiration and the interaction between *Myosotis* and *Typha* becomes competitive. In the field, the overall effect of *Typha* on *Myosotis* was positive, as *Myosotis* plants growing next to transplanted *Typha* were larger and produced more fruit than those isolated from *Typha* plants.

This phenomenon of so-called third parties affecting the competitive abilities of two other species was called mediation of competition (Allen & Allen 1990). Similar to the suggestion of Price *et al.* (1986) in their review on mediation by parasites, parasites act as a third party affecting the interference competition or resource competition between two other species (see Pennings & Callaway 1996). In the 1990s this was termed 'apparent competition', but I prefer the more clear-cut notion of mediation, because it is applicable to both negative and positive interactions. Mediation by parasites is very common in nature and must be regarded as one of the major types of interaction in ecological systems, comparable in importance to direct competition, predation, parasitism or mutualism (Price *et al.* 1986). Experimental and theoretical studies have provided new insight into the mechanisms and conditions that could influence coexistence or exclusion (reviewed by Hudson & Greenman 1998). Parasite-mediated competition is particularly significant, given the rise in emerging diseases and the opportunity that pathogens have to reduce host abundance. Parasite-mediated competition can act when an invading species introduces a parasite to vulnerable, resident species. This might have been the scenario when grey squirrels (*Sciurus carolinensis*) were introduced to Britain, bringing with them a parapox virus that reduced the competitive ability of the native reds (*Sciurus vulgaris*). The reverse can also take place. If a parasite weakens the competitive ability of the resident species, invasion can occur. Interactions between the pathogens of rabbits (*Oryctolagus cuniculus*) and hares (*Lepus europaeus*) provide some field evidence that pathogens can influence a competitive outcome. Studies in the Netherlands found a stomach worm (*Graphidium strigosum*) to be more pathogenic in hares than rabbits. Jefferies (1999) reviewed interactions between herbivores and micro-organisms affecting nutrient fluxes in the ecosystem, and discussed special conditions where increased dominance from herbivores overrides the regulatory controls imposed by other organisms, which leads to trophic cascades and discontinuous vegetation states.

5.5 Concluding remarks

For management purposes, an integrated approach is often required. Therefore, to link up with the previous chapter on ecosystems and biotic communities, and the next one, which deals with populations, one must pay attention to the plea by WallisDeVries *et al.* (2002) for an integrated approach to conservation of diversity in biotic communities at the ecosystem level. They exemplified their views by referring to calcareous grasslands. In these ecosystems, the biological diversity is high and includes a variety of rare species from different taxonomic groups. Among plant communities, chalk grasslands rank as one of the richest in plant species, both at a small spatial scale and on a large scale. They include for instance highly valued orchids. A potentially high species richness is found, as well, for butterflies and various other invertebrates. Calcareous grasslands are predominantly semi-natural communities and thus require some form of management by grazing or mowing, which raises conflicts about the best management regime for particular groups of species. Indeed, managing for plant diversity requires low nutrient levels, which can be arrived by grazing, mowing or even mulching. But the importance of heterogeneity in vegetation structure is emphasized for the invertebrates, some groups of which require a less-intensive management by grazing at low stocking rates or by rotational mowing. Botanically centred management may thus lead to an impoverished vertebrate fauna. This can only be avoided by an integrated approach that considers various taxonomic groups of the biotic community.

What these considerations, with respect to nature conservation and management, imply in the context of restoration ecology is hard to say, apart from the need – to state it in terms of a metaphor – to pay attention to the forest if one aims at studying or even preserving a tree population (see Odum 1983).

6

Populations: intraspecific interactions

Jelte van Andel

6.1 Introduction: island biogeography and metapopulations

Population dynamics deals with numbers of individuals and genotype frequencies in different stages of the life cycle of the local representatives of a species. An introduction to population processes in a textbook on restoration ecology, recognizing the often fragmented structure of habitat or biotope, cannot but explicitly place them in the context of processes at the level of landscapes. This implies a focus on the dynamics of spatially structured (meta)populations in a network of habitat patches. The island biogeography (MacArthur & Wilson 1967), though developed to explain species richness and interactions between species at the community level, also includes such single-species population aspects as migration, colonization and extinction. The metapopulation model of Levins (1969) can be characterized as a single-species version of the island model of MacArthur and Wilson (1967). According to Hanski and Simberloff (1997), the main difference between the metapopulation approach and island biogeography is that in the latter case there is a permanent source population on the 'mainland', whereas habitat patches of metapopulations can mutually be temporary sources and sinks. As soon as patches of varying size and quality are included in a metapopulation study, the difference with the island biogeographical approach becomes even smaller. Migration from one population to another does reduce the number of individuals in the donor population, but this effect becomes negligible if the donor population is relatively large, as is often the case with mainland populations.

A metapopulation in the strict sense is an assemblage of local populations that are connected by mutually dispersing individuals (see Levins 1969, Hanski & Gilpin 1991). The concept was elaborated with mobile animals in mind. For sessile plants, seeds and pollen are the major dispersal units (Kwak *et al.* 1998). Hanski (1997) outlined four conditions needed for a regional population to persist as a metapopulation:

1 suitable habitat occurs in discrete patches which may be occupied by local breeding populations;
2 even the largest local populations have a substantial risk of extinction;
3 habitat patches must not be too isolated to prevent recolonization following local extinctions;
4 local populations do not have completely synchronous dynamics.

Later on, the concept of source/sink populations was incorporated (Pulliam 1988, Eriksson 1996), in which ensembles of individuals in sink habitats are maintained by continuous immigration from source habitats, but the latter is actually an important aspect of MacArthur and Wilson's (1967) theory of island biogeography. Eriksson (1996) discussed concepts of regional-scale dynamics in plants and added a third type, termed 'remnant population dynamics', in local populations that survive long enough to bridge periods of unfavourable successional development, including survival in a long-lived seed bank (Figs 6.1

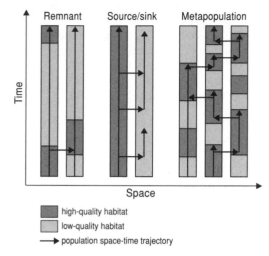

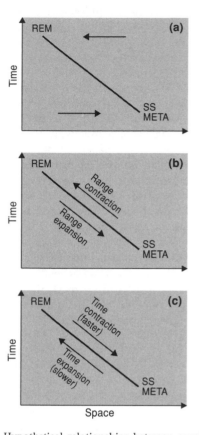

Fig. 6.1 A space-time diagram illustrating the principal difference between three types of regional dynamics in plants: remnant, source/sink and metapopulation dynamics. After Eriksson (1996). Reproduced by permission of Oikos.

and 6.2). The literature reveals evidence, indeed, for the existence of all three kinds of regional-scale dynamics in plants. For example, Van der Meijden *et al.* (1992) distinguished between three types of biennial plant species that form metapopulations: transient species and fugitive species, which are both types of classical, typical metapopulation, and persistent species, with metapopulations within sites. Long-lived seed banks of the species that do not occur in the established vegetation represent remnant populations. Freckleton and Watkinson (2002) proposed a rearrangement of the terminology when dealing with regional plant population dynamics, which I tend to adopt here, namely:

- **metapopulations** *sensu stricto*: classic metapopulations of mainland–island systems with predominantly source/sink populations;
- **regional assemblages**: local processes dominate and the constituent populations may be basically unconnected;
- **spatially extended populations**: a spatially extended form of the local populations, with continuous rather than patchy habitat.

Fig. 6.2 Hypothetical relationships between capabilities of plants to disperse in space and persist (disperse in time), in relation to three types of expected regional dynamics: remnant (REM), source/sink (SS) and metapopulation (META) dynamics. A general trade-off between temporal vs spatial dispersal is assumed. (a) Regional dynamics related to plant features; (b) range contraction and expansion may alter regional dynamics; (c) rate of dynamics may alter regional dynamics. After Eriksson (1996). Reproduced by permission of Oikos.

In his classic study of the migration of terrestrial arthropods, den Boer (1968) introduced the concept of 'spreading of risk', which implies that an investment in dispersal can reduce the risk of extinction of a population as a whole, even if local subpopulations become extinct, provided that suitable sites are within travelling distance. Den Boer (1990) placed 25 carabid beetle species into one of two groups: L-species such as *Calathus melanocephalus*, occurring in stable habitats and having low dispersal capabilities, and T-species such as *Amara plebeja*, occurring in unstable habitats and having good dispersal potentials. The turnover in populations of L-species was 1–3% per year, whereas for T-species this amounted to 3–10% per year. Human-induced changes in the environment resulted in a decline of L-species until extinction, whereas T-species were not affected. Carabids, adapted to extensive stable habitats, usually live there in heterogeneously structured metapopulations, and spreading the risk of extinction over interconnected and differently fluctuating local groups is suggested to result in almost unrestricted survival of the species in the region. In later research, much effort has been put into studies of birds and butterflies. The European nuthatch (*Sitta europea*) is found to show all the characteristics of a Levins-type metapopulation: the distribution is dynamic in space and time, extinction rate appeared to depend on patch size and habitat quality, and colonization rate depended on density of surrounding patches occupied by nuthatches (Verboom *et al.* 1991). Another example is the cranberry fritillary (*Boloria aquilonaris*), a butterfly of peat bogs in western Europe. Baguette and Schtickzelle (2003) have shown that the 14 patches of the metapopulation on the Plateau des Tailles upland (Belgium) are just below the minimum available suitable habitat for the long-term persistence of a viable metapopulation, and that the population network is unstable. There is a huge number of examples showing that the metapopulation approach may be of central importance when dealing with restoration ecology in modern-time landscapes. Although a range of forms of local spatial dynamics exist, these are qualitatively different from the forms of population structure at the regional level. One important characteristic of all metapopulation systems is that the metapopulation itself can be much more stable than the component populations, because migration can buffer individual populations from negative stochastic and genetic events in their local environments.

In the present chapter I will first deal with some background information on the dynamics and genetics of metapopulations. Thereafter I will focus on restoration problems, mainly related to re-introductions, such as the identification of suitable habitat, the determination of connectivity between habitat patches as related to dispersal problems, and the identification of a minimal (meta)population size. I will also pay attention to cascade effects, recognizing that populations do not occur as single monocultures; the individuals are also components of local communities.

6.2 Metapopulation dynamics and genetics

6.2.1 The rescue effect

Though colonization and extinction of island populations, or of the local populations of a metapopulation, are natural phenomena, literature aimed at nature conservation may qualify the risk of extinction of a local population as a negative aspect and start referring to a so-called rescue effect. Defined as 'immigrants from other populations increasing the local population size and therefore limiting the extinction risk' (Brown & Kodric-Brown 1977), the term goes beyond the more neutral term buffer and actually implies a conservation or restoration approach. The rescue effect was originally envisioned to occur in island populations, rescued through input from the larger mainland, but a similar effect may also occur in metapopulations without a mainland. In this case, existing populations provide a kind of mutual aid to each other: a reduced probability of population extinction due to immigration.

Animals

Even a very limited amount of migration can have a profound effect upon the recipient population (Stacey *et al.* 1997). We refer briefly to a few examples these authors provided. All the studies on butterflies suggest a pattern of evidence for important rescue effects in butterfly metapopulations. Many species are clearly predisposed towards a metapopulation structure because they occupy distinct habitat patches as a result of specialized vegetation requirements during some stage of their life cycle. Also, for small mammals such as pikas (*Ochotona princeps*) migration among habitat fragments appeared to be important in maintaining genetic variation within the metapopulation and preventing the loss of heterozygosity, a general measure of genetic diversity, in component populations. Most movements of water voles (*Arvicola terrestris*) in Scotland were confined within a particular social group, but several female voles dispersed long distances and joined new social groups. Populations or colonies of blacktail prairie dog (*Cynomis ludovicianus*) consist of several family groups called coteries. These populations remain spatially distinct from each other because of a rigid social structure. The regular influx of male immigrants should reduce inbreeding within local populations and elevate levels of heterozygosity, potentially rescuing populations from inbreeding depression and possible extinction. White-footed mice (*Peromyscus leucopus*) have persisted in a remnant network of woodlot patches in North America, connected by migration routes. Populations linked by these high levels of migration had higher growth rates than populations linked by lower levels of migration. Since winter populations can be as small as two individuals, rapid population growth in the spring could be critical for the persistence of local populations. There is clear evidence for a similar rescue effect in several other small mammal species. For amphibians, such as red-spotted newts (*Notophtalmus viridescens*) in the eastern United States, there is evidence for a relatively high migration rate between ponds, and no local extinctions were observed. Natterjack toad (*Bufo calamita*) populations in Germany showed low genetic distances between local populations. Most adult female toads and some juveniles migrate between breeding ponds within seasons, while most males leave their first breeding sites. Most sites are occupied continuously, and local populations persist due to immigration and the rescue effect. Along the Baltic coast of Sweden, pool frogs (*Rana lessonae*) occur in natural metapopulations, reproducing only in distinct water bodies. Over a 6-year period, populations isolated by greater than 1 km became extinct, whereas less-isolated populations tended to persist.

Plants

The relationship between genetic diversity and survival in plant populations has been reviewed by Booy *et al.* (2000), who concluded that there is no general relationship between genetic diversity and various fitness components. If a lower level of heterozygosity represents an increased level of inbreeding, a reduction of fitness can be expected. Indeed, Richards

(2000) has shown that the degree of inbreeding depression in populations of *Silene alba* varied with the degree of isolation from other established populations. Crossing experiments confirmed that isolated populations are inbred and had probably been founded by related individuals. If seed movement is rare and pollen movement is common, then the potential for inbreeding to occur soon after establishment is offset by genetic rescue through pollen-mediated gene flow from another population. Crosses among isolated sites restored germination rates to levels comparable to outcrossed lines.

The question now remains as to the relative importance of genetic variation compared to other aspects of habitat fragmentation, such as habitat quality and demographic factors, in determining the fate of local populations. Vergeer *et al.* (2003) applied path analysis to separate and quantify the relatively direct and indirect contributions of genetic variation and soil conditions on population size and plant performance in *Succisa pratensis* (Fig. 6.3). The number of seeds per flowerhead, germination percentage, germination rate, seed weight and seedling survival were all positively correlated with population size. Reduced plant performance was better explained by genetic erosion and habitat deterioration (eutrophication) than by population size in itself.

6.2.2 Effective (meta)population size

Colonization and extinction are determined by the effective population size rather than by the overall population size. The effective population is composed of only those individuals that contribute genes to the next generation. The effective population size is classically defined as the size of an idealized population that results in a given variance in allele frequency or amount of inbreeding (Hedrick & Gilpin 1997). The amount of genetic variation in a population is generally determined using heterozygosity as a measure. In metapopulations, the average levels of heterozygosity of the component populations and the spatial distribution of heterozygosity are important parameters. The loss of genetic variation in a metapopulation can be dramatically lower than that expected from a population the size of an average census number in the system (Fig. 6.4). Both the steady-state levels of

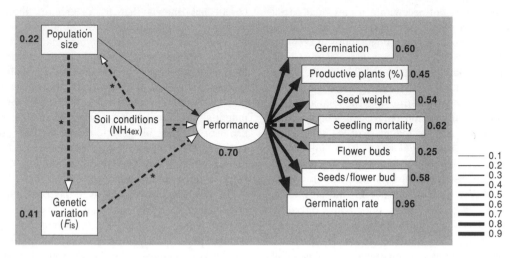

Fig. 6.3 Path model depicting the hypothesized causal relationships between population size, genetic variation, soil conditions and plant performance in 17 populations of *Succisa pratensis* in the Netherlands. Arrow width is proportional to the standardized path coefficients, and dashed lines indicate negative paths. Asterisks indicate values significantly different from 0 ($P < 0.05$). Numbers in bold are estimates of the proportion of the total variance explained (squared multiple correlations) for each independent variable (i.e. all except soil conditions). NH_{4ex}, exchangeable ammonium; F_{is}, inbreeding coefficient. After Vergeer *et al.* (2003). Reproduced by permission of Blackwell Publishing.

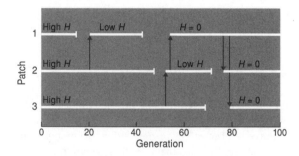

Fig. 6.4 The level of heterozygosity (*H*) over time in a simulation of a population existing in three patches. The short vertical bars on the right-hand ends of horizontal lines indicate extinctions in a patch and the vertical arrows indicate recolonization. After Hedrick & Gilpin (1997). Reproduced by permission of Elsevier.

heterozygosity and changes in heterozygosity are governed by the effective size of the population. Metapopulation structure might affect variance-effective size by reducing total population size, and by increasing or reducing the variance in the reproductive success of the individuals (Hanski 1999). From their modelling

approach, Hedrick and Gilpin (1997) concluded that the rate of extinction of patches and the number and type of founders recolonizing empty patches are the main factors determining the minimum effective metapopulation size. Overall, the effective metapopulation size increased if the rate of extinction (or the rate of turnover, extinction and recolonization) was reduced, while the number of local populations, the numbers of founders and the rate of gene flow increased. In cases where many human impacts are fairly recent, fragmented populations might be in the initial phase of heterozygosity loss, which may proceed much faster than predicted from either census numbers or traditional estimates of effective population size.

These results encourage genetic approaches to conservation and support the importance of preserving genetic variability as a way of increasing the variability of wild populations. However, there does not yet seem to be a general rule as to what the minimum size or the minimum genetic variation of populations should be, in order to survive and safeguard the potential for evolutionary adaptation (see section 6.5).

6.3 Suitable habitat patches

Effects of habitat fragmentation on birds and mammals in landscapes with different proportions of suitable habitat have been reviewed by Andrén (1994). Many studies he referred to have found that communities or populations of single species in small patches were not random samples from large patches. A more fine-tuned analysis showed that studies in which an effect of area and/or isolation on species number or density was found were from landscapes with highly fragmented habitat, whereas those yielding results that were not different from those predicted by the random-sample hypothesis were mainly from landscapes with a larger proportion of suitable habitat. In this respect, there was no difference in principle between mammals and birds, nor between resident birds and migratory birds. The review indicated that there might be a threshold in the proportion of suitable habitat in the landscape, above which habitat fragmentation is pure habitat loss. Indeed, a review by Fahrig (2003) has shown clearly that it is most important to distinguish between effects of habitat loss and effects of habitat fragmentation, because habitat loss has large negative effects on biodiversity, whereas the effects of habitat fragmentation are rather weak and may be positive or negative.

In the metapopulation approach both immigration and local extinction are normal phenomena. This implies that habitat patches that are suitable may be temporarily empty. For restoration purposes there is therefore a need to identify potentially suitable habitat patches, which may or may not be occupied. How can suitable habitat be identified? Habitat models originated in the USA in the 1970s, and were developed further in Australia in the 1980s and in Europe in the 1990s (Kleyer *et al.* 2000). The focus has been on statistical models and on expert-based habitat models. A major contribution to nature conservation planning resulted from relating habitat models to geographic information systems (GISs). Corsi *et al.* (2000) combined deductive and inductive approaches to identify species distributions (see Fig. 6.5). The deductive approach uses known species' ecological requirements to extrapolate suitable areas from the environmental variable layers in the GIS database. The inductive approach is used to derive the ecological requirements of the species from locations in which the species occurs. These two approaches can contribute to habitat suitability mapping. Concerning statistical models, recent publications favour logistic regression as compared to discriminant analysis. Note that statistical habitat models imply the assumption that the observed occurrence is in equilibrium with the environmental factors and their spatial arrangement. Clearly setting the preconditions necessary to transfer habitat models in time and space, which then have to be validated, is seen as a prerequisite for application in

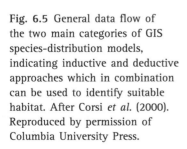

Fig. 6.5 General data flow of the two main categories of GIS species-distribution models, indicating inductive and deductive approaches which in combination can be used to identify suitable habitat. After Corsi *et al.* (2000). Reproduced by permission of Columbia University Press.

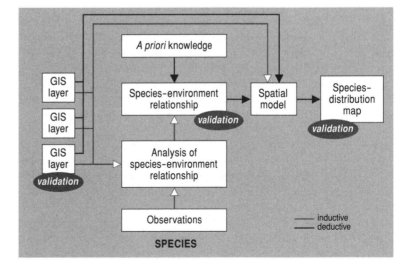

planning for nature conservation (Schröder & Richter 2000).

The blue-winged grasshopper (*Oedipoda caerulescens*) has been used as an example to demonstrate how highly resolved data on the habitat factors can be provided for an entire landscape (Kuhn & Kleyer 2000). The first step was to develop a regionally valid statistical habitat model, in terms of probability of occurrence, which quantified the habitat quality of all single areas in a landscape and predicted potential habitats. This habitat model, derived from presence/absence data and habitat factors (temperature, vegetation cover and structure, all mapped on a landscape scale), produced the potential habitat map of the study area. The relationship between the presence of the species and the spatial configuration of the habitats was then presented in a second model considering those habitats exclusively. In this second model, presence/absence data were recorded from the potential habitats and their isolation and size were computed as independent variables. The authors used iterative statistical habitat models to predict the incidence of *O. caerulescens*.

Strykstra *et al.* (2002) developed a framework for the relative importance of dispersal, seed-bank longevity and life-span spectra as related to an estimated reliability of safe-site dynamics in a number of plant communities. Their analysis brought them to conclude that (i) a higher percentage of species with a specialization in persistent seed-bank formation corresponds with a lower reliability in time, (ii) a higher percentage of species with a specialization in long-range dispersal corresponds with a lower reliability in space and (iii) a higher percentage of species with a long life span corresponds with a lower level of site disturbance. This kind of analysis may result in estimates of the potential establishment of particular types of plant communities, rather than plant species. For individual plant species it is often quite difficult or even impossible to define suitable patches *a priori* (Freckleton & Watkinson 2002). Even if seed banks (dispersal in time) are present and would be taken into account, in addition to dispersal in space, it is not easy to determine whether a site that is occupied by living seeds is suitable for recolonization by seedlings and survival of adults. The site may dramatically change in the meantime, and established plants are not capable of moving.

6.4 Connectivity between habitat patches

A central issue in the analysis of metapopulations is the frequency of migration, or demographic connectivity, among component populations (see also Chapter 3 in this volume). Habitat patches are parts of a landscape mosaic, and the presence of a given species in a patch may be a function not only of patch size and isolation, but also of the kind of neighbouring habitat (Andrén 1994) and of the species composition in the patches (Kwak *et al.* 1998). I refer to three definitions of what is called landscape connectivity as a feature of a landscape:

1 the degree to which the landscape facilitates or impedes movement among resource patches (Taylor *et al.* 1993);
2 the functional linkage among habitat patches, either because habitat is physically adjacent or because the dispersal abilities of the organisms effectively connect patches across the landscape (With *et al.* 1997);
3 a property of locations to maintain spatial or functional relationships with other locations in terms of flows of entities such as materials, energy, information, people and animals (van Langevelde *et al.* 1998).

Based on simulated dispersal across heterogeneous landscapes, Tischendorf and Fahrig (2000) compared the responses of three connectivity measures – dispersal success, search time and cell immigration – to habitat fragmentation. From their analysis they concluded: (i) two common measures of landscape connectivity, dispersal success and search time, both averaged over all patches in the landscape, indicate higher connectivity in more-fragmented landscapes; (ii) landscape connectivity measured as immigration into all habitat cells in the landscape predicts higher connectivity in less-fragmented landscapes and (iii) the three connectivity measurements respond differently to landscape structure and dispersal characteristics. Consistent measurement of landscape connectivity is crucial to ease comparisons across different studies.

Dispersal traits of species are crucial to cope with habitat fragmentation. Krebs (2001) adopted the three modes of dispersal as proposed by Pielou (1979):

1 **diffusion**, the gradual movement of a population across hospitable terrain for a period of several generations;

2 **jump dispersal**, the movement of individual organisms across large distances followed by the successful establishment of a population in the new area;

3 **secular dispersal**, diffusion in evolutionary time, including natural selection and evolutionary change within or beyond the geographic area.

The majority of terrestrial animal species are mobile, whereas the dispersability of terrestrial plant species is largely limited to the reproductive phase (fruits, seeds and pollen). Dispersal mechanisms of plant species and associated hypotheses about their ecological functions have been reviewed by Howe and Smallwood (1982). Aquatic and amphibious vascular plants have developed various specialized dispersal mechanisms, where dispersal by water and by animals and humans play an important part (Cook 1987). Vegetatively developed diaspores contribute as much as seeds to the dispersal patterns of species. The diaspores of some species sink at once whereas those of other species float. Water dispersal interacting with floating time may play a role in the small-scale structuring of riparian plant communities (Nilsson *et al.* 1991). Some diaspores are adapted for endozoic transport, in particular by water birds and aquatic mammals. The seeds of most amphibious plants germinate under water. The seedlings can start floating as soon as they assimilate and develop aerenchyma or remain submerged by anchors or weights before they are rooted and established. In general, the distribution of aquatic macrophytic species of river floodplains does not seem to be limited by dispersal (Skoglund 1990, Cellot *et al.* 1998).

The problem to be discussed here, however, is not so much what dispersal capabilities species have developed, but whether the dispersal traits, which result from evolutionary adaptation in the past, are still suitable to cope with the current rate of changes in the landscape, still apart from changes in climatic conditions. During the postglacial period, before human settlement, important plant-dispersal vectors in the natural landscape were wind, water and animals. Wind could have been an important vector in the open tundra landscape, but not in wooded landscape. Water was probably a more important dispersal

vector during that time, at least for trees and shrubs such as *Salix* and *Alnus* species (Delcourt & Delcourt 1991). Animals are assumed to have been the most important dispersal vectors of plants in the postglacial time, especially over long distances. Most trees and shrubs may have been spread by birds. Agricultural practices also included important processes for the dispersal of plants to reach suitable sites, for example associated with sowing seeds, fertilizing with manure, harvest methods, livestock movements and artificial flooding (see Bakker *et al.* 1996a, Poschlod & Bonn 1998). With the increasing fragmentation of natural and semi-natural habitats, especially in central and western Europe, there has been a dramatic loss and change of dispersal processes and vectors in our man-made landscape since the last century. If dispersal processes cannot be restored it is clear that any efforts in restoration management will be at odds with the goal to provide new habitats for locally or even regionally extinct species, except if they could survive in a diaspore bank. If this is not the case, dispersal processes have to be stimulated or replaced by others which can include the artificial introduction of species (see Chapter 7).

6.5 Minimum viable (meta)population size

Metapopulations represent an interesting meeting point between population biology and landscape ecology (see Chapter 3). The approach of defining a minimally required (meta)population size is in concert with the science-based proposal of Pickett *et al.* (1989) to identify minimal structures at different hierarchical levels as a challenge to make explicit what the key components of a system are. But the issue of identifying a minimum structure has also been raised from an economic interest: which is the part that can be harvested from a population without causing damage below the level of sustainability? A quantification of the minimum structure of a (meta)population is also relevant if we want to define the lower limit to be preserved or to restore a degenerated population: what is the minimum population size to be conserved or to be restored in order to guarantee sustainability? The minimum viable metapopulation size was defined as the minimum number of interacting local populations in a balance between local extinctions and recolonizations, and thus necessary for long-term persistence (Hanski *et al.* 1996). Apart from minimum viable metapopulation size, one has also to consider the minimum amount of suitable habitat necessary for metapopulation persistence (Andrén 1994); not all suitable habitat may be occupied simultaneously. As there is not so much useful information about minimum viable metapopulation sizes, we focus on the concept of the minimum viable population size, which is defined as the minimum number of individuals in a local population that has a good chance of surviving for some relatively long period of time; for instance a 95% chance of surviving for at least 100 years (Soulé 1980, 1987, Lande 1988).

While asking the question, how small is too small?, we should not only take into account the minimum number of individuals of a species under concern, but also the minimum size of the area, biotope or wildlife reserve. Gurd *et al.* (2001) derived estimates of the minimum area requirement from species-area curves; for mammals in parks and nature reserves in the Alleghenian-Illinoian mammal province (Canada) the minimum area would be 5037 km^2 (ranging from 2700 to 13,296 km^2), whereas the majority of reserves are smaller. Similarly, Fritts and Carbyn (1995) estimated the area required for a minimum viable population of 100 grey wolves in North America as 3000 km^2; under otherwise favourable circumstances 500–1000 km^2 could be adequate.

From a genetic point of view, the minimally required number of genotypes, which is in general the minimum amount of genetic variation, can be more important than the number of individuals. However, Lande (1988) argued that demography may usually be of more immediate importance than population genetics in determining the minimum viable sizes of wild populations. In many species, individuals in populations declining in low numbers experience diminished viability and reproduction by non-genetic causes, and there may be a threshold density or number of individuals from below which the population cannot recover. Examples are known of viable populations, despite a very low level of genetic variation. The northern elephant seal (*Mirounga angustirostris*), in a region off the coast of California and Mexico, experienced an extreme population bottleneck of less than 20–30 seals during a period of 20 years due to

exploitation in the 19th century. Legislative protection in the United States and Mexico resulted in a dramatic recovery in number, although their genomic diversity was greatly reduced (Bonnell & Selander 1974, Hoelzel *et al.* 1993). Having mentioned this, it would be wrong to conclude that genetic variability is of minor importance in general. A large amount of literature is available on the effects of genetic erosion (see Vergeer *et al.* 2003), but these results have not been considered in view of a minimum viable (meta)population. A few examples are available of the rescue of depauperate populations by the (re)introduction of new genes, which may point at genetic bottlenecks. Madsen *et al.* (1999) published on the successful restoration of an inbred adder population. Their population data from 1983 to 1995 indicate that the Smygehuk adders (*Vipera berus*) were on the brink of extinction, with falling numbers and negligible recruitment. Introducing new genes from a different population enabled the adders to make a dramatic recovery. Bijlsma *et al.* (2000) have shown experimentally in *Drosophila melanogaster* that populations with a low rate of genetic variation may remain stable as long as they are cultivated under optimal conditions. Imposed stress, however, has a much stronger impact; high temperature stress and ethanol stress upon small vital populations differing in the level of prior inbreeding under optimal conditions revealed increasing extinction probabilities with increasing inbreeding levels, even for low levels of inbreeding. The authors emphasized the need for further research on the interaction between genetic and non-genetic processes. It is recommended that breeding programmes not only avoid inbreeding, but also are done under conditions that mimic future natural situations (Bijlsma *et al.* 1999).

Results of recent empirical studies suggest that whereas genetic variation may decrease with reduced remnant population size, not all fragmentation events lead to genetic losses and different types of genetic variation, as for example allozyme and quantitative variation, may respond differently. In some circumstances, fragmentation actually appears to increase gene flow among remnant populations, breaking down local genetic structure (Young *et al.* 1996). There are many causes of extinction, the fate of a specific population cannot generally be predicted and there is no single answer to the problem of what the minimum viable population size is for a species (Nunney & Campbell 1993). It depends on the biology of the species and on the options that are available regarding size, number and location of the habitat that can be preserved.

6.6 Cascade effects in metapopulations

Habitat fragmentation can be studied at the level of populations of individual species, but can also be expected to play a coincidental role for several species that co-occur in a particular type of habitat. These may be communities of coexisting species or any other type of species assemblage. Tscharntke (1992) provided qualitative evidence for potential effects of fragmentation of *Phragmites australis* habitats on a number of insect and bird species, in view of their dependence on a metapopulation structure. This was shown to differ for each of the species, but also for the relationships between the species. Although the concept of metacommunity (Holt 1997, Hanski 1999) confusingly suggests that all the components of a community are affected by fragmentation in a similar way, it does put emphasis on the notion that habitat fragmentation affects species groupings as a consequence of affecting single species. Habitat fragmentation may change or disrupt interactions such as pollination, parasitism or predation, and interspecific competition (see Steffan-Dewenter & Tscharntke 2002); these authors hypothesized that the effects of habitat fragmentation on species richness and community interactions would be stronger at higher trophic levels, and mentioned different *z* values for the slopes of species-area curves of a number of insects as related to their specialization on food plants: from 0.07 in polyphagous species to 0.11 in oligophagous, 0.16 in strongly oligophagous and 0.22 in monophagous species. Theoretical studies on the consequences of spatial heterogeneity on community structure (Holt 1997) suggest that sparse habitats in a heterogeneous landscape are likely to sustain a biased array of species, including habitat specialists with unusually high colonization or low extinction rates and habitat generalists sustained via spillover from more abundant habitats. Trophic specialization would lead to a kind of magnification of these effects, whereas trophic generalization would lead to an avenue of direct interactions among alternative prey species. If alternative prey species are habitat specialists, but a predator is a habitat generalist, predator colonization can couple the dynamics of these prey species. This gives rise to apparent competition or mediation (see Chapter 5). Indeed, most food chains are part of a complex food web, composed of multiple species on each trophic level and complex linkage patterns across levels. I will elaborate with a selection of examples.

6.6.1 Plants, insect herbivores and parasites/predators

In a study on patches of *Urtica dioica*, of different area and degree of isolation, habitat fragmentation was shown to reduce species richness of phytophagous Heteroptera, Auchenorrhyncha and Coleoptera (Zabel & Tscharntke 1998). Monophagous herbivores had a higher probability of absence from small patches than all (monophagous and polyphagous) herbivore species. Species richness of herbivores correlated positively with habitat area, and species richness of predators correlated negatively with habitat isolation. The authors concluded that increasing habitat connectivity in the agricultural landscape should primarily promote predator populations.

Food chains are useful starting points for examining the implications of metapopulation dynamics for community structure. Van der Meijden and van der Veen-van Wijk (1997) analysed whether and to what extent interactions within a tritrophic system are, indeed, affected by spatial distribution of habitat patches. Specifically, they reviewed their long-term data (two decades) on the relationships between the plant populations of ragwort (*Senecio jacobaea*), its most important herbivore, the monophagous cinnabar moth (*Tyria jacobaeae*), and the specialist parasitoid of the herbivore, *Cotesia popularis*. Populations of ragwort showed tremendous fluctuations in biomass at both local and regional scales, with local fluctuations frequently resulting in extinction. The metapopulation type that fits ragwort best is the source/sink metapopulation consisting of patches with mostly negative population growth rate. Herbivory of cinnabar moth, especially in open habitats, contributes to the local extinction risk of the plants, but the overall metapopulation of ragwort does not seem to be in danger of extinction. On a local scale, the cinnabar moth has an ephemeral existence of 1 or only a few years. Its survival is closely linked with the presence of ragwort. By numerically overshooting the local carrying capacity, larvae are subject to scramble competition for food, which frequently leads to mass starvation and local extinction. Even on the regional scale, the cinnabar metapopulation seems to be in danger of extinction during such events. The positive, and often high, correlations between temporal fluctuations in local plant populations and the metapopulation of cinnabar moth indicated that the searching capacity of the female moth is high and apparently not much hampered by interpatch plant population distances. In this respect, the cinnabar moth may be considered in one, albeit patchy, population and not in a classical metapopulation. Habitat heterogeneity in local ragwort populations even adds to the probability of cinnabar moth survival. The parasitoid *C. popularis*, with its supposed limited power of dispersal between local populations, causes a delay in the local recovery of the cinnabar moth and consequently enables some of the local ragwort populations to grow for one season without herbivory.

6.6.2 Plants and insect pollinators

Habitat fragmentation for pollinators can be characterized by spatial heterogeneity in the number of flowers to be visited. Steffan-Dewenter and Tscharntke (1999) studied experimental patches of *Sinapis arvensis* and *Raphanus sativus*, visited by a number of bee species. Number of seeds per fruit and per plant decreased significantly with increasing distance from the nearest grassland for both mustard and radish. The number of seeds set per plant was positively correlated with the number of flower-visiting bees. In general, changes in the species composition of a plant community may have a great impact on pollination and pollen flow due to the differences in pollination efficiency and flight distances; see Kwak *et al.* (1998) for a review. A reduction in local flower population size of all or several component plant species causes changes in a decrease in the richness of the assemblage of insect pollinators as well, which affects pollination quantity and pollination quality. Indeed, insects must visit several plant species to meet their energy demands, thus increasing the chance of heterospecific pollen deposition on the stigmas. This often results in a reduction of seed set and greater inbreeding in the plant population, which can potentially be counteracted through gene flow between local populations. The fitness consequences of fragmentation depend on the amount of gene flow still possible between local populations, and within populations as well.

6.7 Concluding remarks: the interface between populations and communities

The theory of island biogeography deals primarily with communities, whereas the metapopulation approach focuses on species populations (see section 6.1). Here I will comment on a few aspects dealing with the interface between populations and communities.

6.7.1 Fractal geometry

A recent development stems from the application of fractal geometry (spatial scaling laws) as a useful tool in the quantification of spatial patterns in ecology and for a causal analysis of species interactions in spatially structured habitats (Ritchie & Olff 1999b). For the spatial distribution of large herbivores at a global scale, Olff *et al.* (2002) have shown that all patterns can arise from simple constraints on how organisms acquire resources in space. They used spatial scaling laws to describe how species of different sizes find food in patches of varying size and resource concentration and derived a mathematical rule for the minimum similarity in size of species that share resources. This packing rule yielded a theory of species diversity that predicts relations between diversity and productivity more effectively than previous models. Size and diversity patterns for locally coexisting East African grazing mammals and North American savanna plants strongly support their predictions. In India, however, their predictions about potential herbivore species densities could no longer be checked owing to the current intensive land use. The theory also predicts relations between diversity and area and between diversity and habitat fragmentation, which in a modified form may also be applicable to the scale of the landscape.

6.7.2 Hybridization

An interesting phenomenon, which so far has not been dealt with in the context of migration and invasion in metapopulation research, is the potential of inter-specific hybridization. Stace's (1991) *New Flora of the British Isles* lists 715 hybrids, and of those known to have been produced in the British Isles 70 are deemed to be products of hybridization between native and alien species, and four between an introduced hybrid and a native species. In total, approximately 7% of introduced species have been involved in the production of interspecific hybrids now recognized to be part of the flora of the British Isles, while approximately 4% of the native plants have hybridized with alien species. These data have been provided by Abbott (1992), who also suggested that interspecific hybridization between a native and an invading plant species, or two invading species, sometimes results in a new, sexually reproducing taxon. Several examples of such taxa have been confirmed by molecular and isozyme analyses.

Hybridization is rarely considered among the biotic factors that promote species extinction (Levin *et al.* 1996), but it may have a profound effect on the persistence of a species. Is there a risk of maladaptive genes being introduced into a new population as a result of hybridization? Ellstrand (1992) believes that interspecific hybridization and subsequent introgression or hybrid swarm formation is harmful to an endangered species, because local adaptive differentiation can be disrupted or prevented. Others promote the view that plant hybridization and the formation of hybrid swarms allows for the maintenance of genes from the rare species and may well contribute to the formation of new species (Grootjans *et al.* 1987, Rieseberg 1991). Interspecific hybrids and their derivatives are predominantly observed in disturbed or intermediate habitats (Stace 1975). Although the notion of hybrid habitat has been specified only occasionally after it was launched by Anderson (1948), this concept could be of great value when identifying suitable habitats. Again, cascade effects should be taken into consideration. For example, a number of studies addressed the distribution of herbivores and parasites in hybrid zones of their hosts. Strauss (1994) pointed to the susceptibility of herbivores to parasites, as related to behavioural, ecological and genetic differences between the herbivores, and genetically based resistance to parasites could become disrupted as a result of hybridization. This issue is still calling for original research. In Chapter 7 the risk of hybridization in reintroduction programmes will be dealt with.

7

Populations: re-introductions

Sipke E. van Wieren

7.1 Re-introduction of species and restoration ecology

According to the Society for Ecological Restoration (SER International; SER 2002), ecological restoration is the process of assisting the recovery of an ecosystem that has been degraded, damaged or destroyed. The very first 'attribute of restored ecosystems' mentioned in the *SER Primer on Ecological Restoration* (SER 2002; www.ser.org) is that 'The restored ecosystem contains a characteristic assemblage of the species that occur in the reference ecosystem and that provide appropriate community structure.' In this context, intended introductions of species are an important tool, because dispersal is very often a major constraint, especially in highly fragmented habitats and landscapes. Thus, restoring diversity is a crucial part of ecological restoration, but while the SER (2002) considered it primarily in an ecosystem context, the issue of the re-introduction of species has also frequently been considered at the species, or subspecies, level. For example, Falk *et al.* (1996), in their volume on strategies for the re-introduction of endangered plant species, considered re-introductions also as a **conservation** tool. In this chapter I will examine experiences with re-introductions, independent of whether they have been performed in a strict restoration context or rather as a species-conservation tool. Indeed, re-introductions are nearly always experiments and the science of re-introduction is in its infancy, which urges us to learn from earlier experiences (Falk *et al.* 1996).

In response to the increasing occurrence of re-introduction projects worldwide and to help ensure that the re-introductions achieve their intended conservation benefit, the Re-introduction Specialist Group of the International Union for the Conservation of Nature's (IUCN's) Species Survival Commission has developed guidelines (IUCN 1995), which are implemented in the context of the IUCN's broader policies pertaining to biodiversity conservation and sustainable management of natural resources. According to the IUCN, the principal aim of any re-introduction should be to establish a viable, free-ranging population in the wild, of a species, subspecies or race, which has become globally or locally extinct, or extirpated, in the wild. The population should be re-introduced within the species' former natural habitat and range and should require minimal long-term management. The objectives of a re-introduction may include the enhancement of the long-term survival of a species, the re-establishment of a keystone species in an ecosystem (or an emblematic species from a cultural point of view), or the maintenance and/or restoration of biodiversity in (semi-)natural landscapes. In the literature, both the terms **re-introduction** and **translocation** are being used. Strictly speaking these terms do not mean exactly the same thing. A re-introduction is an attempt to establish a species in an area that was once part of its historical range, but from which it has been extirpated or become extinct. Re-establishment is often used as a synonym, but implies that the re-introduction has been successful. A translocation is a deliberate and mediated movement of wild individuals or populations from one part of their range to another.

Community, ecosystem and landscape changes and transformations carried out in the past may have great consequences for the success of re-introduction attempts because the ecosystem or habitat at issue may have become permanently unsuitable for the species of concern. However, even when habitat suitability is ensured, many re-introductions nevertheless fail. In some cases it has been hypothesized (Law & Morton 1996, Lundberg *et al.* 2000) that this could be caused by 'community closure'; that is, the feasible and persistent community to which the lost species once belonged is no longer 'open' for reinvasion. This approach could, for example, help explain the results of a study on dispersing prairie voles (*Microtus ochrogaster*) by Danielson and Gaines (1987). In this experiment, voles were introduced into enclosed resident populations of the same species, of southern bog lemmings (*Synaptomys cooperi*), of cotton rats (*Sigmodon hispidus*) or into an empty enclosure. The results indicated that colonization by dispersing voles was negatively affected most by resident conspecifics. Introduced female voles were more strongly affected than males during the growing season but not during the non-growing season when reproductive activity was typically low. Resident bog lemmings also negatively affected colonization by dispersing voles, but after the colonization phase coexistence was possible. Cotton rats did not affect colonization by dispersing voles. Further investigation is required to reveal to what extent this kind of interspecific interaction within guilds plays a role in re-introduction attempts.

There are, however, many other factors involved in re-introduction becoming either a success or a failure. Wolf *et al.* (1996) evaluated 80 translocations of birds and mammals in Australia, New Zealand and North America, and compared the results with a similar analysis carried out in 1987 by Griffith *et al.* (1989). The analysis revealed that approximately 58% of all translocations conducted with thousands of individuals of threatened, endangered or sensitive birds and mammals have failed to establish self-sustaining populations. Furthermore, keystone species play a critical role in communities, and their effects are generally much larger than would be predicted from their relative abundance. The importance of keystone species is essentially recognized through removal experiments (Paine 1966; see also Chapter 2 in this volume). A keystone species often referred to is the sea otter (*Enhydra lutris*), living in the north Pacific. Sea otters feed on sea urchins (*Strongylocentrotus franciscanus*), which in turn feed predominantly on kelp (macroalgae; e.g. Mate 1972). If keystone species become threatened or go extinct in their habitat it can be expected that the system changes dramatically and that, next to trying to re-introduce the keystone species into its habitat, the changes may have become so extensive that re-introduction becomes very difficult. Recently it has been shown through modelling work that even (random) removal of species can lead to cascading extinctions far beyond the target one (Borvall *et al.* 2000, Lundberg *et al.* 2000), and that cascading extinctions are positively related to species abundance and connectance (Law 1999). If extinctions are followed by community closure, re-introductions are even more difficult. If ordinary (i.e. non-keystone) species can already have such effects, what can we expect if keystone species become extinct? No clear field data are available at present, but this question stresses the need for the conservation of keystone species while they are still present in their original habitat.

In the remainder of this chapter I will highlight some of the important aspects of the art and science of re-introducing species that largely determine either success or failure.

7.2 Source populations

Individuals to be re-introduced can come from various sources and as a first step a careful assessment should always be made of the taxonomic status of the candidate sources or provenances. Even though the species concept as a basic taxon unit is controversial, individuals should ideally be of the same subspecies as those that were locally extirpated. Genetic studies should be carried out, if possible, to determine the relative degree of taxonomic and genetic similarity between possible substitutes and the pre-existing population. Genetic analyses may also permit prediction of the likelihood of hybridization taking place with other taxa in the target or release area or region. For animals, it is preferable that source animals derive from wild populations. For plants and animals, the source population should ideally be closely related genetically to the original native stock and also show ecological characteristics (morphology, physiology, behaviour, habitat preference, etc.) similar to the original population or subpopulation.

If a subspecies has become extinct in the wild and in captivity, a substitute form may be chosen for possible release. Such substitutions are actually a form of benign introduction. Selection of a suitable substitute should focus on extant subspecies and consider genetic relatedness, phenotype, ecological compatibility and the conservation value of potential candidates. For example, a local population of ibex (*Capra ibex*) that became extinct in Czechoslovakia was replaced by re-introductions of Austrian *C. ibex* and Turkish *Capra hircus aegagrus* and *C. ibex nubiana* from the Sinai desert (reviewed in Stanley-Price 1989). The inevitable hybrid forms dropped their kids in the middle of the winter, 3 months earlier than pure *C. ibex*, resulting in the death of all offspring. This case illustrates the need to assess both hybridization risks and ecological compatibility (Seddon & Soorae 1999). In general, there is a need for information on whether the introduction can literally be considered a re-introduction or whether it entails a risk of effects like those related to unintended invasions by aliens (see Chapter 2 in this volume).

Removal of individuals for re-introduction should not endanger the wild source population, and individuals should only be removed from a wild population after the effects of translocation on the donor population have been assessed and evaluated. When removals from source populations are large relative to its size, problems may arise (Stevens & Goodson 1993). Sometimes a species may become so threatened in the wild that it is taken into captivity, and the loss of wild animals may leave only captive populations. Examples include the Arabian oryx (*Oryx leucoryx*), the Przewalski horse (*Equus przewalskii*) and the Sorocco dove (*Zenaida graysoni*; Stanley-Price 1989). In such cases there is still the potential to breed species in captivity although the results of genetic and phenotypic changes such as genetic drift, inbreeding, domestication, increased tameness and the loss of behavioural traits will tend to preclude the chances for successful re-introduction and subsequent *in situ* conservation. However, many attempts are and should be made to conserve and restore critically threatened species through the re-introduction of captive-bred animals into suitable habitats. Recent examples include programmes for the black-footed ferret (*Mustela nigripes*), the golden lion tamarin (*Leontopithecus rosalia*) and the red wolf (*Canis rufus*). Unfortunately, the success rate of re-introduced captive-bred individuals is highly variable and often very low (James *et al.* 1983).

A special hazard to successful re-introductions of animals is the risk of disease introduction. The guidelines of IUCN (1995) prescribe that prospective release stock should be subjected to a thorough veterinary screening process before transport from the original source. There are many examples of devastating effects of diseases introduced unintentionally. From 1893 to 1906, 332 elk (*Cervus canadensis*) were released in the Adirondack region of New York. Additional animals were released in 1916 and 1932. The releases initially appeared successful, and in 1906 the population was estimated at 350 elk. However, the elk slowly disappeared, and there has not been an authenticated report of elk in the Adirondacks since 1953. The parasitic round worm *Pneumostrongylus tenuis* was the likely cause of the failure of elk to survive in the Adirondacks (Severinghaus & Darrow 1976). However, it is not sure when the round worm appeared for the first time in this area.

7.3 Founding numbers, diversity and population structure

In general, the number of individuals that are released in re-introduction attempts is small. This means that founding groups are susceptible to the same dangers of increased extinction risks as small, natural populations: environmental fluctuations, demographic stochasticity and inbreeding. Therefore, to achieve the highest possible success, a primary goal of re-introduction should be to maximize the initial rate of population increase in order to shorten the period during which the introduced population is exposed to these risks. This can be brought about by releasing a high number of individuals in a high-quality habitat. Komers and Curman (2000) investigated how the rate of increase of more than 30 newly re-introduced populations was affected by various population characteristics such as population size, sex and age structure in Artiodactyla (even-toed ungulates). Their results were in line with the general notion that re-introduction success increases with the number of animals released (Fig. 7.1). The function became asymptotic at about 20 animals. When fewer than 20 animals were released, the variance in growth rate increased substantially and, of a number of factors, only the age structure explained a significant portion of this variance. The population growth increased

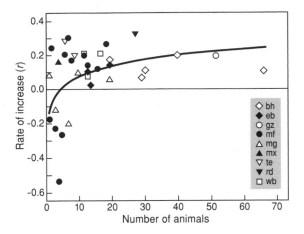

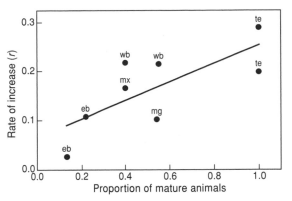

Fig. 7.1 The rate of population increase in relation to the number of animals in re-introduced ungulate populations. Source species are shown in the key: mf, mouflon; mx, muskox; te, Tule elk; rd, red deer; mg, mountain goat; eb, European bison, wb, wood bison; bh, bighorn sheep; gz, mountain gazelle. After Komers and Curman (2000). Reproduced by permission of Elsevier.

Fig. 7.2 The rate of population increase in re-introduced ungulate populations in relation to the proportion of socially mature animals in populations of fewer than 20 animals. Source species: mx, muskox; te, tule elk; mg, mountain goat; eb, European bison; wb, wood bison. After Komers and Curman (2000). Reproduced by permission of Elsevier.

with the proportion of mature animals in the population (Fig. 7.2). This finding could be explained by a higher fecundity of mature females.

Loss of genetic variability, due to genetic drift and/or inbreeding, is especially likely when an effectively small number of individuals is used in founder populations. Because of its importance, many conservation plans call for the maintenance of genetic variability in translocated populations. Stockwell *et al.* (1996) studied the effects of translocation in mosquito fish (*Gambusia affinis* and *Gambusia holbrooki*). These fish have two life-history traits that should minimize the loss of genetic variability; they have high reproductive potential, and females retain sperm and commonly have multiply sired broods, maximizing the ratio of the effective population size N_e to the total population size N (see Chapter 6). Ten translocated populations were examined. These populations had significantly lower levels of heterozygosity than their respective parental source populations. The most striking result was a reduction in allelic diversity in the translocated populations that varied from 24 to 40%. All losses were of relatively rare alleles, and probably due to an undocumented bottleneck in the early

introduction history. The results were surprising because initial translocations involved hundreds of fish and because mosquito fish, as mentioned above, have various reproductive traits that appear to minimize the effects of bottlenecks (on genetic diversity). Similar effects have been found in other introduced populations (e.g. seabream, trout, salmon, Anolis lizards, house sparrow, common myna, reindeer and ibex), as reviewed by Stockwell *et al.* (1996). In 50% of the cases examined, translocated populations had lower heterozygosity than their parental sources. In approximately 75% of the cases, refuge populations had reduced levels of allelic diversity. This pattern agrees with theoretical expectations: founding events should have a stronger effect on allelic diversity than on heterozygosity (Nei *et al.* 1975, Allendorf & Leary 1986). Also, reductions in allelic diversity are often due to loss of rare alleles, which typically have little effect on overall heterozygosity. It is clear that re-introduction programmes should attempt to create populations with high levels of genetic diversity. However it will not be easy to prevent some loss of genetic diversity. Starting with the highest possible number and ensuring a high initial population growth rate will help to maintain high genetic diversity.

Sometimes, insights in metapopulation theory may be used to understand the success rate of translocations.

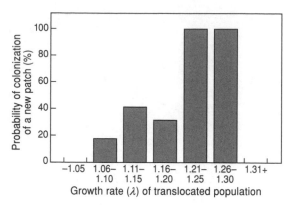

Fig. 7.3 Probability of successful colonizations of new patches is correlated with population growth rates (λ) of 31 translocated populations of bighorn sheep in the western USA in 1946–97. After Singer *et al.* (2000). Reproduced by permission of Blackwell Publishing.

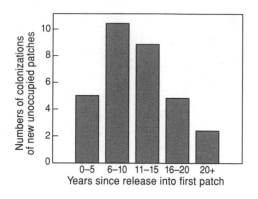

Fig. 7.4 Probability of successful colonizations in relation to the number of years since release in the first patch for 31 translocated populations of bighorn sheep released 1947–91. After Singer *et al.* (2000). Reproduced by permission of Blackwell Publishing.

Singer *et al.* (2000) related the fate of a number of bighorn sheep re-introductions to this species naturally occurring in metapopulations. At present, most extant populations of bighorn sheep (*Ovis canadensis*) consist of fewer than 100 individuals occurring in a fragmented distribution across the landscape, whereas the species formerly occupied a more continuous and wider range. They investigated the correlates for the rate of colonization of 79 suitable, but unoccupied, patches by 31 translocated populations of bighorn sheep released into nearby patches of habitat. Dispersal rates were 100% higher in rams than in ewes. Successful colonizations of unoccupied patches (24 out of 79 patches were colonized) were associated with rapid growth rates of the released population (Fig. 7.3), years since release (Fig. 7.4), larger area of suitable habitat in the release patch, larger population sizes and a seasonal migration tendency in the released population (Fig. 7.5). In this study area, colonization rates were much higher than other studies have reported and this could be attributed to the presence of larger regions of unoccupied suitable habitat with a greater probability for detection than the other studies. It is possible that bighorn sheep existed mostly in metapopulations but that human disturbance has accelerated extinction rates in these metapopulations, and that bighorn sheep now occur in a non-equilibrium state. The results of this study also

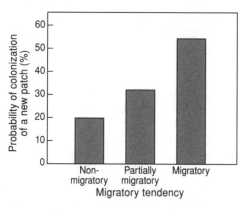

Fig. 7.5 Probability of successful colonizations in relation to migratory tendency in the release patch for 31 translocated populations of bighorn sheep. Migratory, > 75% of the population uses distinct seasonal changes; partially migratory, part of the population migrates; non-migratory, year-round use of the same ranges. After Singer *et al.* (2000). Reproduced by permission of Blackwell Publishing.

indicate that many restoration projects in the past probably suffered from poor procedures. Many prior translocations consisted of small founder groups (typically fewer than 25 animals) released into small, isolated patches of habitat, probably representing a near-perfect prescription for failure.

7.4 The re-introduction site

Re-introduction in the core of the historic range is sometimes indicated to be better than along the periphery (Griffith *et al.* 1989, Wolf *et al.* 1996). However, Lomolino and Channell (1995, 1998) found that 23 out of 31 species of endangered mammals persisted along the periphery, not in the core or central portion of their historic range. In addition, persistence was greater for insular than for continental populations. According to Lomolino and Channell, the range periphery, in comparison with core sites, encompasses a much more diverse collection of habitats and environmental conditions. They referred to the California condor (*Gymnogyps californianus*) as an important case in point. Which range should investigators adopt as the raptor's historic range? Recent efforts include release to a site in northern Arizona, well outside the condor's present range, but also providing protection from anthropogenic threats. From their review of re-introductions of Marsupialia in Australia, Short *et al.* (1992) came to the conclusion that the success rate of island (re-)introductions (60%) was far greater than those in mainland Australia (11%), even though the successful island (re-)introductions were all to islands with no historic record of the occurrence of the (re-)introduced species. Success of (re-)introduction of these macropods appeared to depend critically on control or exclusion of exotic terrestrial predators such as foxes and cats. Peripheral sites should thus not automatically be discarded as suitable re-introduction sites.

A crucial aspect of any re-introduction plan is an assessment of the availability and quality of the re-introduction site. Re-introductions can have a chance of success only if the habitat and landscape requirements of the species are or could be satisfied, and are likely to be sustainable. The area should have sufficient carrying capacity to sustain growth of the re-introduced population and support a viable self-sustaining population over time. Identification and elimination, or reduction to a sufficient level, of previous causes of population decline and/or habitat transformation should take top priority. In a habitat suitability study for an otter re-introduction project in Utah, it was found that 94% of the studied streams were unacceptable for re-introductions. Escape cover

was the most limited habitat attribute, whereas food for otters appeared to be available in adequate quantities (Bich 1988). This study, therefore, recommended that no otter re-introductions should be made until riparian zones were rehabilitated and protected, since re-establishment of stream-bank vegetation was deemed essential to provide escape cover for re-introduced otters. Similarly, Howells and Edwards-Jones (1997) studied the feasibility of re-introducing wild boar (*Sus scrofa*) to Scotland through an assessment of suitable woodland habitat that could support a minimum viable population of the target animal. This species has been the focus of early attempts to re-introduce it into Britain. Based on a review of *S. scrofa* ecology, the authors identified woodland habitats suitable for supporting wild boar. Only long-established woodlands containing some stands of semi-natural origin and larger than 500 ha in size were considered. None of the woodlands could be considered optimum habitat for wild boar and none was large enough to support a minimum viable population of 300 animals. The study concluded that the goal of (re-)establishing a self-sustaining population of wild boar in Scotland was unrealistic in the short term.

Habitat destruction and modification can also be brought about in the form of invasive species. Such invasions often result in dramatic changes in ecosystem structure or function (Gordon 1998, Hobbs & Mooney 1998). Invasive species may not only lead to changes in ecosystem properties but can also hamper re-introductions through predation. Recent attempts to recover razorback suckers (*Xyrauchen texanus*), an endangered piscivorous fish species, by re-introducing them into their native range of mainstream Colorado River have not been successful because of predation on the native young suckers by non-native fishes (Johnson *et al.* 1993). In another study, Bergerud and Mercer (1989) reviewed 33 (re-)introductions of caribou that took place in eastern North America between 1924 and 1985. Twenty introductions resulted in sustained populations and 13 failed, the majority as a result of predation by wolves. The fate of these 33 introductions is consistent with the view that predation (natural and hunting) is a major factor in the decline of caribou in eastern North America following European settlement. In Europe, meanwhile, attempts to re-introduce black grouse (*Lyrurus tetrix*) and capercaillie (*Tetrao urogallus*) have been hampered, in part by predation by goshawk (*Accipiter gentilis*) and pine marten (*Martes martes*; Kalchreuter & Wagner 1982). Predation may have particularly severe impacts on very small populations, especially if a more common primary prey species is present (prey switching), while at the same time the number of re-introduced individuals is almost always small or very small. Only a sufficiently large re-introduction might overcome predation and succeed where a smaller one would fail. The minimum viable population would then, however, be much larger than that predicted by standard population-viability analysis. When McCallum *et al.* (1995) used a simple stochastic model based upon the bridled nailtail wallaby (*Onychogalea fraenata*) to explore this possibility, even very small amounts of predation (two to four individuals per 6 months) were sufficient to cause large re-introductions of up to 50 animals to fail. No clear threshold population size was found beyond which re-introductions would succeed. The moral is that if predation is a serious risk, a single re-introduction of a given size is preferable to multiple re-introductions of the same total number of individuals.

7.5 Re-introduction techniques

In the majority of cases of animal re-introductions or translocations, the focus is on populations, rather than communities, whereas for plants the focus is on communities. Many techniques are applied to help new animal populations to get established. Use can be made of individuals caught in the wild or of individuals kept and bred in captivity. Choices have to be made concerning which stages in the life cycle of species are most suitable for translocation activities. Should it be eggs/seeds, subadults/seedlings, or adults? For plants, individual plants or whole turfs can be transplanted, or seed mixtures can be harvested from hay and spread over the introduction site. A few commonly applied techniques will be discussed below.

7.5.1 Plants

Community translocation involves the wholesale removal of an assemblage of species from a site and the attempt to establish it as a functioning community at a receptor site. The translocation of species assemblages is used to move communities that would otherwise be completely destroyed by a change in land use at the donor site (e.g. civil engineering and excavation projects; Klötzli 1987). Bullock (1998) reviewed, among several others, 10 translocation projects in Britain. Four different techniques were used: hand turfing, machine turfing, macroturfing (1 m × 2 m) and spreading (of excavated soil and vegetation). In most projects, post-translocation management was similar to the original management at the donor site. All communities, except the species-poor heath, showed both losses and gains of species. At some sites, all translocated communities were becoming more similar to the original communities at the receptor sites. Rare plant species were lost on a regular basis. The associated invertebrate communities showed larger and more obvious changes than did the plant communities, and often showed losses in rare species of conservation importance.

The restoration of a former plant community *in situ* is quite another issue. If seeds of the target species (characteristic of the original plant community) are no longer available in the soil seed bank, they have to immigrate from elsewhere, for example attached to hay-making machinery or after deliberate re-introduction. Somerford Mead is an old flood-meadow along the River Thames near Oxford, UK, which harboured a *Alopecurus pratensis/Sanguisorba officinalis* plant community in the 1950s. From 1960 to 1982, however, it was used as grassland for haymaking or silage cutting and received artificial fertilizers. From 1982 to 1985 it was ploughed and used for barley. In 1985 it was agreed to take Somerford Mead out of this high productivity and set in motion regimes to create an *Alopecurus/Sanguisorba* flood-meadow community again. The possible benefits of removing the surface soil to reduce fertility was set against the disadvantage of synchronously removing much of the seed bank. Therefore, in 1986 the last crop of barley was grown without any fertilizer in order to start the reduction of nutrient availability. Further restoration efforts have been described by McDonald (1992, 1993, 2001) and McDonald *et al.* (1996). In July 1986 a seed mixture was harvested from the reference site Oxey Mead, an ancient flood meadow, 2 km downstream. Its exploitation has not changed since at least the 13th century (Baker 1937). It features, therefore, a notably low-fertility grassland community, the *A. pratensis*/*S. officinalis* association (MG4, according to Rodwell 1992). The seed mixture was broadcast over prepared soil on Somerford Mead the following October. Management included cutting for hay in early July and grazing the aftermath by cows and/or sheep. During the first 3 years target species such as *Bromus* spp., *Cynosurus cristatus*, *Festuca pratensis*, *Leucanthemum vulgare*, *Ranunculus* spp., *Rhinanthus minor* and *Trisetum flavescens* had become established. After 6 years, 20 target species were found in the established vegetation that had not been re-introduced from the reference site. They must have spread spontaneously or by haymaking machinery. The position of many re-introduced target species became critical. *Silaum silaus* and *Leontodon hispidus* occurred in the seed bank and only rarely in the established vegetation. *A. pratensis*, *Briza media*, *Hordeum secalinum* and *S. officinalis* were not found in the seed bank and were rare in the established vegetation. Species with short-lived seeds cannot form a seed bank, and hence cannot survive years when they are absent from the established vegetation.

In 1989 a management experiment began in Somerford Mead consisting of an annual hay cut at the end of June followed by 4 weeks of grazing in October – by sheep or cattle – in comparison to a control, non-grazed treatment. From 1990 onwards, the differences between grazed and ungrazed treatments increased. The ungrazed plots became dominated by tall grasses such as *Arrhenatherum elatius*, *Dactylis glomerata*, *Festuca rubra*, *Holcus lanatus* and *Lolium perenne*. At the same time, the frequency of *Bromus hordeaceus*, *Cirsium arvense*, *C. cristatus*, *Ranunculus bulbosus*, *Trifolium pratense* and *Trifolium repens* decreased. The ungrazed plots changed in composition towards *Arrhenatherum elatius* grasslands common on road verges in Britain. Both the cattle- and sheep-grazed plots became more similar to the community in the reference site, but were still far from the species composition of Oxey Mead, even 15 years after the re-introduction of target species.

7.5.2 Fish and herpetofauna

The restoration of historical spawning areas, or the provision of new, suitable spawning habitat, are important for successful re-introduction of fish and amphibians. Both translocation from the wild and the release of captive-bred individuals are commonly applied techniques. Stocking appropriate life stages of target species is clearly important for successful introductions or re-introductions. For fish, using older/larger individuals has been more successful than using spawn (Noakes & Curry 1995), whereas the reverse seems to have been the case for amphibians. Cooke and Oldham (1995) monitored the establishment of large populations of common frogs (*Rana temporaria*) and common toads (*Bufo bufo*) for 6 years in a newly created reserve, following stocking with spawn of both species and with toads rescued from a site to be destroyed. Transfer of spawn proved to be more effective as a means of establishing a new population of toads than transfer of adults.

The Great Lakes ecosystem has changed dramatically in the past 50 years. A review of historical changes reveals complex interactions of overexploitation of fishery resources, invasion of non-indigenous species, eutrophication, extensive habitat modification and toxic contamination. Native fish species that required tributary or near-shore habitat for spawning and nursery areas have declined markedly. Among surviving native species, such as walleye (*Stizostedion vitreum*), stock diversity declined with the loss of tributary-spawning stocks and lake-spawning stocks became dominant. With the rarefaction of native species, the abundance of formerly subdominant species increased. Species such as smelt (*Osmerus mordax*), gizzard shad (*Dorosoma cepedianum*) and white perch (*Morone americana*) depend less on critical tributary and near-shore habitat (Koonce *et al.* 1996). Invasive species pose a special problem. The Great Lakes ecosystem is home to at least 139 non-indigenous species of fauna and flora that have become established following invasions or intentional introductions. About 10% of the exotic species have caused economic or ecological damage to the system. Despite activities to reduce the causes of decline, most problems have not yet been solved adequately. Nevertheless, several re-introduction attempts have been made with various species. Much attention has been given to the rehabilitation of the lake trout (*Salvelinus namaycush*). It seems that a complete restoration of the Great Lakes is unlikely, due to naturalization of exotic species, habitat degradation and destruction, heavy fishing mortality, lack of native gene pools and complicated political jurisdictions that rarely work towards a common vision. Meffe (1995) proposes that a more realistic goal would be rehabilitation, a movement along the trajectory towards complete restoration.

Until now, most re-introduction projects involving amphibians and reptiles have not been very successful (Dodd & Seigel 1991), but efforts undertaken for the natterjack toad (*Bufo calamita*) represent an interesting exception. The species is endangered in Britain and has been legally protected since 1975. This amphibian suffered a major decline during the first half of the 20th century, due partly to habitat destruction but mostly to successional changes in its specialized biotopes and anthropogenic acidification of breeding sites. Extensive autecological research over the past 25 years has provided the foundations for an intensive, 3-year species-recovery programme funded by statutory nature-conservation organizations. Management of heath and dune habitats focused on restoration and maintenance of early stages of serial succession, initially through physical clearance of invasive scrub and woodland vegetation, followed by applying grazing regimes similar to those prevalent in earlier centuries. In some cases extra breeding pools were constructed to either increase or stabilize natterjack toad populations that had become reliant on one or very few pools at small sites, or to promote range expansion within large habitat areas. Re-introductions also had been attempted. At least six out of 20 re-introductions resulted in the foundation of expanding new populations, and an additional eight have shown initial signs of success. Conservation methods developed for *B. calamita* provided a useful precedent for long-term conservation of early successional habitats and species (Denton *et al.* 1997).

7.5.3 Birds

In birds, making use of captive-produced eggs that are fostered or cross-fostered is a common and viable re-introduction technique (Derrickson & Carpenter 1983). Sometimes eggs are collected from wild populations. Fostering has proved to be a much better technique than the release of hand-reared individuals as they are much more prone to all sorts of danger (e.g. predation). This has been found, among others, in whooping cranes (*Grus americana*), hand-reared capercaillie (*T. urogallus*), white storks (*Ciconia ciconia*) and raven (*Corvus corax*). Releasing individuals straight into the wild (hard release) is not recommended by Bright and Morris (1994). Most species of birds (and mammals) rely heavily on individual experience and learning as juveniles for their survival. They should be given the opportunity to acquire the necessary information to enable survival in the wild. Therefore, soft-release techniques have been developed whereby the animals are kept in pens or other holding devices and slowly are made acquainted with their new environment.

A commonly used soft-release technique for the introduction of birds of prey is called hacking. Hacking is the release of free-flying young birds at a site where food is provided until independence. Hacking was used in the re-introduction of Montagu's harrier (*Circus pygargus*; Pomarol 1994). It took place in an enclosure measuring $3-4 \text{ m} \times 2 \text{ m} \times 1 \text{ m}$ high. The re-introduced harriers were between 20 and 30 days of age. After 5–8 days the enclosure was opened. The young birds became independent on average 34 days after their first flight (at 70 days of age). Over a 5-year period 87 birds were (re-)introduced with a success rate of 83%. Only three birds had been seen returning to the area in subsequent years. Hacking has also been applied very successfully in the many re-introduction projects of the peregrine falcon (*Falco peregrinus*). Over the past 25 years more than 1000 birds have been re-introduced in this way in many parts of the USA.

More than 1670 attempts have been made to establish several hundred avian species worldwide. Among them are many raptors. At least six species of owls and 15 species of diurnal raptors have been established successfully. Examples of raptors that have been re-introduced or newly introduced are little owl (*Athene noctua*) in Britain, eagle owl (*Bubo bubo*) in Sweden and Germany, goshawk (*A. gentilis*) in Britain, white-tailed sea eagle (*Haliaeetus albicilla*) in Scotland and other parts of Europe, bald eagle (*Haliaeetus leucocephalus*) in New York and California, Seychelles kestrel (*Falco araea*) on Praslin (Seychelles) and the peregrine falcon (*F. peregrinus*) in the USA, Canada and Germany. A raptor that almost became extinct is the Mauritius kestrel (*Falco punctatus*). By 1974, the species had declined to only four known wild birds, including one breeding pair, as a result of habitat loss and pesticide contamination. A conservation project begun in 1973 has used many management techniques including captive breeding, supplemental feeding of wild birds, provision of nestboxes, multiple clutching, egg pulling, artificial incubation, hand rearing and release of captive-bred and captive-reared birds by hacking, fostering and predator control. A total of 331 kestrels were released in the 10 years up to the end of the 1993–4 breeding season; one-third of these were captive bred and the rest were derived from eggs harvested from the wild. By the 1993–4 season, an estimated 56–68 pairs had established territories in the wild with a postbreeding population, including floating birds and independent young, of 222–86. Since the pesticides responsible for their decline are no longer used, the number of Mauritius kestrels should continue to rise through natural recruitment. The distribution of suitable habitat suggests that an eventual population of 500–600 kestrels on Mauritius is possible. Due to its outstanding success, the release programme for the Mauritius kestrel ended after the 1993–4 breeding season (Jones *et al.* 1995).

7.5.4 Mammals

Mammals can be taken from wild source populations or from captive breeding stock. Catching animals from the wild can be a costly and time-consuming operation, and is not without risk. Like birds, also mammals should be given the opportunity to acquire the necessary information to enable survival in the wild, and soft releases are therefore recommended. Mammals propagated in an enclosure tend to develop an affinity for their immediate surroundings and therefore, upon release, exhibit a slow dispersal rate. This behaviour generally enhances survival. An example of a successful soft release is the case of the scimitar-horned oryx (*Oryx dammah*) in Tunisia which disappeared from that country in 1902 due to desertification, competition with domestic livestock, disturbance and hunting. Ten young scimitar-horned oryx (five males, five females) from Britain were re-introduced into the Bou-Hedma National Park in Tunisia in December 1985. They were acclimatized in a 600-m² pen for 4.5 months and then released into

a 10-ha pre-release enclosure. Social organization was established peacefully, and the oryx adjusted to the new climate and natural foods. In July 1987, the oryx were released from the enclosure into the total-protection zone of the park. This zone is a 2400-ha area that has been protected from domestic livestock since 1977 (Bertram 1988).

Sometimes a more spectacular technique is indeed the only solution. In a translocation project for beaver (*Castor canadensis*) in Idaho, the mountains, heavy forests and lack of roads in Idaho made transplanting a labour-intensive, expensive and time-consuming task. In addition, it resulted in high beaver mortality. The use of planes and parachutes with animal holding boxes proved to be a much more efficient and much less expensive method of transportation. In 1948, 76 live beavers were dropped with only one casualty. Observations made in 1949 showed that the beavers that had participated in the airborne transplantation had settled and were well on their way to producing colonies (Heter 1950).

7.6 Socio-economic aspects and concerns

The California population of sea otter (*E. lutris*), in the 1970s introduced from Amchitka, Alaska, is considered vulnerable and therefore a Fish and Wildlife Service Recovery Plan for the sea otter has been made which calls for the establishment of a second California population as a hedge against devastation by a possible oil spill. During the 1970s transplant operations, 86 animals were captured, 24 of which died in the nets or holding pens. Of the 79 otters caught in 1971, 15 died due to capture complications (Mate 1972). Also the second translocation was not without its problems, including emotional reactions of various groups with widely different interests (Booth 1988). This example demonstrates that re-introductions can be highly controversial, especially with species that combine a high cuddling status with potential negative interaction with economic interests. The European habitats directive 92/43/EEC demands a proper consultation of the public in case of re-introduction of species listed in Annex IV of the directives (EC 1992a).

Re-introductions are generally long-term projects that require the commitment of long-term financial and political support. It is important that socio-economic studies should be made to assess impacts, costs and benefits of the re-introduction programme to local people. According to the IUCN (1995) guidelines, a thorough assessment of attitudes of local people to the proposed project is necessary to ensure long-term protection of the re-introduced population, especially if the cause of a species' decline was due to human factors (e.g. over-hunting, over-collection or loss or alteration of habitat). The relevance of these guidelines should not be underrated because there are many examples of failures due to not paying sufficient attention to the attitudes of local communities. From an extensive literature review on exclosures, afforestation, reafforestation, rehabilitation and other regeneration operations over several million hectares in Mediterranean bioclimatic areas from the Atlantic Ocean to the Aral Sea, combined with 50 years of personal field experience, Le Houérou (2000) concluded that, while the main constraint for success is the restoration of habitat factors that have caused degradation, the most difficult constraints to overcome are usually of a socio-economic and/or sociocultural nature. Poaching can also be a problem, for example in the relocation of 22 Tule elk (*Cervus elaphus nannodes*) from the Tupman Tule Elk Reserve near Buttonwillow to Fort Hunter Liggett (both in California) in 1978. Factors conducive to the high poaching rate were tameness of the relocated elk, location of release site, lack of monitoring and resentment by locals to changing policies at Fort Hunter Liggett (Hanson & Willison 1983).

Resentment can be especially strong against predators. Thus, when nine European lynx (*Lynx lynx*) were released in central Austria in 1975, 100 years after the last native lynx had been killed, there was strong local opposition from hunters, especially in Carynthia. Carynthia has few federal forest estates, but many large private forest estates pursuing trophy hunting by tourists as a source of income (Gossow & Honsig-Erlenburg 1986). Similar problems are encountered with wolves. In response to popular resistance, red wolves (*Canis rufus*) re-introduced to the Alligator River National Wildlife Refuge in North Carolina were classified as a 'non-essential experimental population' and did not have the full protection of the Endangered

Species Act when released. Proposed re-introduction of grey wolves (*Canis lupus*) to Yellowstone National Park met similar opposition from livestock interests, hunters and state agencies (Wilcove 1987).

Some species have a much more positive press. Especially, the release of high-profile flagship species may raise public awareness of conservation issues and generate funding for wider programmes. In Saudi Arabia the first wildlife conservation project targeted the Houbara bustard (*Chlamydotis macqueenii*), which is threatened as a resident. Programmes directed towards the re-introduction of this large, appealing bird have attracted wide public attention owing to the emblematic status of the bird throughout the Middle East as the premier quarry for falconry, and thus these programmes have helped generate support for other, lower-profile species in need of protection. The aesthetic value or economic benefits of an animal may also be tied to the generation of public support and the means to raise public awareness of conservation issues. In Latvia the re-introduction of the beaver (*Castor fiber*) resulted in the creation and conservation of wetlands; their value in water purification has been estimated at up to £1.3 billion sterling, and beavers re-introduced into France and Sweden have become tourist attractions (Seddon & Soorae 1999).

In conclusion it seems to be clear that the idea of re-introducing species within their former habitat has gained quite some acceptance within the context of the restoration paradigm. An important incentive is that in most cases species are not able to colonize these areas by themselves and need a little help. Nevertheless, as has been amply demonstrated, much can go wrong and indeed has gone wrong in the many thousands of re-introduction attempts already set in motion. The ones that were successful, however, also teach us that it can be done and that success cannot be attributed to sheer luck alone. If re-introduction programmes take into account that the habitat is suitable (or can be made suitable again), the founding population is sufficiently large, the population structure is right, a high level of genetic diversity is ensured, the proper techniques are applied, careful planning has been applied and the public has been consulted properly, then the chances for a successful re-introduction are enhanced considerably.

PART 3

Restoration perspectives

8

Restoration of dry grasslands and heathlands

Jan P. Bakker and Rudy van Diggelen

8.1 Introduction: the historical context

Dry grasslands and heathlands can be characterized from a hydrological point of view as infiltration areas, fed by rainwater. Hence, for these types of habitat restoration does not have to deal with complicated groundwater-management practices as is the case for mires (see Chapter 9 in this volume). After the last glacial period natural communities of dry grasslands and heathlands are supposed to have occurred on rocky inland outcrops, for example on chalk grasslands, along coasts, in river floodplains and at the edge of bogs. Their area was restricted to marginal habitats with extreme abiotic conditions such as shallow soil with subsequent low nutrient availability. This traditional view has recently received much attention. Several authors suggest that large herbivores maintained a mosaic of forest and grassland/heathland in a wooded meadow system in the pre-agricultural landscape (Gerken & Meyer 1996, Vera 2000). However, wooded meadows may have been restricted to small stretches along rivers, being the places with relatively open forest on nutrient-poor soil where herbivores are able to maintain patches of short plant communities (Olff *et al.* 1999). Archaeologists doubt whether the density of large herbivores was high enough to maintain wooded meadows. Svenning (2002) reviewed the literature and he concluded from findings of pollen, macro-fossils (seeds, leaves) and dung beetles that large herbivores may have played a role (together with fire) in pre-agricultural locally open landscapes in chalk-

lands, floodplains and nutrient-poor soils, whereas uplands were probably covered by dense forest. It is not clear in this context whether the megafauna in North America might have been able to maintain extensive prairies. When they were extirpated within 1000 years about 11,000 BP, climatic conditions were different from the current ones. It is assumed that human impact strongly increased the area of dry grasslands and heathlands as a result of agricultural exploitation, in what is nowadays referred to as low-intensity farming (Bignal & McCracken 1996). Hence, they can be classified as semi-natural landscapes (Westhoff 1983). They can only be maintained by this type of farming; changes in land use both by intensification and abandonment will change the communities. The question for restoration is whether such changes are reversible or irreversible.

In Europe, this agricultural farming system was introduced from the Middle East, and it spread quickly after about 7000 BP. In wetlands, people of the Swifterbant Culture started keeping livestock at about 5800 BP and crop cultivation at about 5300 BP. In uplands they started agriculture at about 5000 BP in the northern part of the Netherlands (Bakker 2003). Apart from arable fields, common grazing lands were exploited and resulted in the extension of heathland in the Atlantic part of Europe, whereas dry grasslands developed elsewhere. Dry grasslands can be subdivided in acid grasslands and calcareous grasslands depending on the occurrence of a limestone subsoil. Dry grasslands and heathlands developed under a common grazing system with mainly cattle and sheep. Often dung was collected in overnight stables to fertilize the arable fields in spring (Bonn & Poschlod 1998). In north-west Europe, sods were also cut to enhance the production of arable fields. Cutting for hay only developed during the last 2000 years, when animals needed for traction power (oxen, horses) were used intensively; they needed hay for winter forage. The removal of biomass resulted in oligotrophication of the soil; that is, lowering of the nutrient availability. This is a characteristic phenomenon of low-intensity farming systems.

North American tallgrass prairies expanded greatly beginning about 8000 BP. From that time until about 3000 BP the region's climate was hotter and drier than at present, favouring grassland development as woodlands retreated. The origins of tallgrass prairies and the roles of climate, grazing animals and human activity are still not completely understood. Humans have been part of the prairie environment since before the last glaciers retreated from the Midwest. Prairies were home to Native Americans for thousands of years prior to European settlement. They managed them according to their needs and abilities. Deliberate fires, along with occasional lightning, were major forces in shaping prairie vegetation. Landscape fires were used to facilitate travelling and hunting, to stimulate new growth for game, to reduce fuel loads near habitations. Large areas of former prairie that have not been destroyed through conversion to agriculture, intensive grazing or development, have been degraded through fire suppression, drainage or other alterations (Knapp et al. 1998).

Grazing by domestic livestock has most likely dispersed many plant species from the Middle East into Europe (Bonn & Poschlod 1998). Transhumance must have played an important role in the long-distance dispersal of diaspores (Poschlod & Bonn 1998). For example, the species composition of calcareous grasslands has changed during the past millennia. Archeological finds of macro-fossils (seeds, leaves, stems) indicated that the current dominant grass *Bromus erectus* was not found in sites occupied 500–1000 years ago (Poschlod & WallisDeVries 2002). The invasion of species from the Middle East during the past millennia means that in Europe only a few invasive species are currently considered a problem in conservation. In contrast, in North America the invasion of plant species from other continents over the past few centuries is considered one of the problems for biodiversity (D'Antonio & Meyerson 2002).

Both intermediate disturbance and intermediate nutrient supply are believed to support species richness in grasslands. Stability of vegetation dynamics in dry acidic grasslands is preserved and promoted by sufficient small-scale disturbance under conditions of low resource availability (Jentsch & Beyschlag 2003). Disturbances in grasslands are often caused by herbivores. Both small and large herbivores have a major impact on dispersal and colonization of plant species. Cattle are identified as most important for seed dispersal, whereas rabbits have a main effect as creators of disturbances (Bakker & Olff 2003).

8.2 Threats for biodiversity due to changes in land use

Changes in land use have resulted in reclamation of grassland and heathland. Intensification of agricultural practices has led to a strong decline of these areas. On the other hand, many marginal agricultural areas have nowadays been abandoned, with subsequent bush encroachment and hence a decline in the area of grassland and heathland. Examples of the decrease of habitats are heathland (Webb 1997), calcareous grasslands in north-west Europe (WallisDeVries et al. 2002), wooded meadows in Sweden (Mitlacher et al. 2002), tallgrass prairies in the USA (Samson & Knopf 1994) and changes within habitats, such as bush encroachment in alvars in Sweden (Rejmánek & Rosén 1992). Abandonment coincides with bush encroachment accompanied by a decrease in plant species richness. For instance, on the alvar grasslands in Öland, Sweden, the number of characteristic alvar species has declined continuously with an increasing cover percentage of *Juniperus communis* scrub and a dramatic drop in species number between 75 and 100% of shrub-cover species (Rejmánek & Rosén 1992). As a result of previous agricultural practices, dry grasslands and heathlands occur on nutrient-poor soils. This makes them vulnerable to impacts from adjacent fertilized agricultural areas, rivers and coastal seas, and atmospheric deposition resulting in eutrophication, especially by nitrogen. It coincides with the dominance of indigenous plant species such as *Deschampsia flexuosa* in dry heathland, *Molinia caerulea* in wet heathland, *Calamagrostis epigejos* in dunes, *Brachypodium pinnatum* in calcareous grasslands and *Elymus athericus* in salt marshes on the continent. In the UK an increase of the indigenous bracken (*Pteridium aquilinum*) and the alien *Rhododendron ponticum*, planted in 1763, is found in heathlands (Cross 1975). In North America invasive species can take over. They pose a serious threat in at least 194 of the 368 National Park Units in the USA (D'Antonio & Meyerson 2002). Many invasive species are perennial plants or persistent annuals that set up positive-feedback mechanisms that perpetuate their own existence. The list of 'One hundred of the world's worst invasive alien species', including all groups of organisms, published in the Global Invasive Species database (www.issg.org/database) reveals that plant species make up 60% of invasive aliens in grasslands and shrublands. One of the most troublesome species is the perennial shrub *Chromolaena odorata* which is invasive in Australia, Africa and Oceania (McFadyen & Skarratt 1996).

Apart from the aforementioned changes in plant diversity, dramatic shifts in birdlife have been monitored as a result of changes in land use. A comparison between data in the new breeding bird atlas of the Netherlands (1998–2000; SOVON Vogelonderzoek Nederland 2002) and the previous atlas (1973–7) revealed a decline of species groups breeding in open grassland, and in dunes and heathland, with ten and five species, respectively (Saris *et al.* 2002). At the same time six species of dense scrub showed an increase. The encroachment of scrub has strongly affected the breeding bird population in dunes. Typically scrub-dependent species such as lesser whitethroat (*Sylvia curruca*) and nightingale (*Luscinia luscinia*) have increased. However, birds characteristic of more-open dunes have declined dramatically, such as Montagu's harrier (*Circus pygargus*), wryneck (*Jynx torquilla*), nightjar (*Caprimulgus europaeus*) and red-backed shrike (*Lanius collurio*) (Sierdsema & Bonte 2002).

Many insect species depend on grasslands with a typical structure of the sward or a mosaic of short canopy and scrub. Up to 20% of European butterflies

are threatened or near threatened (Bourn & Thomas 2002). Calcareous grasslands are extremely important for butterflies in Europe: 48% (274) of all native species (576) occur in calcareous grasslands (van Swaay 2002). Only for the UK and the Netherlands are historical data available on changes over the past century. In Suffolk, UK, 42% of the butterfly species became extinct (Bourn & Thomas 2002). The Netherlands harbour a total of 71 native species of butterflies, of which 17 (24%) have become extinct and 30 (43%) are endangered (van Ommering *et al.* 1995).

Data on the decline of biodiversity as a result of changes in land use should be treated with caution. We will elucidate some plant data collected before 1950 and after 1970 in the well-monitored province of Drenthe, The Netherlands. This province showed a dramatic decline in the area covered by heathlands and bogs between 1850 and 1990 (Plate 8.1). For the whole of the Netherlands the area decreased from about 600,000 ha in 1833 to 36,000 ha in 1990. Despite a strong intensification of agricultural practices, expansion of urban areas and other significant land-use changes, the biodiversity expressed as the average frequency of 126 indicator species of nine plant communities hardly declined (van Diggelen *et al.* 2005).

The first reason for this phenomenon is that the data on the occurrence of plant species are presence/absence data, and do not take into account the abundance of species, as shown by the following example. The number of marked grid cells (cell size 1 km^2) after 1970 shows a nearly complete coincidence with the former heathland area of the dominant species of dry heathland, *Calluna vulgaris* (Fig. 8.1a). At the national level the number of marked grid cells decreased from more than 10,000 in 1930 to 3000–10,000 in 1995 (Tamis & van 't Zelfde 2003). Small fragments of heathland harbouring *C. vulgaris* still exist in most grid cells. Moreover, after reclamation of heathland, the seeds of *C. vulgaris* may survive for many decades in the soil (Thompson *et al.* 1997). Intensively exploited arable fields still harbour viable seeds 70 years after reclamation from heathland (ter Heerdt *et al.* 1997). The intensification of infrastructures such as roads and ditches often brings viable seeds to the surface and hence allows the re-establishment of *C. vulgaris* in road verges. In contrast, *Arnica*

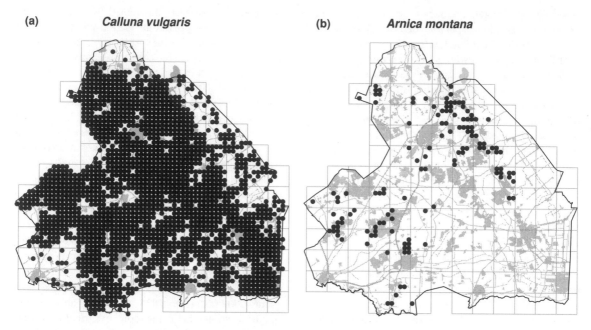

Fig. 8.1 Number of grid cells (1 km²) with the presence of *Calluna vulgaris* (a) and *Arnica montana* (b) in 1990 in Drenthe, The Netherlands. After Werkgroep Florakartering Drenthe (1999). Reproduced by permission of the author.

montana, characteristic of the same plant community, produces seeds with short longevity (Thompson *et al.* 1997), which prevents re-establishment. Indeed, the pattern of *A. montana* features a strong decline in the number of grid cells (Fig. 8.1b). At the national level it decreased from 1000–3000 grid cells in 1930 to 31–100 grid cells in 1995 (Tamis & van 't Zelfde 2003).

A second reason why biodiversity does not always reflect significant land-use changes is that characteristic ecosystems are sometimes very species-poor whereas their degradation phases are much more species-rich. Typical examples are bogs and the afore-mentioned heathlands. Bog drainage leads to a decrease in characteristic species (Moore 2002) but at the same time to an increase in species richness (Vepsäläinen *et al.* 2000). The third reason has to do with changes in landscape structure. Whereas most characteristic ecosystems used to cover comparatively large and uniform areas this is no longer the case today. The landscape has become increasingly fragmented and the large bogs, heathlands, forests, etc. of the past have developed into a mosaic of small woodlands, meadows, arable fields, hedgerows, ponds and so on.

If such landscapes developed from species-poor landscapes such as large bogs, biodiversity can actually have increased.

8.3 Constraints for restoration

8.3.1 Seed longevity

Plant species richness may not change during the first years of bush encroachment (abandonment) or of establishing the dominance of grasses (eutrophication) or invasive species. As a result of decreasing light availability at soil level, fewer species will flower and set seed. This implies a depletion of the soil seed bank, especially of species with a transient or short-term persistent seed bank (Thompson *et al.* 1997). After bush encroachment of several decades on the alvar of Öland, Sweden, both the species richness of the soil seed bank and the established vegetation have decreased (Bakker *et al.* 1996a). Spectra of plant communities in dry grassland and heathland in north-west Europe show that these communities can

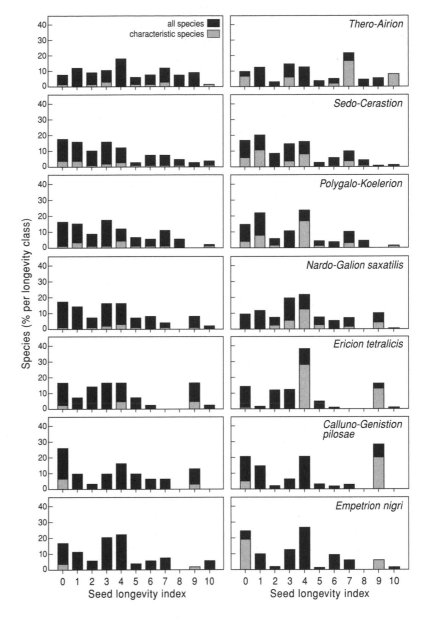

Fig. 8.2 Seed-bank longevity spectra of heathland (*Ericion tetralicis, Calluno-Genistion pilosae* and *Empetrion nigri*) and dry grassland (*Thero-Airion, Sedo-Cerastion, Polygalo-Koelerion* and *Nardo-Galion saxatilis*). On the *x* axis longevity classes are indicated by the seed longevity index ranging from 0 (transient) to 10 (long-term persistent) according to data from Thompson *et al.* (1997). The *y* axis gives the percentage per longevity class of the species in the community. The left-hand panels indicate the proportion of the longevity class within the community for sheer presence of the species, and the right-hand panels indicate the proportion taking into account the frequency of species in reference relevées of the database of the *Vegetatie van Nederland*. After Bekker *et al.* (2002). Reproduced by permission of the author.

only partly rely on the soil seed bank for restoration (Fig. 8.2; Bekker *et al.* 1998, 2002).

8.3.2 Seed dispersal

Low-intensity farming implied an enormous dispersal of seeds (Poschlod & Bonn 1998) as a result of graz-ing, cutting and the application of manure. Losvik and Austad (2002) reported that seeds from barns were brought into the field. Current intensive farming in north-west Europe strongly reduced the rate of dis-persal of seeds (Poschlod & Bonn 1998). Remnants of extensive oligotrophic dry grassland and heathland are separated in the currently fragmented landscape. They are no longer connected by the 'moving ecological infrastructure', i.e. large-scale tending of flocks of live-stock (Poschlod *et al.* 1996). Hence, dispersal is supposed to become an important constraint in restoration. This does not only hold for seeds, but also for gene exchange by pollen between populations (Kwak *et al.* 1998). A complication here is the finding that species of different trophic levels (plants, herbivores/pollin-ators and predators/parasitoids) show an increasing positive correlation between fragment size and spe-cies richness (Steffan-Dewenter & Tscharntke 2002). This implies that common insects will become more and more dominant (Kwak *et al.* 1998). Therefore, specialized relationships between trophic levels may be disturbed in small fragments with subsequent problems in restoration.

8.3.3 Soil organisms

Many plant species depend on mycorrhizal fungi. However, soil organisms disperse very slowly, and hence are a constraint for restoration (Baar 1996, Van der Heijden *et al.* 1998) as indicated above for seed dis-persal. Grasslands dominated by *Sporobolus wrightii* once covered riparian floodplains in the south-western USA, but were reclaimed for agriculture. Inoculation of the grass with arbuscular mycorrhizal fungi can benefit restoration efforts in abandoned agri-cultural fields in semi-arid regions (Richter & Stutz 2002). Also, other soil communities may be important. Disturbed sites that have been planted with native perennial species to facilitate restoration revealed that the soil microbial community composition differed from that of relic native stands: it was closer to that of grasslands composed on invasive, non-native species (Steenwerth *et al.* 2002). In a microcosm experiment, soil herbivores such as nematodes forage particularly on grass species dominant in early eutro-phic successional stages of restoration, and hence may accelerate the establishment of later successional target species in restoration of species-rich meso-trophic grasslands (de Deyn *et al.* 2003).

8.3.4 Eutrophication

The atmospheric deposition and transformation of oligotrophic dry grassland and heathland into intens-

ively exploited arable fields and pastures created mesotrophic and eutrophic soil conditions. Authorit-ies in charge of management cannot reduce the atmospheric deposition from precipitation. However, dry deposition can be reduced by maintaining a low canopy. In the Netherlands, atmospheric deposition of nitrogen compounds ranges from between 50 and 170 kg of N $ha^{-1} yr^{-1}$ in forests and between 30 and 50 kg of N $ha^{-1} yr^{-1}$ in grassland and heathland. Critical nitrogen loads for heathland and grassland are below 20 kg of N $ha^{-1} yr^{-1}$ (Bobbink *et al.* 1998). Reducing the nutrient availability in the soil is a challenge.

8.3.5 Invasive species

Even when removed as adult plants, invasive species may leave a legacy that makes long-term restoration of the site difficult. This legacy may be in the form of a long-term persistent soil seed bank that is often larger in a species' new home than in their native hab-itat. Another legacy may be chemical alteration to the habitat, such as elevated nitrogen mineralization by nitrogen fixation (D'Antonio & Meyerson 2002). Atmo-spheric deposition may enhance this latter problem.

Box 8.1 When maintenance of grassland or heathland fails: can forests be restored easily?

Ancient forest species cannot rely on a long-term persistent seed bank for recovery after reclamation to arable field or pasture (Bossuyt *et al.* 2002). Dispersal is another constraint. Although a large number of forest species will be able to colonize new forest, some typical forest species have low rates of colonization. Hence the complete restoration of the understorey requires a time period of over a century (Bossuyt & Hermy 2000). A third point may be competition. Recently established forests on former agricultural land with a relatively high nutrient status and a relatively open canopy provide excellent habitat conditions for competitive species such as *Urtica dioica* and *Rubus fruticosus*. Therefore, competitive exclusion can be an explanation for poor establishment of characteristic forest species (Butaye *et al.* 2002). Restoration of forest will be dealt with further in Chapter 10 in this volume.

8.4 Restoration of degraded ecosystems

Dry grasslands and heathlands with most or at least part of the characteristic species still present can be managed by the continuation or re-instalment of practices inherent to the former low-intensity farming systems, such as mowing, cutting sods, grazing and burning. In Europe, this implies continuation of the still-existing low-intensity farming system in countries such as Greece, Spain, parts of France and the UK (Bignal & McCracken 1996). In other countries existing dry grasslands and heathlands are often nature reserves. Here the management in charge of nature conservation has more or less taken over the former farming practices. However, this is not always successful.

8.4.1 Mowing

Mowing regimes should be applied in a proper way as can be deduced from the following example. Calcareous grasslands in Dutch nature reserves were cut in late autumn to stimulate seed rain of the target species present. However, the grass *B. pinnatum* became dominant, and this resulted in a decline of characteristic species, mainly forbs. Experiments revealed that early cutting (July) prevented the reallocation of nutrients to below-ground storage organs of the grass. The early cutting regime showed a decrease of *B. pinnatum* and an increase in species richness (Bobbink & Willems 1991).

8.4.2 Sod cutting

Sod cutting in heathlands overgrown by grasses is successful for matrix plant species such as *C. vulgaris* and *Erica tetralix*. Initially, cutting was carried out by hand, but later machinery became involved. The latter was also thought to be economically feasible. The sods were stored and after a freezing period sold for horticultural purposes. This type of industrial sod cutting on a large scale has the positive benefit of removing nutrients. However, it is detrimental for animals such as amphibians and invertebrates. These animals favour a mosaic of different structures of the vegetation in their habitat. Therefore, mechanical sod cutting should be a small-scale practice.

Target plant species such as *A. montana* do not spread or re-establish after sod cutting in overgrown heathland. Immediately after sod cutting, the level of ammonia increases sharply (Dorland *et al.* 2003). This turns out to be harmful for *Arnica* seedlings (De Graaf *et al.* 1998a). The addition of chalk, to increase the buffering capacity of the soil, may compensate for this negative effect (van den Berg *et al.* 2003).

8.4.3 Grazing

The success of grazing in heathlands overgrown by grasses depends on the species of herbivore introduced. Sheep grazing is less successful than cattle grazing. The introduction of sheep seems appropriate, as sheep grazing and heathland exploitation are strongly related. However, it has become clear that sheep were only used during the past two centuries. Before 1800, cattle grazing was practised more in heathlands in the Netherlands. Cattle grazing was successful in reducing the grass *Deschampsia flexuosa* (Bokdam & Gleichman 2000), and a cyclic successional pattern has been suggested (Bokdam 2003).

The introduction of livestock grazing in prairie management is heavily debated in North America. This has partly to do with the wilderness concept of nature conservation, which ignores the concept of semi-natural landscapes. Wedin (1992) has compared the approaches of management of tallgrass prairie in the USA and of the semi-natural landscapes in Europe, and he advocates introducing the European practices to North America. Grazing by large herbivores in tallgrass prairie generally increases plant species richness as a result of the creation of a mosaic of patches (Collins *et al.* 1998, Knapp *et al.* 1999). Gazing by bison resulted in high light levels at the soil surface and heterogeneity of the high light levels in the vegetation. Moreover, with bison grazing a positive correlation was found between species turnover (extinction and colonization) and species richness (C. Bakker *et al.* 2003).

8.4.4 Burning

Burning of heathland overgrown by grasses was practised in the Netherlands until 20 years ago. As the dominant grasses spread again rapidly after burning, this practice is no longer applied. The afore-mentioned sod-cutting and grazing practices are nowadays widely adopted. Fire is adopted as a management practice in North American tallgrass prairie (Knapp *et al.* 1998)

8.5 Removal of late-successional invasive species

Local biodiversity can be reduced when a single plant species becomes dominant, and monopolizes resources such as light and nutrients. The invasive species can be an alien introduced from another continent, but it can also be a locally existing species taking over in the process of succession after a change in land use such as abandonment. Remarkably, a strong separation seems to exist in the literature on invasive species and succession. The examples given below include both. Irrespective of the threats for biodiversity, successional invasive species may replace native species and take over their functional role. In that case we may have to 'live with exotics' (D'Antonio & Meyerson 2002). The removal of invasive species may change ecosystems completely. Ideally, there should be (i) pre-eradication assessment, to tailor removal and to avoid unwanted ecological effects, and (ii) post-removal assessment of eradication effects, on both the target organism and the invaded ecosystem (Zavaleta *et al.* 2001).

A variety of techniques can be adopted to remove invasive alien species from reserves or restoration sites. These most commonly include hand removal, mechanical removal, herbicides, fire or some combination of the above. In the USA system-wide management plans called for more than 535 species to be managed between 1996 and 2000 at a cost of over $80 million, but less than 10% of this money was made available (D'Antonio & Meyerson 2002).

8.5.1 Reducing single species' dominance

First we deal with examples concerning the dominance of a single species. Many lowland heaths in the UK have been invaded by *Betula* spp., *Pinus sylvestris*, *P. aquilinum* and the alien species *R. ponticum*. Litter stripping is effective in reducing the nutrients available, and it accelerates the succession of reversion (Mitchell *et al.* 1998). *P. aquilinum* encroachment can be reduced in terms of removal of fronds to allow persistence of other vegetation by both cutting and herbicide application. The success after some years of *Pteridium* control treatments depends on the rhizome reserves. In this respect cutting seemed to be more successful than herbicide application (Pakeman *et al.* 2002). However, cutting will favour grass species over ericaceous species, and so is not appropriate where the intention is to restore a *C. vulgaris* community. Herbicide application in combination with grazing revealed different results depending on the stocking density of sheep. At low stocking rate and sufficient shelter to prevent rapid *Pteridium* litter loss, *Calluna* could become established. At high stocking rate and minimal shelter, the vegetation became dominated by *Rumex acetosella* (Pakeman *et al.* 2000). Addition of a source of *Calluna* propagules is necessary. Experiments with cutting and herbicide application lasting 18 years demonstrated that after an initial increase of *Calluna* a transition took place towards a typical Breck grass heath. This was probably caused by increased atmospheric nitrogen deposition and extreme weather conditions, factors beyond the reach of managers (Marrs & Le Duc 2000).

R. ponticum is being controlled by cutting down bushes, and burning and excavating the stumps. When herbicide use is allowed, cut stumps are treated with amcide or glyphosphate. Re-invading seedlings can be removed by hand. Cleared areas do not possess a persistent seed bank of *Rhododendron*, since the viability of seeds lasts only 7 months. The control is extremely labour-intensive and hence expensive (Gritten 1995).

The encroachment of *J. communis* at the alvar of Öland, Sweden, is nowadays counteracted by cutting of the scrub and subsequent re-introduction of livestock grazing in the framework of subsidies to farmers. The subsidies are provided by the European Union in the LIFE programme (Rosén & Bakker 2005). Dunes in the Netherlands get overgrown by *Hippophaë rhamnoides* and *Salix repens*. In order to re-establish the communities of pioneer habitats, the shrubs are cut and the remaining vegetation is removed by shovels to give the wind free access (van Til & van Mourik 2001).

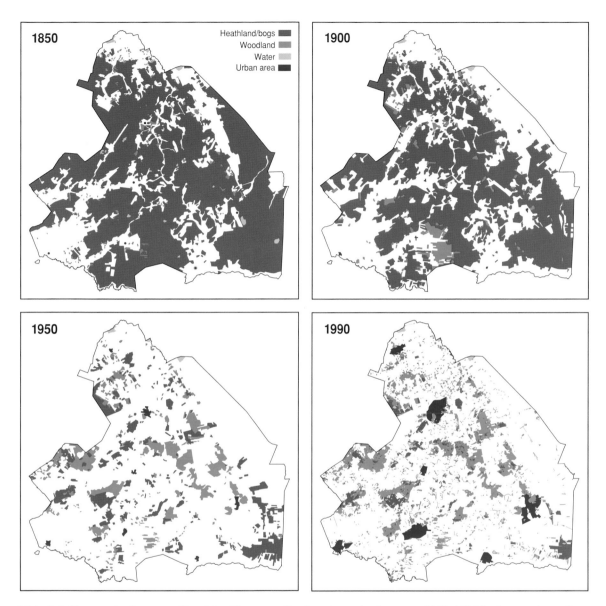

Plate 8.1 Changes in the area of heathland/bog, woodland, water and urban area 1850–1990 in the province of Drenthe, The Netherlands. After Werkgroep Florakartering Drenthe (1999). Reproduced by permission of the authors.

Plate 9.3 Distribution of groundwater types in relation to vegetation pattern in a drained fen meadow complex. Base-poor groundwater (blue) from decalcified valley flanks sustains the development of nutrient-poor bog vegetation, while base-poor groundwater (red) from a calcareous aquifer sustains a species-rich fen meadow vegetation (*Calthion palustris* alliance). In the foreground an acid rainwater lens has developed in the drained fen meadow.

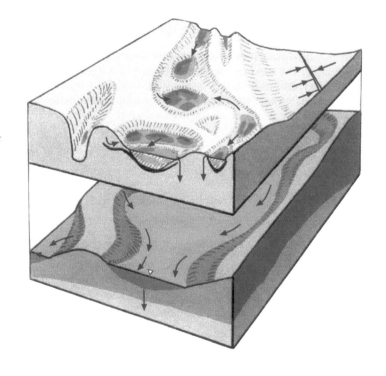

Plate 9.4 Landscape model of small bogs situated in old erosion gullies in the eastern part of the Netherlands. In wet periods the bogs are completely filled with water and the water surplus may flow from one bog to another via the podzolic layers in the subsoil with a high resistance to water flow. Some infiltration through the organic layers occurs, and this water percolates to the groundwater table above the boulder clay. Depressions in the boulder clay layer may increase the water level in wet periods, but may also be windows of water losses during the summer if former erosion processes have left only very thin clay layers. Due to a complex geological history of the area, various groundwater levels are present in the subsoil, and all of them are relevant to sustaining bog growth at the surface. After Grootjans *et al.* (2003).

8.5.2 Reversion to earlier successional stages

The following examples deal with restoration succession of forest after abandonment. Calcareous grasslands that have been overgrown by scrub and later forest lose the majority of their characteristic plant species. The 170 characteristic species of central European calcareous grasslands were classified according to life-history traits such as seed longevity, wind dispersal and dispersal by herbivores. After removal of the forest, initially a small number re-established, most likely from the long-living soil seed bank. Another small number re-established later and was supposed to originate from elsewhere by wind dispersal. Finally, a relatively large number re-established, most likely dispersed by sheep that connected still-existing fragments of calcareous grasslands with the area where the forest was cut (Poschlod *et al.* 1998, Knevel *et al.* 2003).

A wooded meadow in Sweden was partly overgrown by forest. Many of the characteristic wooded meadow species did not occur when compared with an adjacent, still-present wooded meadow. A third previously wooded meadow overgrown by forest was cut, and after 36 years a great deal of the wooded-meadow species were present. The cut part and the forest were connected by grazing livestock. Assessment of the contents of dung revealed that dispersal of the characteristic wooded-meadow species seems unimportant: the majority of the species found in the dung were trivial grassland species growing on more-fertile soil, such as *Poa trivialis* (Mitlacher *et al.* 2002). It is likely that the animals prefer these species, having a better forage quality. Hence species of eutrophic habitats have more chance of being dispersed by herbivores than the target species of mesotrophic or oligotrophic habitats.

8.6 Restoration after reclamation and intensification

An important point in the restoration of communities on mesotrophic or oligotrophic soils from intensively exploited agricultural fields is removal of the surplus of nutrients, especially nitrogen. Several techniques will be discussed.

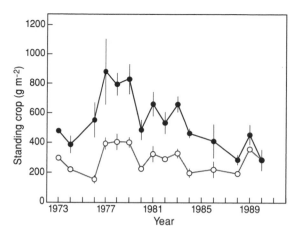

Fig. 8.3 Dry weights (means ± S.E.M., $n = 10$) of standing crop in July, from 1973 to 1990, in N-fertilized (●; 50 kg of N ha^{-1} yr^{-1}) and unfertilized (○) control plots that were cut for hay annually in early August. After van der Woude *et al.* (1994).

8.6.1 Growing crops

An experiment was carried out with cultivating barley in pots with soil from a previously intensively used arable field or oligotrophic soil. The nitrogen content of the crop grown on fertile soil was higher than that of the crop grown on oligotrophic soil. This suggests that growing crops can help to enhance the process of nutrient removal after abandonment of intensively exploited agricultural fields (Marrs 1993).

8.6.2 Adding nutrients and carbon sources

After abandoning intensively exploited agricultural fields, nitrogen will be depleted first (Marrs 1993). This will lead to low production and hence little removal of phosphorus and potassium. Adding only nitrogen annually in a July cut grassland system on sand resulted in a higher standing crop than the control without nitrogen addition. After 17 years the standing crop of both treatments was similar (Fig. 8.3). A factorial design with nitrogen, phosphorus and potassium added individually revealed that potassium had become depleted to such an extent that it became the limiting factor for plant production (van der Woude *et al.* 1994). However, the plant communities did not transform into the target communities of heathland or oligotrophic grassland (Bakker 1989).

Adding carbon sources could be practised to immobilize nitrogen as a means for accelerating the restoration of grasslands on oligotrophic soils (Zink & Allen 1998). Indeed, carbon addition decreased nitrogen availability in laboratory microcosms. Short-term field experiments revealed that immobilization is possible in sites with relatively low organic matter content and soil moisture. The method needs to be evaluated over a period of several years (Török *et al.* 2000).

8.6.3 Mowing

A haymaking regime in July in the sandy uppercourse of a brook valley in the Netherlands after intensive agricultural exploitation showed transformation of a community dominated by *Agrostis stolonifera* and *P. trivialis* into dominance of *Holcus lanatus* and *Anthoxanthum odoratum* after 25 years. The comparison of different cutting regimes revealed that the best result towards the target communities of oligotrophic grassland was by haymaking twice a year as compared to haymaking only once a year. Mulching annually or once every 2 years showed a deviation into tall forb communities (Bakker *et al.* 2002a). An adjacent field that was acquired as a nature reserve some years earlier had a shorter history of fertilizer application (Bakker & Olff 1995), and hence a better starting posi-

tion. After 25 years of haymaking in July, *Festuca rubra* and *Agrostis capillaris* were dominant and several target species of oligotrophic grassland had established. The latter were already present in the established vegetation or in the soil seed bank. The long period of haymaking did not result in the establishment of target species present in a field nearby (500 m). Dispersal for reaching regional target communities seems to be a problem (Bakker *et al.* 2002a).

The decrease of standing crop from 800 to 200 g of dry weight m^{-2} is not enough to create oligotrophic grassland. The nitrogen output by haymaking is too low to compensate for the input by atmospheric deposition (Table 8.1; Bakker *et al.* 2002a). The spread of the moss *Rhytidiadelphus squarrosus* in dry grasslands in the Netherlands is supposed to be related to atmospheric deposition (Londo 2002). A thick moss carpet is disadvantageous for the establishment of target species (van Tooren *et al.* 1987).

In the Netherlands, mulching seems not to be a good practice to create oligotrophic communities. In contrast, it revealed similar results to haymaking in a 25-year experiment in southern Germany (Fig. 8.4). The higher temperature in late summer is supposed to better decompose the litter than in the colder Netherlands (Kahmen *et al.* 2002, Moog *et al.* 2002).

8.6.4 Burning

Burning experiments are practised as a cheap way to remove above-ground biomass and nutrients. Initial results from calcareous grasslands in Germany

Table 8.1 Average nitrogen content at peak standing crop (yield; mean ± S.D.) and nitrogen removal in the first half of July in different cutting regimes and different years. Nitrogen content: 1975, *n* = 1; 1983, *n* = 3; 1999, *n* = 5; yield all years, *n* = 10. After Bakker *et al.* (2002a).

Year	Cutting regime	N concentration (%)	Yield (g of dw m^{-2})	N removal (kg ha^{-1})
1975	July	1.31	625 ± 92	82
	July and September	1.39	810 ± 97	113
1983	July	1.49	723 ± 64	108
	July and September	1.77	641 ± 69	113
1999	July	1.89	269 ± 23	51
	July and September	1.41	291 ± 43	41

dw, dry weight.

Fig. 8.4 Development of species composition under five different management regimes from 1975 to 1999 presented in an ordination diagram of a detrended correspondence analysis (DCA). Total number of species is 84. Eigenvalues: Axis 1, 0.3; Axis 2, 0.1. Treatments: 1, grazing; 2, haymaking; 3, mulching; 4, burning; 5, laissez faire. Years are given in parentheses. After Kahmen *et al.* (2002). Reproduced by permission of Elsevier.

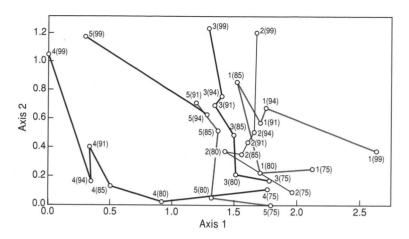

(Schiefer 1981) and France (Chabrerie 2002) seem promising. However, after 25 years the grass *B. pinnatum* became so dominant that the plant community in Germany resembled that of abandonment (Fig. 8.4; Kahmen *et al.* 2002, Moog *et al.* 2002).

8.6.5 Grazing

Grazing by livestock is mainly practised for cattle and sheep. After abandonment of intensively exploited fields, the stocking density adopted is often that used in mesotrophic or oligotrophic communities. Such a grazing regime is more or less successful in areas not too heavily fertilized in the past. Examples show that the development of mesotrophic communities is possible, but it seems that the transformation stops at a dominance of the grasses *F. rubra* and *A. capillaris* (Bakker 1989). Grazing at low stocking rates implies differentiation within the fence. Indeed, locally mesotrophic species establish. In other places, where the grazers tend to rest and to ruminate, accumulation of dung takes place and stands with tall forbs develop (Bakker & Grootjans 1991). The higher the previous fertilizer application, the larger the area with no grazing at all in large areas within the fence, which leads to tall forb stands including *Urtica dioica*. Apart from the lower costs, grazing is often preferred above cutting because the latter produces a uniform effect across the whole field. Extensive grazing, when plant production exceeds utilization by the herbivores,

results in patterns on different scales (Bakker 1998) that cannot be mimicked by cutting. Only when the starting position is good (not very intensive previous exploitation), and the soil seed bank harbours target species, will the establishment of such species be recorded.

Grazed areas often include a mosaic of eutrophic/mesotrophic sites to be restored and still-existing dry grassland and heathland communities on oligotrophic soils. This is done to replace the sharp artificial boundaries by more natural boundaries and allow dispersal of seeds into the area abandoned from former agricultural exploitation. Unfortunately, herbivores may enrich the sites with oligotrophic soil by depositing their dung, while the seeds they introduce in this way seem to originate mainly from species in the sites with the higher nutrient levels (better forage quality for the herbivores; Bakker 1989, Mitlacher *et al.* 2002, Bokdam 2003).

Spontaneous succession in a former debris-disposal site with bare soil resulted in ruderal communities including tall forbs after 3 years. Grazing by sheep could not stop this trend. Target species of oligotrophic dry grasslands were not available in the soil seed bank, and hence did not establish. Inoculation with target species by the introduction of mown material, raked material or intact sods from reference sites turned out to be successful for the establishment of target species. Despite their suppression by sheep, many ruderal species maintained and represented a latent ruderalization potential (Stroh *et al.* 2002).

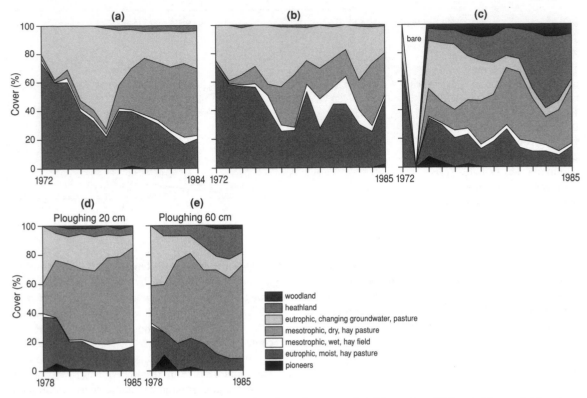

Fig. 8.5 Relative changes of cover percentages of ecological groups after (a) grazing, (b) haymaking and (c) sod cutting from 1972 onwards, and ploughing up to (d) 20 cm and (e) 60 cm from 1978 onwards.

8.6.6 Sod cutting

To enhance the restoration of oligotrophic grassland and heathland, sod cutting (up to 10 cm) is practised. It does remove nutrients and enhances the establishment of target species from the soil seed bank if present (Fig. 8.5). The disadvantage of sod cutting is that it removes part of the soil seed bank including target species. However, shallow sod cutting of up to 10 cm can also activate the seed bank of non-target species. A headache for many authorities in charge of conservation is the establishment of *Juncus effusus*. This rush species is represented in most soil samples in (former) agricultural fields in north-west Europe (Bekker *et al.* 1997). However, sometimes the dominance of *J. effusus* is judged to be positive, as the stands harbour many reptiles (Donker 1999) and birds such as

the grasshopper warbler *Locustella naevia* (Bijlsma 2001). It is under discussion whether managers should change targets depending on unexpected events.

8.6.7 Topsoil removal

In the early 1990s waste deposits had to be covered by soil in the Netherlands. This was for sanitary reasons as birds dispersed waste over the countryside. The process resulted in topsoil removal (up to 50 cm) in several nature reserves. Topsoil removal gives the opportunity to get rid of the whole former agricultural history by removing the ploughed and fertilized soil until the C horizon (the undisturbed parent material) and to start succession with oligotrophic soil conditions. Of course, the seed bank of target species is

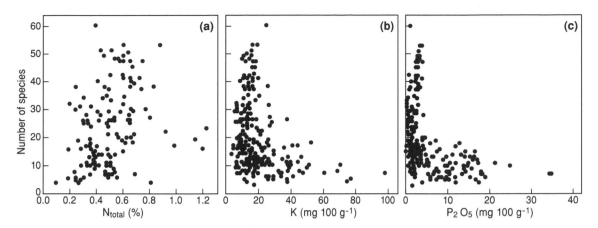

Fig. 8.6 Relation between total nitrogen, percentage of soil-extractable potassium, soil-extractable phosphorus and the number of species per 100 m² in 281 grassland sites in western Europe. After Janssens *et al.* (1998).

depleted, but so is that of non-target species. The process is also mimicked by local topsoil removal and local accumulation of soil within a site. The results so far show that nutrient removal is successful. However, in some sites a considerable amount of phosphorus is found in the C horizon as a result of heavy fertilizer input in the past. Together with the continuous input of nitrogen from atmospheric deposition this will cause problems for the future development of oligotrohic communities. The highest species diversity in grasslands over north-west Europe is more related to low phosphorus levels in the soil than to the levels of potassium and nitrogen (Fig. 8.6; Janssens *et al.* 1998). The phosphate saturation in agricultural soils in the Netherlands increased strongly during the last few decades (www.minlenv.nl).

After topsoil removal, vegetation studied by Verhagen *et al.* (2001) revealed a range of 0–30% similarity with target plant communities of oligotrophic grassland and heathland. Many target species occurred in the same grid cell (1 km²) or adjacent grid cells (Table 8.2). However, after nearly 10 years many target species were still absent. This again indicates that dispersal is a limiting factor for restoration. As mentioned above, seed dispersal by the large herbivores is not always successful when still-existing oligotrophic grassland and heathland are fenced in together with locally more productive communities.

8.6.8 Turning the soil profile

When sod cutting or topsoil removal are too expensive or otherwise not feasible, grasslands can be ploughed. This will activate the persistent seed bank of target species when present. A site that was reclaimed from heathland in the 1950s by ploughing up to 40 cm was ploughed again up to 20 cm and up to 60 cm. In both cases the seed bank of target heathland species was activated. However, after 10 years the cover of heathland species was higher in the area ploughed up to 60 cm than in that ploughed up to 20 cm (Fig. 8.5). This must be due to bringing the oligotrophic C horizon to the surface. After ploughing to 20 cm the topsoil remains productive, which enables competition from non-target species such as the grass *A. capillaris* (Fig. 8.5).

8.7 Concluding remarks: key issues

8.7.1 Nutrient reduction

As dry grassland and heathland depend on low nutrient availability, the removal of nutrients is a key issue. Not only the absolute amount, but also the balance between the limiting nutrients phosphorus, potassium and nitrogen is important. Pastor *et al.* (1984)

Table 8.2 The total number of target species per site, recorded during the first 9 years after top-soil removal, in the local species pool (in parentheses is shown the number of target species within the same block as the site/number of target species within the same or an adjacent block as the site) and the mean longevity index (LI) per target alliance of the seed bank formed by the target species (in parentheses is shown the number of species used to calculate the mean longevity index). After Verhagen et al. (2001).

Alliance	Total target species	Number of target species per site									Mean LI
		Aekingerbroek	Bakkeveensterduinen	Dellebuursterheide	Eemboerveld	Eexterveld	Ennemaborg	Hullenzand	Labbegat	Tichelberg	
Calluno-Genistion pilosae	6							0 (4/4)			0.42 (2)
Caricion nigrae	14			1 (13/14)	1 (7/12)				7 (10/13)		0.23 (11)
Ericion tetralicis	8	1 (8/8)	2 (7/8)	0 (8/8)	0 (2/6)	0 (6/8)	1 (1/2)	1 (5/7)	3 (3/6)	2 (1/5)	0.71 (2)
Hydrocotylo-Baldellion	11	0 (4/7)	0 (3/5)	1 (6/9)	1 (0/3)	0 (1/5)	0 (0/0)	0 (1/6)	1 (2/10)		Unknown
Junco-Molinion	12								5 (9/11)		0.23 (9)
Nardo-Galion saxatilis	14		2 (13/14)	4 (14/14)		5 (14/14)			4 (7/13)		0.27 (10)
Thero-Airion	14							1 (9/11)			0.45 (4)

showed that N mineralization rates in humus-rich woodland soils were better correlated with the P content of fallen leaves than with the amount of N. This suggests that raised P levels will also result in increased N availability. Moreover, when two or more nutrients are limiting productivity, the system is less sensitive to temporary rises in the availability of a single nutrient. Sod cutting and topsoil removal, however, are mainly effective for the removal of nitrogen but much less for phosphorus (e.g. Verhagen *et al.* 2004). Consequently, the productivity is only limited by nitrogen availability and these systems are more susceptible to small fluctuations in N input. This is especially true in the framework of atmospheric deposition that amounts up to 30–40 kg of N ha^{-1} yr^{-1} in the Netherlands. Critical loads for the target communities are less than 20 kg of N ha^{-1} yr^{-1} (Bobbink *et al.* 1998). Modelling the effects of management and atmospheric deposition on *C. vulgaris* revealed an interaction between the two. More-frequent cutting was required with high rainfall when atmospheric deposition is high. This variation should be taken into account when defining critical loads or designing management prescriptions (Britton *et al.* 2001). Long-term haymaking can remove nutrients up to the level of mesotrophic communities. Further oligotrophication is only likely to be achieved after sod cutting or topsoil removal. It certainly enhances the transformation of plant communities towards the targets in habitats that sufficiently meet the other abiotic conditions. Grazing can only be successful when the herbivores are introduced after several years of frequent haymaking or after sod cutting or topsoil removal. Fencing of sites is only feasible when dispersal of nutrients and non-target species into the oligotrophic target sites can be prevented.

8.7.2 Dispersal and establishment of species

Even when the right abiotic conditions are achieved, the transformation into dry grassland and heathland communities is limited by dispersal of target species (Bakker & Berendse 1999). The importance of the 'moving ecological infrastructure' such as animals and humans is underestimated (Strykstra *et al.* 1996b). Experiments with the introduction of seeds or hay showed that poor dispersal can be overcome. A solu-

tion could be the introduction of hay from nearby reference communities (see Chapter 7), which enhances the chance that seeds with similar genetic backgrounds are introduced from the local species pool. Another advantage is that managers in charge of conservation do not have to get rid of poor-quality hay from reference sites.

Introduction experiments suggest that especially rare plant species do not establish, even after the introduction of seeds. Maybe they depend on special soil organisms. These can be symbiotic mycorrhizal fungi, or nematodes grazing on roots of competing species. These soil organisms are supposed to disperse much more slowly than the seeds of plants. Hence, the limiting factor for establishment could be the absence of these soil organisms. Experiments are advocated to introduce soil from reference sites together with seeds of target species.

8.7.3 Conflicting targets

Sites under restoration management should be monitored for the establishment of target species of plants, invertebrates – such as butterflies, grasshoppers and carabid beetles – and vertebrates, including their presence in adjacent grid cells (1 km^2). This will reveal data on the rates of dispersal and re-establishment of different groups of organisms. Many invertebrates and species of birds, mammals and reptiles strongly depend on the structure of the vegetation, especially the spatial arrangement of bare soil, short turf, tall forbs, scrub and forest (van Wieren 1998). This means that conflicts between different nature-conservation interests can arise. Large-scale industrial sod cutting and haymaking are positive for a number of plants – although not all – in view of the removal of nutrients, but negative for many animal species. The latter need tall stands or litter accumulated for shelter, oviposition and the development of young animals. Restoration efforts should aim at the development of complete communities and not just plant communities of dry grassland and heathland (Mortimer *et al.* 1998). A design of mosaics including different successional stages and different scales should be part of restoration management.

Setting ecological targets is necessary to prevent conflicts. It should also be clear what the targets are and what the means necessary to reach them are. For instance, it is clear that grazing management by a farmer even without fertilizer application (organic farming) does not result in the establishment of oligo-trophic grassland communities (Bakker & ter Heerdt 2005). Agro-environment schemes are sometimes considered a goal in themselves without properly evaluating the ecological target (Kleijn *et al.* 2001).

9

Restoration of mires and wet grasslands

Ab P. Grootjans, Rudy van Diggelen and Jan P. Bakker

9.1 Introduction

Mires are peatlands where peat is currently being formed (Sjörs 1948). In Europe loss of mire area has been particularly great in north-western and southern Europe since c.1900. Here less than 1% of the original mire area has remained in most countries (Joosten & Clarke 2002). In eastern European countries, such as Poland, Belarus, Estonia, Latvia and Ukraine, between 10 and 50% of the former mire area has remained, while in European Russia, Sweden, Norway and Romania the mires have been best preserved (> 50% of the mires remaining). On a world scale mire losses are highest in Europe due to its high population pressure on nature and the climatic suitability for agriculture and forestry. In the non-tropical world 80% of mire losses are attributed to agriculture and forestry. Peat extraction in the world is responsible for c.10% of mire losses, but on an annual basis new peat extractions commence in some 10 km^2 (Joosten & Clarke 2002). In North America and Asia mire losses are much lower (5 and 8% respectively).

Mires can be roughly divided into bogs and fens. Bogs are generally considered as mires raised above the surrounding landscape and fed by precipitation (rainwater), whereas fens are usually situated in depressions and, additionally to precipitation, they are also fed by ground or surface water. Bogs are always acidic (pH < 5.5), whereas fens are usually base-rich and slightly acidic or neutral (pH > 5.5; Succow 1988, Wheeler *et al.* 2002). Most fens have been slightly drained in Europe and the vegetation has changed to that of a fen meadow. The fen then shifts from a peat-accumulating system to a peat-degrading system. Fens and fen meadows have many species in common. Fen meadows are usually more rich in species, because after slight drainage many typical fen species remain and many typical grassland species establish themselves in the area. While bogs and fens are peat-accumulating systems, semi-natural fen meadows do not accumulate peat. There has been a strong decline in semi-natural fen meadows and related species-rich grassland in north-west Europe as a consequence of changes in agricultural practices during the past decades. Estimates from the UK are that 95–98% of species-rich hay meadows that were present before 1940 have been lost due to intensification of agricultural exploitation or due to abandonment and subsequent development of forest (Garcia 1992, Muller *et al.* 1998).

We use the notion **wetland** as a general term, to describe an area with vegetation adapted to very wet conditions. A wetland may or may not be a peatland; it can also occur on mineral soil. A peatland is a soil type consisting of dead organic material that has been formed on the spot (sedentarily) and which consists of at least 30% dry mass of organic material.

In this chapter we will mainly discuss two types of restoration project: (i) projects that try to restore damage in existing nature reserves, and (ii) projects that try to increase the biodiversity in peat areas which are no longer destined for intensive agricultural production and where new nature is being developed by increasing water tables and flooding frequencies and sometimes by removing the eutrophic topsoil entirely.

9.2 Description of natural mire systems

Mire systems are by definition peat-forming systems. They accumulate organic matter, but mineral sediments – such as sand, clay or chalk – can also be deposited in a mire system. Mire systems are basically natural systems that can develop without human interference. Present-day mire systems, however, are almost all influenced by humans in one way or another, particularly in densely populated areas. The description of the following mire types is, therefore, based mainly on stratigraphical and palynological studies, describing natural mire systems without human influences (Kulczynski 1949, Brinson 1993, Grünig 1994, Dierssen & Dierssen 2001). Succow (1988) integrated vegetation composition of mires based on geomorphological and hydrological features with palynological records and nutrient status of the mire systems. He distinguished c.10 different mire systems. In the present study we will only distinguish between the following five mire types: rainwater-fed bogs, groundwater-fed fens, terrestrialization mires, spring mires and floodplain mires.

9.2.1 Rainwater-fed bogs

Bogs are fed by rainwater although some peripheral parts of the bog system can be influenced by water that has been in contact with the mineral soil (Plate 9.1a). Normally they consist of a peat body, which has grown above the immediate surroundings. This rather firm and decayed peat layer (catotelm) is covered by a loosely structured layer (acrotelm) of mainly living mosses (*Sphagnum* species), intermingled with phanerogams which have adapted to very wet and nutrient-poor conditions. The acrotelm, consisting of mainly living plants, maintains the required wet conditions for the typical bog species. During very wet periods the acrotelm expands and the resistance to water flow becomes very low. Excess water can flow through the acrotelm to the periphery of the bog. In dry periods the acrotelm shrinks and becomes more firm, thus increasing the resistance to water flow. This feedback mechanism prevents a rapid loss of water from the bog during dry periods (Succow & Joosten 2001).

Various types of bog have been distinguished on the basis of their geomorphology (Vasander 1996, Dierssen & Dierssen 2001). Shallow-layered blanket bogs are common in Ireland and England. Plateau bogs, large lowland bogs with a flat top, used to be common in north-west Europe, but are now almost unknown. Kettle-hole mires are relatively small but usually very deep bogs, since they grow in Pleistocene areas where, after the retreat of the glaciers, ice blocks have melted, forming small lakes, which later terrestrialized as bogs. Kettle-hole mires are common in eastern Germany and Poland (Dierssen & Dierssen 2001, Succow & Joosten 2001). In northern Germany and in the Netherlands, with much precipitation during the summer, even smaller bogs of c.1–2 m deep occur in wet heathlands (Tüxen 1983, Grootjans *et al.* 2003). These bogs are usually found in former riverbeds, which have been cut off from groundwater discharge.

9.2.2 Groundwater-fed fens

Groundwater-fed fens are usually large peat-forming systems almost without trees and dominated by bryophytes (Plate 9.1b). Usually these fens are gently sloping, in particular in mountain areas, but they have also been known to occur in almost flat lowland river valleys, where the groundwater was supplied by surrounding hills. As in natural bogs, natural fens have loosely structured topsoils that permit a rather fast flow of water during wet periods. Such mires are, therefore, referred to as percolating mires (Succow 1988). Large natural fen systems of hundreds of hectares have been described from Canada (Glaser *et al.* 1990), Scandinavia (Dierssen 1996, Vasander 1996) and Russia (Sirin & Minaeva 2001). Such large peat-forming groundwater-fed mires were once widespread in north-west Europe too, but they do not occur any more under natural conditions in Europe (Dierssen & Dierssen 2001, Succow & Joosten 2001). In the best-preserved fen systems the typical fen species co-occur with a large number of fen meadow species. Locally some peat forming can occur, but many of such fen systems do not accumulate peat any more and in most systems that are regarded as fens the peat is actually degrading (Kotowski 2002).

9.2.3 Terrestrialization mires

Terrestrialization mires (see Plates 9.1c and 9.2a) are peat-forming systems, which in western Europe are usually found in human-made environments. Relatively natural terrestrialization mires may be found in eastern Europe and Scandinavia. Most terrestrializing mires are relatively small. Their size ranges from less than 1 ha to hundreds of hectares, such as the Norfolk Broads in England and the Weerribben in the Netherlands (van Wirdum *et al.* 1992). Sasser and Gosselink (1984) described a very large, floating freshwater system of *c.*3000 ha in the Mississippi River delta plain in Louisiana, USA. In Europe the best-developed terrestrialization mires occur in extensive peat-cut areas, where regeneration of the fen vegetation occurs after the original peat layers have been removed. Terrestrialization mires are ecologically very diverse. Eutrophic plant communities, such as reeds, form floating rafts on which in later stages mesotrophic or even oligotrophic plant communities can develop. When the floating rafts are large enough and the peat layer is sufficiently thick, rainwater lenses are formed which sustain nutrient-poor acid bog vegetation (van Wirdum 1995). Later in the succession the terrestrialization mires tend to turn into willow and alder shrub. In order to keep such mires open they are mown regularly. This can be done in summer with small mowing machines or in winter over the ice.

9.2.4 Spring mires

Spring mires are almost exclusively fed by discharging groundwater (Plates 9.1c and 9.2c). They occur in landscapes with much relief under rather complex geological and hydrological conditions. Usually they are situated where large groundwater aquifers are forced to discharge, due to geological phenomena that have changed the original stratification of sediments. At this geological breach a clay layer that has been pushed upwards may then block a sandy aquifer. Spring systems can be found in lowlands as elevated spots, in the middle of a sloping landscape or even close to a hilltop. In calcareous areas in particular, they can accumulate large amounts of $CaCO_3$ (travertine) up to a height of 30 m, but only under the condition of a regular discharge of groundwater. When groundwater, supersaturated for $CaCO_3$, comes into contact with the atmosphere, CO_2 escapes into the atmosphere and $CaCO_3$ precipitates. Spring mire species, in par-

ticular mosses, stimulate this process by using the high concentration of CO_2 in the discharging groundwater as a carbon source. The result is that $CaCO_3$ precipitates on the leaves and when these leaves die off and decompose small amounts of $CaCO_3$ are added to the soil (van Breemen & Buurman 2002). Under favourable climatic conditions peat forming may occur. Usually peat-forming processes in spring mires do not last very long. Small changes in the landscape or in climatological conditions trigger changes in the groundwater discharge and peat layers start to decompose. Strongly decomposed peat has a high resistance to water flow, thus blocking water transport in the spring cupola itself. The discharging groundwater then will force its way out somewhere else in the spring system, creating new opportunities for peat forming or travertine building. A spring mire, therefore, is a rather dynamic mire system that can shift rapidly from an accumulating to an erosive system where sediments are washed away or decomposed (Wolejko *et al.* 1994). The remaining spring mires in Europe are very small (often less than 1 ha). Some of the best-preserved systems can be found at the base of the Alps and the Slovakian Tatra Mountains.

9.2.5 Floodplain mires

Floodplain mires (Plate 9.1d) are also hydrologically very dynamic systems. In winter and spring intensive flooding may occur, which can deposit large amounts of sand, silt or clay. In summer water tables may drop to over 1 m below the surface. Under such conditions the availability of nutrients can be very high (Loucks 1992, Olde Venterink *et al.* 2003). The productivity of floodplain vegetation consequently can also be very high. On a geological timescale downstream rivers frequently change their course. As a result flooding frequencies and sedimentation rates also change in a wide range of mire ecosystems. Eutrophic floodplains may change into mesotrophic fens and existing fens or even bogs may become flooded with surface water. Examples of still existing large floodplain mires in Europe are the Danube floodplain in Romania, the Oder floodplains on the border of Poland and Germany, the Narev floodplain in Poland and the Rhone delta in France (see also Chapter 11 in this volume).

9.3 Consequences of land-use changes

The change from natural mires to semi-natural eco-systems is depicted in Plate 9.2, showing typical remnants of natural mire types after large changes in land use during the past centuries. The time period in which such land-use changes began may differ greatly between various parts of Europe, but the outcome is practically always the same. Important changes in land use are excavations of mires for fuel, intensification of agriculture and afforestation (see also Chapter 10). All these changes affected hydrological conditions, including the composition of groundwater and the direction of water flows. Most fens and bogs have been excavated for fuel and the peat structure of the remaining peat layers has been irreversibly changed due to deep drainage in adjacent agricultural fields. Moreover, the natural fluctuation pattern has been reversed in many European lowlands: low water levels in winter and spring and high water levels in summer. This is done to increase agricultural produc-tion, but the effects are also noticeable in nature reserves, causing direct or indirect eutrophication.

Before the introduction of artificial fertilizer, farmers depended on the natural soil fertility within a landscape and manure from livestock. Before the Second World War, almost every bit of landscape was used for food production in densely populated areas. This led to large losses of forested areas and natural mires. Fens were transformed into fen meadows on a large scale, but they had gained in species richness due to shallow drainage, resulting in spreading of meadow species into the fen without a total replacement of the original mire vegetation. In order to increase the soil fertility of the fen meadows artificial flooding was applied on a large scale in north-west Europe (Girel 1994, Hassler *et al.* 1995). Extensive networks of ditches and small dams guided base-rich surface water to and from the meadows. Soil fertility became less dependent on landscape processes after the intro-duction of artificial fertilizer. Many peat areas were drained deeply for food production. Towards the end of the 20th century the semi-natural landscape had almost disappeared in most of Europe. In western Europe small remnants of mires and species-rich meadows had been preserved as nature reserves, but they were usually influenced in a negative way by deeply drained adjacent agricultural fields. In eastern Europe, however, most of the remaining fen meadows declined in diversity due to abandonment (Ilnicki 2002, Kotowski 2002).

9.3.1 Rainwater-fed bogs

Large bog complexes (Plate 9.2a) that have been excavated commercially are left with a totally bare peat surface, consisting of remnants of the catotelm, the firm and decayed peat layer underneath the flexible and little-decomposed top layer. Hydrological and ecological conditions are very extreme after strip cutting. In particular in the summer water tables drop deep and desiccation can be severe. The bog peat is very acidic with pH values between 3 and 4. Under such conditions *Sphagnum* species cannot form a new vegetation layer. Grasses such as *Molinia caerulea* and *Deschampsia flexuosa* and trees such as *Betula pub-escens* soon cover the whole area. In western Europe most of the former bog areas have been reclaimed for agriculture.

Kettle-hole mires and other small bog systems can still be found in large numbers in afforested areas in the Netherlands, northern Germany and Poland. Remnants of such mires can also be found in agricul-tural areas, but most often they have been severely eutrophicated due to inflow of surface water with high amounts of nutrients. In afforested areas peat cutting may have occurred and drainage ditches are usually present to drain the direct surroundings of the mire.

9.3.2 Fens and fen meadows

From a hydrological point of view fens and fen meadows (Plate 9.2b) are very different systems (Grootjans & van Diggelen 1995). The distinction between the two systems, however, cannot easily be made in the field and many ecologists do not make a distinction between fens and fen meadows (Wheeler & Proctor 2000). Fen vegetation can still be found in numerous ecosystems ranging from small terrestrial-izing peat pits to spring mires and fen meadows, and even in calcareous dune slacks on almost bare mineral soil. Well-preserved fen and fen meadow complexes are still present in the mountain areas in

Scandinavia on sloping soils (Moen 1990, Dierssen 1996).

Changes in land use have most affected groundwater-fed fen systems. In the lowlands in particular, dense networks of canals and drainage systems were built to quickly transport water out of the area in winter, while in summer water was brought into the area from large rivers. Such practices had a major effect on the integrity of hydrological systems in the area and consequently also on the nutrient cycling in the wetlands. Ecohydrological research in the Biebrza catchment in eastern Poland, for instance, showed that the hydrochemical and associated trophic gradients were very smooth over large distances (Wassen *et al.* 1990, 1996). This spatial uniformity was related to the absence of drainage works resulting in an almost stagnant water body in the mire that kept it saturated with mineral-rich groundwater. Compared to Poland the fen systems in the Netherlands were much more varied on the landscape level. The former large hydrological gradients had been split up in numerous small ones due to large-scale changes in the regional hydrology. This resulted in significant hydrochemical differences even within a single small fen (van Wirdum 1995, Wassen *et al.* 1996). Despite these large differences in hydrological systems, the species composition and the site conditions in the fens are very similar. The groundwater table never dropped more than a few tens of centimetres below the surface and the water in the rooting zone was rich in base cations, especially Ca^{2+} and Fe^{2+} (Wassen *et al.* 1996).

In most of north-west Europe natural fens are almost non-existing. Extensive fens and fen meadow areas in Europe still occur in the Biebrza Valley, Poland, in the Baltic states and in Scandinavia. In other parts the fen meadow reserves are numerous, but small. Some examples of relatively well-preserved fen meadow reserves are the Drentsche Aa in the Netherlands, the Zwarte Beek in Belgium and the Wümme and Peene Valleys in Germany.

9.3.3 Terrestrialization mires

In flatlands, in particular, the original mires (fens or bogs) are long gone and terrestrialization has taken place under quite different hydrological conditions (Plate 9.2a). Once such areas were the lowest areas in

the landscape but, due to deep drainage of surrounding agricultural peat areas, they are now elevated wetlands that have turned from exfiltration areas into infiltration areas. Their hydrological function is mainly to store water, which is pumped into the wetlands from agricultural areas. Nowadays the later successional vegetation stages (shrubs of *Betula*, *Alnus* and *Salix*) prevail in most areas. The initial stages with mesotrophic vegetation have become rare. Those fens that have not been overgrown by shrubs acidify rather quickly (Kooijman & Bakker 1995, van Diggelen *et al.* 1996). The mires growing in these wetlands are also threatened by eutrophication, because they are influenced by surface water from polluted rivers during most of the summer (Barendregt *et al.* 1995).

9.3.4 Spring mires

Most spring mires (Plate 9.2c) in western Europe have disappeared due to a decreased inflow of groundwater from surrounding infiltration areas as a result of changes in land use. Intensive agricultural crop growing on the former heathlands and forests have decreased infiltration of rainwater considerably. In many areas, the spring systems themselves have been completely removed. From the Netherlands, for instance, documentation exists that iron-rich spring cupolas were exploited as a source of iron ore in the 19th century. In north-west Poland many spring systems are still present and their infiltration areas have not been hydrologically disturbed. Yet most of them have been damaged by drainage in downstream lake areas (Wolejko *et al.* 1994). Due to decreasing lake levels in the 19th century, erosion processes have been triggered that have washed away most of the peat that had accumulated around the spring systems.

9.3.5 Floodplain mires

Floodplain mires (Plate 9.2d) usually have very fertile soils and most of them have been reclaimed for agriculture when financial means and technical skills were available to drain the areas. Remnants of former floodplain mires can still be found in very low-lying areas that could not be drained easily and were left as nature areas. Examples are sections of the Biebrza and Narev Rivers in Poland, the Peene Haffmoor in Germany and also the Odra National Park on the border between Poland and Germany. Very often large canals have replaced the original streams. Dikes were built to prevent flooding of agricultural and urban areas, which led to increased fluctuations in the water table in the remnants of floodplain areas: high water levels in the wet periods and low water levels in the summer. Floodplain systems that have vegetation types which resemble natural peat-forming vegetation types are either groundwater-fed in addition to regular flooding, or they are supplied with additional water from surface-water systems (Wassen *et al.* 1996).

9.4 Causes of further deterioration in semi-natural wetlands

Remnants of semi-natural vegetation are now situated in a landscape that has completely different properties compared to the time in which these ecosystems evolved. Deep drainage in agricultural areas and abstraction of groundwater for the public water supply and for industrial purposes has left wet nature areas with considerable water shortages, especially during the summer. This process is sometimes called desiccation, but in areas with a large precipitation surplus lack of groundwater does not so much lead to water shortage for plant growth, but to increased mineralization of the peat and replacement of groundwater by rainwater (Grootjans *et al.* 1996). In order to compensate for these water losses additional surface water was sometimes supplied to these wetlands. The nutrient content of surface water usually is much higher than that of groundwater and it also has different concentrations of iron, sulphate and chloride (Roelofs 1991, Barendregt *et al.* 1995). Apart from that, the intensive agricultural land use emits high amounts of NH_3 into the atmosphere, which also reaches nature areas through dust particles and precipitation. In some regions in Europe this atmospheric N deposition can be very high ($25-40$ kg of N ha^{-1} yr^{-1}).

9.4.1 Acidification

Acidification of the topsoil occurs when calcareous and iron-rich groundwater in the topsoil is replaced by rain-

water (Plate 9.3). This process is generally promoted by drainage activities in the landscape (Grootjans *et al.* 1988), but it can also be a natural phenomenon, when floating mats of vegetation in terrestrializing mires increase in thickness and retain rainwater in the topsoil (Schot *et al.* 2004). Such processes mark the beginning of bog formation. In autumn and winter a rainwater lens develops in the profile and can reach a depth of over 1 m. In areas with a large precipitation surplus this leads to acidification of the topsoil, because practically all base cations are being exchanged for H^+ and Al^{3+} (Steinberg & Wright 1994, Brady & Weil 1999). In the long run the soil may even become iron depleted. Over a period of several decades, this acidification leads to a drop in soil fertility. Moist sub-neutral grassland types, in particular, are very susceptible to acidification, because buffer mechanisms are weak in such ecosystems. Species such as *Arnica montana*, *Antennaria dioica*, *Botrychium lunaria* and many others have declined in north-west Europe, even in well-managed nature reserves, due to drainage and atmospheric deposition. This acid rain has often depleted the buffer mechanisms in the topsoil, thus increasing the availability of Al^{3+} and also NH_4^+ at the cost of Ca^{2+} and NO_3^-. De Graaf *et al.* (1998b) have shown that *A. montana* and *A. dioica* are very vulnerable to high concentrations of NH_4^+ and Al^{3+} in the topsoil. They found that the absolute values of Al^{3+} were not important, but that the $Ca^{2+} : Al^{3+}$ ratio in the soil solution should be above 1.

9.4.2 Eutrophication: increased influence of surface water systems

In natural fens nitrogen is often the primary limiting nutrient, while harvesting eventually results in phosphorus limitation (Verhoeven *et al.* 1996). Experimental research had shown that Ca^{2+} can limit P availability in very calcareous fens (Boyer & Wheeler 1989) but in less calcareous fens Fe^{2+}/Fe^{3+} and Al^{3+} play a more important role (Boeye & Verheyen 1994, Wassen *et al.* 1996).

The decline of groundwater discharge in wetlands has lead to widespread eutrophication in most European wetlands, in north-west Europe in particular. Eutrophication means an increase in availability of nutrients and is expressed in the vegetation by a shift from low-productivity fen vegetation to high-productivity reeds (Olde Venterink *et al.* 2002, Wheeler *et al.* 2002). This process is often associated with an influx of nutrient-rich surface water into the fens (Harding 1993).

An increase in the sulphate concentration in the surface water can also increase nutrient cycling in the peat soils due to microbial sulphate reduction under anoxic conditions (Smolders & Roelofs 1993, Koerselman & Verhoeven 1995). During sulphate reduction organic material is used as a carbon source and sulphide is produced, which is toxic for most vascular plants in relatively high concentrations (Lamers *et al.* 2002). Also an increase in alkalinity can be observed during sulphate reduction (Stumm & Morgan 1981). In most groundwater-fed fen and fen meadow soils sulphide will not reach toxic concentrations since it is chemically bound by iron. However, when sulphide production exceeds the availability of iron, free sulphide concentrations increase (Caraco *et al.* 1989). A side effect of increased sulphide concentrations is that phosphate concentrations can also increase considerably in the soil pore water. In soils phosphate is normally bound to organic matter, clay, but also to iron, calcium or aluminium (Stumm & Morgan 1981, Scheffer & Schachtschabel 1992, Brady & Weil 1999). Phosphate sorption to iron hydroxides is particularly important in wetland soils that are not very acid or very calcareous. Sulphide can interfere with phosphate binding to iron hydroxide. It can bind with iron complexes forming iron sulphides such as FeS and FeS_2 (pyrite). Under limited iron availability phosphates are released and become available for plant growth (Koerselman *et al.* 1990, Smolders *et al.* 1995, Lucassen *et al.* 2004). Such changes in nutrient cycling, called **internal eutrophication**, can be very harmful for fens and fen meadows that are phosphorus-limited (Verhoeven *et al.* 1996, Lamers *et al.* 1998). In summary, it can be concluded that anoxic iron-rich groundwater tends to slow down nutrient cycling, while oxic surface water tends to stimulate nutrient cycling.

9.4.3 Atmospheric deposition

Atmospheric nitrogen and sulphur deposition have a very negative effect on almost all low-productivity wetland ecosystems. Between 1930 and 1980 the annual atmospheric N deposition in the Netherlands increased from 8 to 40 kg of N $ha^{-1} yr^{-1}$. In areas with intensive cattle farming up to 80 kg were measured. Increased N loads tend to increase the speed of succession. Fast-growing species start to dominate the vegetation and replace many characteristic and protected plant and animal species, even in nature reserves. The concept of critical N load has been developed to indicate the annual amount of N deposition where the ecosystem involved is stable with respect to essential ecosystem functions. Critical loads for bogs are 5–10 kg of N $ha^{-1} yr^{-1}$, for moist acid grasslands 10–15 kg, for mesotrophic terrestrialization mires 20–35 kg (unmanaged) and for wet heathlands 17–20 kg (Bobbink *et al.* 1998). When the deposition of N increases the *Sphagnum* species start to accumulate N in amino acids or mineral N is not taken up and becomes available for vascular plants such as *M. caerulea* and tree species such as *B. pubescens* (Lamers *et al.* 2000, Tomassen *et al.* 2004a).

9.5 Restoration approaches, successes and failures

Apart from aesthetic reasons, economic reasons have become important in initiating restoration projects in peatland areas. Most peatlands have become unsuitable for modern agricultural production (Pfadenhauer & Grootjans 1999), and in many peat areas in Europe the maintenance costs for keeping such areas suitable for agricultural production are simply too high to sustain agricultural production in an economic way. Environmental considerations are also important. Deeply drained peatlands which are used heavily by agriculture suffer badly from shrinking and peat loss through deflation and mineralization. This may amount to up to 2 cm yr^{-1} and results in the release of enormous amounts of CO_2 into the atmosphere (release of up to 40 kg of N_2O $ha^{-1} yr^{-1}$ and up to 10 t of CO_2; Armentano & Menges 1986, Augustin *et al.* 1996). So there is a high societal need to restore high water levels in degrading peat areas, which for biodiversity reasons should be fed again with clean groundwater from restored hydrological systems, and which are capable of reducing nutrient loads of polluted surface-water systems.

Most traditional nature-conservation organizations aim to rehabilitate damaged mires or closely related systems. Wheeler and Shaw (1995b) described such activities as repairing a system. Such activities consist of raising the water tables by removing drainage ditches, terminating groundwater-abstraction facilities, reinstalling traditional management, etc. Sometimes part of the threatened vegetation type is transplanted to another site. When a wetland system is damaged beyond repair, restoration may end in rebuilding the system to a state that probably has never existed before. Most large nature-development projects are actually close to (re)building a new wetland ecosystem in areas that have been abandoned by agriculture, industries or other land users. Rehabilitation of a former stage of ecosystem development is often not an option any more, since nothing of the former stages has remained.

9.5.1 Restoration of rainwater-fed bogs

In north-west Europe and in North America restoration efforts include attempts to restart peat growth in the large bog complexes where peat has been extracted for commercial use. Restoration of such flat surfaces has proven to be very difficult. Simply rewetting remnants of large bog complexes usually does not result in renewed *Sphagnum* growth (Vermeer & Joosten 1992). Shallow flooding often results in much fluctuation in water levels, and the topsoil dries out again in dry years, unless renewed *Sphagnum* growth is able to start from floating rafts of vegetation (Tomassen *et al.* 2004b). Wheeler and Shaw (1995b) and Wheeler *et al.* (2002) present an overview of techniques and approaches to bog restoration in north-west Europe and North America. Often we can see a rapid growth of *Sphagnum* species in early stages of bog formation, particularly in small peat-cut pits (Lütke-Twenhöven 1992), but the development of late-successional, bog-building sphagna is often very poor or non-existent. Several studies showed that in the first stages of peat regeneration *Sphagnum* plants need support of tussock-forming species, such as *Eriophorum vaginatum* or from perennial species such as relatively tall *Carex* species that can form many erect shoots, without shading the *Sphagnum* plant too much (Pfadenhauer & Klötzli 1996). Smolders

et al. (2001) and Lamers *et al.* (2002) showed that *Sphagnum* growth is stimulated by addition of small amounts of HCO_3^-. The HCO_3^- dissociates into CO_2 and provides the *Sphagnum* with an additional C source. Concentrations between 0 and 1 mM stimulated the growth of *Sphagnum*, while concentrations above 1 mM restricted *Sphagnum* growth. Limpens *et al.* (2003) showed that high atmospheric N deposition (30–35 kg of N ha^{-1} yr^{-1}) reduces the growth of ombrotrophic *Sphagnum* species. Tomassen *et al.* (2004a) also showed that elevated atmospheric N deposition could favour the growth of *M. caerulea* and *B. pubescens*. Above 18 kg of N ha^{-1} yr^{-1} *Sphagnum* species cannot take up all the NH_4^+ from the soil solution and the nitrogen becomes available for vascular plants, which start to shade the *Sphagnum* plants, thus reducing their growth considerably. Water-table fluctuations in large bog remnants are usually sub-optimal for *Sphagnum* growth, due to the absence of a flexible acrotelm. Furthermore, water tables around the bog remnants are usually very low due to agricultural drainage. Building large dams in and around the bog remnants can decrease the water-table fluctuations.

To summarize, the restoration in large bog remnants in north-west Europe is usually slow due to high water-table fluctuations (> 25 cm), high atmospheric N deposition, a very low pH and low CO_2 concentrations in the pore water.

Kettle-hole restoration was quite successful in north-east Germany (Jeschke & Paulson 2001). After felling woods and closing drainage ditches, *Sphagnum* growth proceeded rapidly and within 5–10 years extensive *Sphagnum* carpets developed. Rapid regeneration of *Sphagnum* growth was also observed in small bogs within a Dutch heathland area (Grootjans *et al.* 2003). Here the closing of drainage ditches not only raised the water tables in the small bogs, but it also stimulated water flow above still-existing water-impervious organic B horizons (consisting of a layer of fine, dispersed organic material that has been transported by water movement to deeper layers). These thin, organic layers connected several small bogs (Plate 9.4), providing them with additional water with a high CO_2 concentration. These shallow flows over and through organic soil layers not only reduced water-table fluctuations and increased the pH, but also provided the sphagna with an additional C source (elevated CO_2 concentrations in the flowing water).

9.5.2 Restoration of fens and fen meadows

It is generally believed that under the present high rates of N deposition fens and fen meadows cannot be maintained without a regular mowing regime. The first thing to do in abandoned hay meadows is to resume the traditional haymaking. In some cases a combination with extensive grazing is also possible. Sod cutting is sometimes applied when the topsoil of a fen meadow has been severely acidified by drainage or when the vegetation has shifted to a grass-dominated stage after long-term exposure to high N deposition from the air.

Resume traditional management

Resuming the traditional management of haymaking without fertilization has proven a very suitable measure to restore species-rich meadows in Europe. This type of management is widely applied in nature reserves in western Europe where intensive agricultural production is not allowed. In some countries private or state nature-conservation organizations carry out the mowing themselves, but more often local farmers are paid to do the mowing. A success story is, for instance, the creation in 1965 of the largest meadow reserve in the Netherlands, the Drentsche Aa catchment area, where most of the meadows on peat soils had never been used intensively and where deep drainage had not been applied on a large scale. Most of such meadows responded very well to restoration management after a short period of abandonment. The increase of target species depended on the amount of nutrients accumulated during former fertilizer application (Bakker & Olff 1995). It may take 5–15 years before the nutrient output has reduced the nutrients stocks to a level in which the productivity of dominant species no longer prevents the establishment of target species (see Prach *et al.* 1996). Even more target species established in former agricultural areas on mineral soils after topsoil removal, but only when former hydrological conditions were restored, seed banks had not been depleted and where dispersal mechanisms (flooding) from adjacent nature reserves were effective (Moen 1990, Jansen *et al.* 2000).

Restoring the hydrological regime

Restoring the hydrological regime is more than just increasing water levels (rewetting). In groundwater-fed mires, in particular, increasing water levels may lead to acidification when the discharge of base-rich groundwater cannot be restored (van Wirdum *et al.* 1992). Restoring groundwater-fed mires includes increasing water levels in the whole surrounding catchment area, which in most cases is not possible any more, due to other hydrological claims (farmers, cities). When groundwater pressure in a mire is still high, digging of small shallow ditches can be recommended (Grootjans *et al.* 2002), which can transport most of the infiltrating rainwater out of the mire. This prevents the influence of acid rainwater lenses dominating in the mire (Schot *et al.* 2004).

Nutrient removal

Resuming the traditional management in fen meadows that have been fertilized intensively in former days will not lead to rapid success without rewetting and topsoil removal. This is clearly illustrated for the *Veenkampen* experiment (Oomes *et al.* 1996) where mowing without fertilization was very unsuccessful, even after 10 years, despite several target species being present in the seed bank. Mowing in combination with rewetting was much more successful and combined with sod cutting it has lead to a reappearance of many Red List species with a long persistent seed bank (Matus *et al.* 2003). These species had survived in the soil seed bank for up to 40 years. Under appropriate site conditions, therefore, activation of seed banks can be beneficial for the survival of endangered plant species, since the rapid population expansions after restoration measures raise the opportunity to produce a new seed bank (Jansen & Roelofs 1996). Tallowin and Smith (2001) reported on successful restoration of a species-rich fen meadow after topsoil removal combined with seeding and planting of seedlings of target species.

Shallow sod cutting in former agricultural fields, drained and fertilized for a long time, is usually ineffective (Pfadenhauer *et al.* 2001, Grootjans *et al.* 2002), since (i) soil degradation has initiated irreversible changes in the topsoil and (ii) seed banks are depleted while dispersal mechanisms are ineffective (van

Diggelen 1998). Soil degradation consists of irreversible changes in the structure of organic matter, which in extreme cases can become water repellent in dry periods and after rewetting does not have the ability to retain water as much as less-decomposed peat types. The reduced capillary rise from the groundwater can cause severe drought in the topsoil even when the groundwater tables are relatively high (Schmidt 1995, Schrautzer *et al.* 1996, Succow & Joosten 2001).

Transplantation

Transplantation of threatened (meadow) vegetation to botanical gardens already has a long tradition in Germany. A well-documented example of fen meadow transplantation in Switzerland was given by Klötzli (1987), who monitored vegetation changes in transplanted turfs at the Zürich national airport, where nutrient-poor fen meadow vegetation had to make way for a new runway. The turfs were transplanted to a newly prepared site next to the runway. Lessons to be learned from this project were that in wetlands it can take several decades before the species composition of a transplanted vegetation changes, but that eventually such an ecosystem has to change, because the hydrological conditions of the newly constructed site are always different from the original site. Hutchings and Stewart (2002) also question the success of transplantations of vegetation turfs in dry calcareous grasslands. Transplantation of individual species to restoration sites has been applied in restoration projects on a large scale (Perrow & Davy 2002b) and is generally accepted as a tool to speed up the restoration process or to test whether or not the ecological conditions of a restoration site meet the requirements of the target species (van Duren *et al.* 1998; see also Chapter 7).

9.5.3 Restoration of terrestrialization mires

Restoring terrestrialization mires often starts with improving the water quality of the lakes. Technical solutions, such as sewage-treatment plants, can restore the oxygen balance in surface water and can decrease the nutrient concentrations. In most cases this is not enough, since a large amount of nutrients originates from diffuse sources that can enter a nature area from canals and agricultural drainage ditches. A proper way to deal with the complex relationships in the water systems and human exploitation (drinking water, navigation and tourism) is to practise integral water management. This includes management of a chain of problems in a specified watershed, such as pollution of underwater sediments, local and regional input of pollutants, shoreline management, management of water levels in the whole catchment, abstraction of drinking water as well as industrial use of groundwater.

Not all terrestrialization mires need management. In oligotrophic lakes terrestrialization soon leads to the development of bog or fen vegetation. In eutrophic lakes, however, terrestrialization can lead to monocultures of reed (*Phragmites australis*), especially when flooding with eutrophic surface water occurs. In such cases, reed cutting, either in summer or during the winter, can be a practical measure to increase biodiversity in such mires.

Beltman *et al.* (1995) reported on attempts to regenerate rich fen vegetation by sod cutting to remove the existing vegetation or digging of shallow ditches to remove the acid-precipitation surplus. They showed that rich fen vegetation survived after 2 years only in the areas where both measures had been applied and even then this was only successful in areas with upward groundwater movement (Beltman *et al.* 2001).

Excavating new turf ponds is an expensive way of starting terrestrialization anew, and should only be practised when the new turf ponds can be supplied with very clean water, preferably groundwater (Beltman *et al.* 1996). However, not all plant species can reach such restoration areas, since they are usually isolated from the surrounding watercourses. Connecting peat-cut areas with larger surface-water systems may be positive for dispersal of propagules, but is often not an option when the surface-water system is influenced by intensive agricultural production areas.

9.5.4 Restoration of spring mires

Most spring-mire systems in Europe cannot be restored anymore since the original peats and sediments have often disappeared and the water discharge sites have shifted to lower sections of the spring system and cannot be directed upwards. The replacement communities, consisting of plant and animal species that are adapted to low temperatures and constantly flowing water, are also highly endangered in western Europe (Wolejko *et al.* 1994). Restoration measures should be aimed at stabilizing the water outflow in the springs and preventing pollution in infiltration areas.

When the discharge of groundwater has been diminished or if the spring water has been polluted in surrounding infiltration areas, the most obvious measure is to protect the direct catchment areas and abolish drainage and fertilization practices.

Koska and Stegmann (2001) reported on the restoration of a severely damaged spring mire in north-east Germany. Several measures were taken to raise the water tables in a degraded spring-mire complex (6 ha) in the Sernitz source area. These measures included (i) construction of a series of wooden dams in the largest drainage ditches, (ii) complete filling of ditches with peat, (iii) reflooding parts of the mire with spring water and (iv) perforating the impervious gyttja layers at the base of the mire. The best results were obtained with complete filling of drainage ditches. Flooding parts of the mire with spring water that had accumulated behind wooden dams in the main drainage could not prevent severe desiccation during the summer. The series of wooden dams in the main drainage ditch were also ineffective, since it had only a very local effect and these measures only slow the drainage down, but do not prevent it. Perforating the impervious layers to enable spring water to discharge at the top of the spring mire had also a very local effect and did not contribute to the rewetting of the mire. Perforating impervious layers in spring systems is, in general, not recommended since it can damage mire systems further down the hydrological gradient. Yet in the case where shortcuts have been made at the base of still active springs, the construction of an artificial outlet in the top of the system could be a measure for repairing the damage at the base of the spring. If at all possible, the outlets at the base of the springs should be closed after installation of the new outlet closer to the top.

9.5.5 Restoration of flood plains: stimulate traditional management

Stimulating flooding in relatively eutrophic flood mires can be a good measure to combat acidification,

and it can under certain conditions compensate for the influence of drainage. Negative effects of deep drainage in agricultural areas can sometimes be compensated for by using existing hydrological structures, such as a large canal with a high water level. Connecting the old ditch system and the remnants of river meanders with the canal enables the continuation of flooding in winter, while groundwater losses to the adjoining agricultural fields are largely compensated by a lateral (groundwater) flow from the canal and by high water levels in the old meanders. The composition of the surface water is, however, crucial in such artificial hydrological systems. Nutrient levels and sulphate levels should be very low to prevent rapid eutrophication of the soil. More often remnants of former extensive flood mires are now completely surrounded by agricultural areas with much lower water levels. Former discharge areas have turned into infiltration areas and acidification of the topsoil has degraded the fen meadows to the extent that, for instance, iron levels have become critically low. Reflooding with surface water, even when pre-purified by helophyte filters (reed swamps that reduce the nutrient content of the surface water), does not always restore the iron contents to higher levels, since surface water usually contains very low concentrations of iron. If iron is fixed as FeS_2 (pyrite), flooding with surface water may lead to a higher pH during the winter and spring period (van Breemen & Buurman 2002). However, when discharge of calcareous and iron-rich groundwater is no longer available during summer, the groundwater levels will drop and the topsoil acidifies again due to oxidation of FeS_2 (Lamers *et al.* 2002). Permanent flooding could prevent acidification, but it will also increase the availability of nutrients in the soil, leading to the very productive plant communities (Lucassen *et al.* 2004). Such restored wetlands, however, have regained an increased capability to capture nutrients from the surface-water system (Hoffmann 1998, Olde Venterink *et al.* 2003).

Stimulated flooding in former agricultural areas usually results in very eutrophic soil conditions (Richert *et al.* 2000, Roth *et al.* 2001). Most floodplain species on eutrophic soil are very persistent in the cultural landscape. They have remained in wet spots and in ditches and after rewetting they can expand very rapidly (*Glyceria maxima*, *Acorus calamus* and *Phalaris arundinacea*). Although most of these species are not peat-forming species, they can rapidly form a highly productive marsh vegetation, suitable for sustaining large populations of waterfowl. In order to restore the peat-forming function of the mire the vegetation should consist of marsh species with less easily degradable tissue (*Phragmites australis*, tall *Carex* species; Hartman 1999, Richert *et al.* 2000). Flooding is also important for redistribution of seeds within restoration sites (Danvid & Nilsson 1997, Jansen *et al.* 2004, Wheeler *et al.* 2002). Middleton (1999) advocates the need for a 'flood pulse' water regime, which is the natural regime; flooding in winter and lower water levels in summer. Such a regime stimulates seed dispersal in winter, allows germination in spring and prevents desorption of phosphates and possibly sulphide toxicity in summer (Lamers *et al.* 2002).

9.6 Generalization and concluding remarks

9.6.1 Successes and failures

In Europe, the restoration of natural mires in a landscape that has been used intensively by humans is an almost impossible task, since hydrological conditions of natural mires are practically always connected to large hydrological systems of the surrounding landscape. In densely populated areas support from society will be lacking entirely when such goals are formulated. Targets should be formulated very clearly in order to obtain political support for restoration projects (Swart *et al.* 2001). In practice, natural mire ecosystems can only be restored on a very local scale, and although aspects of natural mires may return in restored wetlands, the new ecosystems will differ from the ones that have been destroyed in former times. Restoring semi-natural ecosystems may gain considerable popular support, particularly in western Europe, where very few elements of the semi-natural landscape, shaped by past generations, have remained. Large sums of money are now being spent to conserve and restore species-rich grasslands or wetlands with large numbers of waterfowl. Restoration projects aimed at restoring semi-natural ecosystems are usually considered a success if rare and endangered species (Red List species) return. It is not surprising

that resuming the traditional management in predominantly cultural landscapes produces the best effects in mires and wetlands that have not been drained or fertilized. Successful projects have been executed at sites that have been least affected by intensive agriculture and drainage. Successful restoration in western Europe is often the result of repairing damage at high costs. From a European perspective the conservation of still-existing semi-natural ecosystems is much more cost-effective. Failures to repair damaged elements of the semi-natural landscape are as numerous as the successes, but they are usually not well documented. Failures are often caused by an incorrect diagnosis of the restoration prospects of the site, lack of knowledge on ecological processes affecting the site negatively, and expectations that are too high (lack of knowledge on appropriate references). The availability of reference communities is, indeed, an acute problem in target areas where intensive fertilization has taken place for a long time, soil degradation has occurred and where practically all target species have disappeared in both the established vegetation and the seed bank.

9.6.2 Prospects for the future

Reduction of the amount of N deposition remains a prerequisite for successful restoration of many nutrient-poor ecosystems, such as bogs, fens and several types of fen meadow. In the Netherlands the atmospheric N deposition has declined (*c.*40%) since 1985 and the atmospheric deposition of SO_2 has also decreased strongly (*c.*70%) during the past 20 years. Freshwater ecosystems have benefited considerably from drastic reductions of the phosphorus emissions in the surface water, but in many streams and rivers the amount of sulphate is still much too high and causes eutrophication in many terrestrial mire systems when they are flooded.

The prospects of restoration of damaged ecosystems are relatively good in most western European countries, since (i) much experience is now available to repair or even rebuild damaged (semi-natural) ecosystems, (ii) some of the environmental stress factors such as sulphate and nitrogen deposition are decreasing and (iii) public and political support for restoration is increasing. In eastern and southern parts of Europe the situation is less clear. Since environmental laws are less strict or less strictly controlled, environmental problems may be imported from the west. Public and political support for restoration activities is largely lacking. However, the quality and the number of relatively undisturbed semi-natural wetlands is still very high in many east European countries and the costs to repair the damage in such systems are still low. It could be very beneficial to develop projects that combine economic development and maintenance of semi-natural mires and wetlands (Wichtman & Succow 2001). It makes little sense to follow the example of western Europe, where large sums of money have been spent to destroy semi-natural landscapes and shortly afterwards restore small relic areas at all costs and usually with little success.

10

Restoration of forests

Anton Fischer and Holger Fischer

10.1 Introduction

In 2000 the world's forests covered about 3.87 billion ha, or about 29.6% of the world's total land area. Between 1990 and 2000 there was a net loss of forest area of about 9.4 million ha yr^{-1} (−0.22% yr^{-1}; FAO 2001). These figures demonstrate clearly that forests are an important natural resource that is decreasing rapidly.

The concept of potential natural vegetation (PNV; Faber 1937, Tüxen 1956) is a tool to indicate and map sets of site conditions relevant for plant growth in terms of vegetation types. The *Map of the Natural Vegetation of Europe* (BfN 2000; scale 1 : 2.5 million) is the most general PNV map for Europe. Its analysis (Table 10.1) shows that – according to the site conditions – forests represent the European natural large-scale ecosystem, among which the boreal coniferous and the temperate deciduous forests predominate. In central Europe (taking Germany as an example; see Table 10.1) there are only very limited areas which, according to recent abiotical site conditions, cannot support forests, for example some coastal and alpine areas as well as raised bogs. In Europe, large-scale, natural open land only occurs in the high north (arctic tundras) and the far east (steppes). For Bavaria, situated in the core of central Europe, Walentowski and Gulder (2001) calculated that the three PNV units, namely *Luzulo-Fagetum*, *Galio odorati-Fagetum* and *Hordelymo-Fagetum*, together cover about 70% of the state forests.

Over two thirds of the potential (and former) forest area, however, has been converted into either agricultural land or land used for settlement and infrastructural purposes (59.5% and 11.3%, respectively, in Germany; BfN 1996). Compared with the PNV the tree species composition of about two thirds of the remaining forest stands changed fundamentally due to the impact of humans (in central Europe needle trees replaced beech). There are, however, remarkable numbers of forest stands existing which, according to their floristical composition, can be interpreted as near-natural. Nevertheless, the stand structure is usually intensively influenced by forest management, and many groups of organisms depend especially on stand structure; for example, many insects and fungi live in dead wood.

The conversion of forests to open land implied the establishment and expansion of vegetation types characteristic of the historic cultural landscape, such as unfertilized grasslands and heathlands, both dry and wet, and many types of sedge wetland as well as weed and ruderal communities. These vegetation types, which are discussed in other chapters of this book, essentially depend on the management of humans. Forest is the only (potentially) widespread ecosystem type in large parts of Europe that is capable of persisting without anthropogenic impact.

This chapter on **forest restoration** covers aspects of (i) forest **improvement** in the sense of optimizing close-to-nature forestry, (ii) **conversion** of tree plantations, mostly Scots pine and Norway spruce, into (mixed) deciduous forests, (iii) forest **stand regeneration** after wind-throw, (iv) **reforestation** of areas used in the past for agricultural purposes and (v) **afforestation** of completely sterile new substrates in post-mining landscapes. We focus on central Europe but reflect also boreal and Mediterranean aspects.

Table 10.1 Potential natural vegetation (PNV) in central Europe and Germany. Data shown are the percentage of land area, without lakes, rivers or glaciers. A value of 0 means that the vegetation type is present, but covers less than 0.1%. Calculations are based on BfN data from the *Map of the Natural Vegetation of Europe* (BfN 2000). Data for Germany are for June 2001 and those for central Europe are for May 2002.

Code	PNV unit	Land area (%)	
		Europe	Germany
A	Polar deserts and subnival–nival vegetation of high mountains	0.4	–
B	Arctic tundras and alpine vegetation	5.3	0.1
C	Subarctic, boreal and nemoral-montane open woodlands	3.3	0
D	(Hygro)mesophytic coniferous and mixed broad-leaved/coniferous forests	30.7	1.3
E	Atlantic dwarf shrub heaths	0.2	–
F	Mesophytic deciduous broad-leaved and mixed coniferous/broad-leaved forests	24.6	87.4
G	Thermophilous mixed deciduous broad-leaved forests	5.9	0
H	Hygro-thermophilous mixed deciduous broad-leaved forests	0.1	–
J	Mediterranean sclerophyllous forests and shrub	5.4	–
K	Xerophytic coniferous forests and shrub	0.4	0.1
L	Forest steppes and dry grasslands	4.1	–
M	Steppes	9.7	–
N	Oroxerophytic vegetation (thorn-cushion communities, mountain steppes)	0.1	–
O	Deserts	1.6	–
P	Coastal vegetation and inland halophytic vegetation	0.6	0.1
R	Tall reed vegetation and tall sedge swamps, aquatic vegetation	0.4	–
S	Mires	2.8	0.3
T	Swamp and fen forests	0.5	1.7
U	Vegetation of floodplains, estuaries and freshwater polders	4.3	8.8

10.2 A history of European forests and forest utilization

10.2.1 History

The European landscape and ecosystems changed drastically during the Pleistocene era. During the time of maximum extent of the glacial ice masses, about 15,000 years ago, Scandinavia was completely covered by ice, as were northern Germany and northern Poland. Large areas of the North Sea belonged to the mainland, and the Alps were covered by glaciers. The world mean temperature was about 4 K lower than today, in Europe about 8 K. The climate was cold and dry (continental). Tundra and steppe were not separated by a forest belt at that time; therefore, the species of these different vegetation types had the chance to intermix. Parts of Europe not covered by ice were settled by the so-called tundra-steppe. Trees and forests survived the ice age in southern Europe, with a few needle trees (e.g. *Larix*) in the eastern Alps.

According to our current knowledge (IPCC 2001) the end of the ice age was a sudden event. The temperature increased rapidly within a few centuries (or faster), and as a consequence large areas of Europe became capable of tree growth within a rather short period of time. Hundreds of pollen diagrams, summarized by Lang (1994), reveal that in central Europe pine (*Pinus* spp.) and birch (*Betula* spp.) occupied most of the area at first. Later on, trees that had survived the ice age in more southern parts of Europe invaded, and at last beech (*Fagus sylvatica*) occurred, occupying large areas of Europe. Fig. 10.1 presents the comparably late but triumphal post-glacial procession of beech across Europe. Trees other than beech as well as forest types

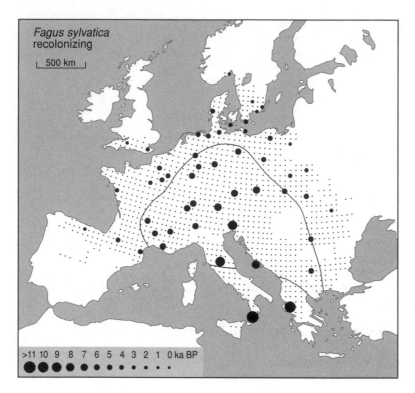

Fig. 10.1 Beech (*Fagus sylvatica*) recolonizing Europe after the end of the ice age. Dot size characterizes the time period since the first record; figures are in kilo-years before present (ka BP). Spotted area: recent distribution. Lines: nothern border of distribution in 10,000 BC (lower line) and 5000 BC (upper line). Drawn from Lang (1994), with the permission of the author.

other than beech-dominated forests are restricted to very limited and rather extreme parts of the recent site spectrum in central Europe. As a result we have to realize that Europe as a potentially forested area is a relatively recent phenomenon, and the same is true for the main mosaic of forest types: (i) pine and spruce forests in boreal northern Europe, (ii) deciduous forests in temperate Europe (in western Europe mainly beech, in eastern Europe mainly hornbeam and lime tree) and (iii) evergreen sclerophyllous forests in southern Europe.

This patch mosaic never had a chance to establish undisturbed, because after the ice age humans also started to settle in Europe. On the Mediterranean islands (e.g. Mallorca, Malta) and in western Europe the remains of the megalith culture show that during the Neolithicum, about 5000 years ago, humans already had relatively high technical standards, and this implied the utilization of natural resources or land consumption. Also in southern central Europe people started to build permanent settlements during the Neolithic age; they cut down the (oak) forest and

carried out agriculture (e.g. Iversen 1941, Bürger 1995, Küster 1997). Forests were pushed back more and more, and after small-scale devastations new sites for pioneer trees like birch and pine arose. Beginning at that time, beech started to invade, but many places had already been occupied by humans.

Up to the 19th century wood and timber were an essential resource for the survival of humans and for the economic development of the human societies in Europe. Timber was used for the construction of houses, bridges and ships. Fire wood (charcoal) was needed for heating and cooking as well as for industrial purposes in the beginning stage of industrialization (e.g. mining industry). Beyond that wood was used for furniture, instruments, utensils, fences and so on; oak bark delivered tanning agent (clothes, leather industry). As a consequence the utilization pressure on forests was extremely high; all these demands were realized in an unregulated way without any care. Thus the forest area decreased (in central Europe less than 20% remained), and the remaining forest stands became more and more devastated.

Fig. 10.2 Tree species composition according to PNV (left-hand diagram) and recent vegetation (right-hand diagram) for Bavaria. Diagrams are based on data from Walentowski and Gulder (2001). Deciduous trees: Fs, *Fagus sylvatica*; Qrp, *Quercus robur* and *Quercus petraea*; Ap, *Acer pseudoplatanus*; Cb, *Carpinus betulus*; Fe, *Fraxinus excelsior*. Coniferous trees: Pa, *Picea abies*; Aa, *Abies alba*; Ps, *Pinus sylvestris*; Ld, *Larix decidua*; Pseudo, *Pseudotsuga menziesii*. Misc., miscellaneous.

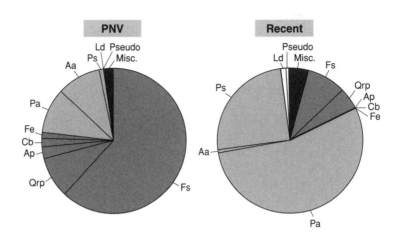

10.2.2 300 years of forest management

Although the first attempts at forest restoration were carried out in the 14th century (the oldest managed forest of the world is the Nürnberger Reichswald; Sperber 1968), forest utilization meant forest plundering for several centuries. It was only in the early 18th century that Hans Carl von Carlowitz (1645–1714), member of the administration of the mining industry in Freiberg/Saxony and responsible for timber production, realized that only with a strong forest management could timber production be maintained for a longer time. In his book *Sylvicultura oeconomica*, published in 1713 (the first textbook on silviculture!) he wrote that 'We have to make every effort in science and forest practice to find out methods for protection and production of timber in a sense that a permanent sustainable utilization can be realized' (translated). This is the first time that sustainability is presented explicitly in a textbook. The idea developed quickly, and since the middle of the 18th century regulated forest management has been introduced into many parts of Europe. Pine and spruce produce a lot of seeds, and the seeds are easy to harvest, transport and handle, and therefore pine and spruce were used intensively in forest restoration after the phase of degradation. This is the reason that today these two tree species make up about three quarters of the temperate European forests, while the natural potential is for about three quarters deciduous trees, mainly beech (Fig. 10.2).

The economic value of today's managed pine and spruce forests is much higher than that of the devastated forests 300 years ago, and a lot of characteristics, like canopy closure, stand structure and microclimate, are closer to the natural forest stands than during that time. Nevertheless, soil quality is changing under the influence of decomposing needle material: decomposing needles force soil acidification, thus reducing the biological activity of the soil. As a consequence essential nutrients such as Ca^{2+}, Mg^{2+} and K^+ are stored in the upper organic soil layer (see section 10.5.2; Wolff & Riek 1997), no longer being available for plant growth. Additionally the storms Vivian and Wiebke in February 1990, throwing down mainly pine and spruce trees (König *et al.* 1995), showed that the recent forest situation should be improved. A conversion from needle-tree dominance to mixed forests is a major goal of forestry for the near future in central Europe.

Today the concept of forest sustainability includes much more than the production of the valuable, regenerating resource timber; it includes also protection of drinking-water resources, protection against soil erosion and noise, use of forests as an element in landscape architecture and a place for recreation and inspiration of humans, conservation of genetic resources and carriers of natural diversity (genotypes, species and ecosystems). The capacity of new established forests to store CO_2 has been given high priority in the discussion of global climate change; CO_2

storage in biomass and soil, storage in short- and long-lived products and substitution of fossil fuels are the most important aspects (Kohlmaier *et al.* 1998, Burschel & Weber 2001).

10.3 Main European forest types

Recent climatic and edaphic site conditions in most parts of Europe – from the coast of the Atlantic Ocean in France to the Ural Mountains in Russia and from southern Sweden to the south-facing slopes of the Alps in Italy – allow summer-green trees to grow and deciduous forests to establish. Evergreen needle trees are predominant in the boreal coniferous forests of Scandinavia and northern Russia. In central Europe they are restricted to the high mountain areas (e.g. the Harz from 800 m a.s.l. upwards, the northern mountain range of the northern Alps from about 1700 m a.s.l. upwards). Evergreen sclerophyllous trees are characteristic of the Mediterranean parts of Europe. The common types are described briefly below.

10.3.1 Beech forests

Beech (*F. sylvatica*) today is the most important natural tree species in central Europe (section 10.2.1). Beech forests represent the zonal vegetation from southern Sweden to the Alps and from central France to western Poland and the Carpathians. Beech is a powerful competitor against other tree species such as oaks (Leuschner 1998). Beech forests cover the whole site spectrum from acidic to calcareous soils and from the sub-mountain to the high-mountain zone. Limitations of optimal beech growth and competition are dry soils, wet soils (in particular stagnant moisture), late frosts and short vegetation period.

Several main types of beech forest can be distinguished, as follows.

- **Acidic** soil type on geological substrates like gneiss, granite, Triassic sandstone, quartz sand and loess loam. Because these substrates are widespread all over Europe this is the most widespread beech forest type. It is very poor in plant species; *Luzula luzuloides* is the most characteristic one in the south and *Deschampsia flexuosa* in the north.

- **Intermediate** soil type on deep soils with a good water supply; *Galium odoratum* is a characteristic species.
- **Calcareous** soil type on shell-limestone, Keuper marls, calcareous moraine material, primary calcareous loess and volcanic material of high alkalinity; this type is usually rather rich in species (e.g. *Cephalanthera* div. spp., *Carex alba*, *Carex montana*).
- **Mixed mountain** type: beech accompanied by fir (*Abies alba*) and/or Norway spruce (*Picea abies*). On sites with very high nutrient and water supply beech is mixed with maple (*Acer pseudoplatanus*) and accompanied by tall perennial herbs; very rich in plant species.

10.3.2 Oak and oak–hornbeam forests

On dry sites beech can be replaced by oak. On such sites in central Europe *Quercus petraea* is mostly becoming dominant (oak forests on dry, acidic soils) and in south-ern central Europe (south-eastern France, Switzerland) *Quercus pubescens* (sub-Mediterranean oak forests) predominates. From western Poland up to the Urals, as a consequence of frequently occurring late frosts, beech is replaced mainly by oak (*Quercus*), hornbeam (*Carpinus*) and lime tree (*Tilia*): continental oak–hornbeam forests.

10.3.3 Elder swamp forests

In flat lowland areas with clay in the subsoil swamp forests developed with *Alnus glutinosa* as the dominating tree species. Most of these sites were converted into productive arable fields by draining. As a consequence of drainage peat material starts to decompose, and nitrogen-indicating species such as *Urtica dioica* invade and spread. Therefore, today there are only few remaining stands of this forest type, and the floristic composition of most of these stands reflects disturbed site conditions.

10.3.4 Riparian forests

Close to rivers, *Salix* species are capable of coping with frequent and intense flooding. They have the capacity to re-sprout quickly after damage and to pro-duce a high number of propagules, and their seeds are able to germinate on mineral substrate. These softwood riparian forests are accompanied by the higher-elevated, hardwood riparian forests. Flooding is less frequent and there is sedimentation of loam, but no destruct-ive power of the running water. Hardwood species such as oak (*Quercus robur*), elm (*Ulmus laevis* and *Ulmus minor*) and ash (*Fraxinus excelsior*) are the most frequent tree species. In general riparian forests are characterized by a very high structural and species diversity. As a consequence of riverbed corrections and of the use of land for agriculture, infrastructure, industrial complexes and buildings, this forest type is today very rare in Europe and the remaining stands are usually eutrophied.

10.3.5 Bolder field forests

This forest type of very limited extent is restricted to natural bolder fields of steep mountain slopes. Maple (*A. pseudoplatanus, Acer platanoides*), elm (*Ulmus montana*) and linden (*Tilia platyphyllos, Tilia cordata*) dominate the tree layer. Depending on geological substrate, substrate structure, elevation and exposition a high number of subtypes can be distinguished. Because of the unfavourable position this forest type can hardly be used and a remarkable number of (small) stands still exists.

10.3.6 Pine forests

While Scots pine (*Pinus sylvestris*) was very common in Europe at the beginning of the post-glacial time period (section 10.2.1), natural pine stands in central Europe nowadays are limited to (i) wet areas in a land-scape with mires and raised bogs, (ii) nutrient-poor, acidic soils, usually quartz sands, and (iii) steep, south-facing calcareous rock slopes with very shallow soil and there-fore very limited water availability. Whereas in cent-ral Europe Scots pine would only account for about 1% in the natural vegetation, it actually covers about 25% of the forest area (Fig. 10.2), due to more than two centuries of intensive forest management.

10.3.7 Spruce forests

Outside the boreal forest zone spruce forests are restricted mainly to the tops of high mountains of the European highlands (e.g. Harz, Bavarian Forest, Ore Mountains, Carpathian Mountains) and of the Alps. Norway spruce (*P. abies*) settles on both acidic and calcareous bedrock; nevertheless, it is usually accom-panied by indicators of low soil pH, either because of the acidic subsoil or because of the acidic organic mater-ial that accumulates beneath the canopy of spruce trees.

10.3.8 Mediterranean evergreen sclerophyllous forests

In the Mediterranean part of southern Europe the nat-ural vegetation is formed by evergreen sclerophyllous trees (mainly *Quercus ilex* or *Quercus suber*). Since the intensive impact of humans (beginning with the Greeks and Romans and including the Middle Ages) most of the forests have disappeared, and intensive soil degradation has taken place (see Chapter 14 in this volume); forests were replaced by evergreen bush-land (macchia) or evergreen shrub-land (garrigue).

10.4 Causes of recent changes in forests

10.4.1 Acidification and eutrophication

The chemical composition of the atmosphere changed remarkably due to industrialization. As a conse-quence of acid rain significant soil acidification is proven to occur in large areas of Europe (e.g. Wolff & Riek 1997). According to calculations of the Euro-pean surveying programme EMEP (1999) in most parts of western Europe the immissions (wet and dry) of oxidized nitrogen (N_{ox}) amounts on average to 5–10 kg ha^{-1} yr^{-1}, in addition to at least 5–10 kg ha^{-1} yr^{-1} of reduced nitrogen (N_{red}) all over Europe.

The change of the floristic composition can be detected by repeated recording of permanent plots (e.g. A. Fischer 1993, Röder *et al.* 1996) or by comparing sets of phytosociological relevées from the same region and the same vegetation type in the course of time (e.g. Wilmanns & Bogenrieder 1986). Two main

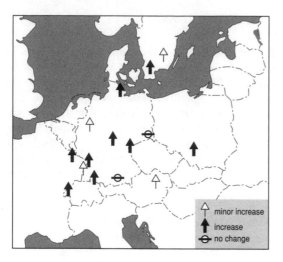

Fig. 10.3 Change in the representation of nitrogen-indicating plant species in central Europe during the last four decades. Modified from A. Fischer (1999).

general trends in European forest vegetation during the past four decades can be seen (A. Fischer 1999).

1 As the most homogeneous trend in vegetation dynamics in central European forests the representation (species number, frequency, cover/abundance and mean ecological indicator value) of **nitrogen-indicating species** is increasing (Fig. 10.3). This seems to be a very strong reaction to the deposition of nitrogen compounds like NO_x and NH_x as mentioned above.

2 Although indicators of acidic soils were expected to increase, such an increase could only be found in a few areas with very strong local industrial immissions. Usually, however, the effect of soil acidification is overcompensated for by other ecological factors such as N input.

Parallel to soil acidification, a tree decline (*Waldsterben*) has been observed. While at first soil acidification was discussed intensively as the main driving factor for tree decline (Ulrich 1986), later on evidence was found that *Waldsterben* is a multicausal phenomenon (Schulze *et al.* 1989). Although the crowns of many broad-leaved as well as needle trees (e.g. beech, oak, spruce and pine) in Europe

appeared to be in bad condition, tree increment has often been increasing for more than two decades and currently surpasses all expectations (Röhle 1995); therefore, other effects seem to overcompensate for the damaging effects (see section 10.4.2).

Altogether it can be stated that the floristic composition of European forest vegetation is indeed changing, but up to now this is only working on very low levels of phytosociological hierarchy (subtype/sub-association, not association). Nevertheless, it is an effect working on a global scale since it could be shown that intercontinental transport of NO_x is possible and significant (Splichtinger *et al.* 2001).

10.4.2 Global climate change

Anthropogenic increases in CO_2 from 280 ppm in 1750 to 370 ppm in 2000 are well documented, and the processes forcing temperature increase (in the 21st century the mean global surface temperature may increase by 1.4–5.8 K; IPCC 2001) seem to be rather well understood. Indeed, it has been shown that – all over Europe from northern Scandinavia to Spain and Greece – the growing season has lengthened by about 11 days since the early 1960s (Menzel & Fabian 1999). In northern and central Europe a floristic response to such temperature increase and growing-season extension up to now cannot be detected by using current methods of vegetation ecology. In forests on the southern edges of the Alps, however, an increase of evergreen broad-leaved woody species has been reported for the past few decades (Klötzli & Walther 1999: 'laurophyllization'), but although there is a close correlation with climatic parameters, other causes such as changes in land-use practices may also play an important role in this change in species composition.

Altogether it seems evident that species composition in European forests will change within the 21st century as a consequence of global climate change, but beech will continue to be the dominating natural tree species in central Europe (A. Fischer 1997).

10.4.3 Forest management

Forest management influences the forest ecosystem. As shown in section 10.2.2 **tree species composition** changed drastically in central Europe as a consequence of human impact. Additionally, species have been introduced that did not establish naturally after the last ice age; for example larch (*Larix decidua*), douglas fir (*Pseudotsuga menziesii*) and red oak (*Quercus rubra*), and in some Mediterranean countries eucalypt (*Eucalyptus globulus*). Some of the introduced species are spreading now intensively: invasive neophytes (e.g. *Prunus serotina*).

Harvesting procedures influence the **forest structure**. Tree cutting over an area up to several hectares followed by seeding or planting of new trees results in even-aged forest patches in a mosaic of different age classes. In some countries selective cutting systems are nowadays common, creating a stand structure that is not so far from a natural forest structure. Nevertheless, timber harvesting reduces the proportion of dead wood in forests drastically.

To improve the soil quality – especially to compensate for former devastating utilizations like litter harvesting and for soil acidification – carbonate fertilization is sometimes used (1500–4000 kg of calcium/ magnesium carbonate ha^{-1}).

Harvesting operations lead to mechanical soil disturbances. Soil compression reduces the growing potential of the site (water availability, gas exchange); the influence on biodiversity is discussed in section 10.5.1. Forestry has the option to manage the forests in a way that is, in principle, not so far from the natural situation (see section 10.5.1).

10.5 Forest restoration: between sterile substrate and primeval forest

In the following sections, we move gradually from forest conservation to forest restoration and rehabilitation, starting with sustainable use of near-natural forest as a point of reference, followed by conversions and transitions back towards formerly existing forests, regeneration after windthrow, reforestation of agricultural land and finally afforestation of post-mining landscapes (sometimes called reclamation).

10.5.1 Close-to-nature forest management

Seen from a global point of view forest utilization in central Europe is a must if sustainability is set as an overall goal in land-use management. Growing conditions for forests as well as quality of timber are excellent in large parts of Europe. Under the term close-to-nature forestry (*naturgemäße* or *naturnahe Waldwirtschaft*) a set of guidelines have been developed to combine utilization and long-term preservation of forest ecosystems. Main topics are (i) preferring native, site-adapted species, (ii) timber harvesting in (group) selective cutting systems, (iii) using natural tree reproduction instead of seeding or planting trees and (iv) leaving dead wood (thick timber, standing as well as lying) in the forest as an important carrier of species diversity (a habitat for fungi, insects, and vertebrates, including birds).

In a recent study on biodiversity in managed and unmanaged forest stands in the Bavarian Forest, south-eastern Germany, species diversity (plants, soil-living beetles and symbiontic as well as soil-saprophytic fungi) besides dead wood was shown to depend mainly on the disturbance regime. In forests a high degree of naturalness is usually not connected with a high number of species of these organism groups. Under harvesting that is typical of selective-cutting systems, plant species number at the plot level was not different from that of the near-primeval situation. Activation of the soil seed bank by mechanical soil disturbance associated with clear-cutting is a very important process to increase plant species diversity significantly at the plot level (Mayer *et al.* 2004). Soil-dwelling beetles, too, react with increasing species number after clear-cutting, invading from surrounding open land. Methods like selective-cutting combined with the acceptance of leaving a limited volume of dead timber are able to combine economical and ecological demands on forests within the same stand.

Forests are the natural habitat of several ungulates: in central Europe this means especially wild boar (*Sus scrofa*) as well as the grazing roe deer (*Capreolus capreolus*) and red deer (*Cervus elaphus*). The cultural landscape often provides much better feeding conditions for grazing ungulates than forests do (and during winter time these animals may be fed by humans). Therefore their populations today are usually much larger than expected under conditions without human impact. A balance has to be found between the aspects of guaranteeing both survival of native animal species and natural regeneration of native tree species. That large animals contribute to micro-site diversity of forest stands is obvious (e.g. digging by wild boars activating the soil seed bank). The ecological role of large grazers like aurochs (*Bos primigenius*) and tarpan (*Equus przewalskii gmelini*), which died out centuries ago, are not exactly known. Although there are current examples of open woodland resulting from several thousands years of domestic cattle grazing in forests (Pott & Hüppe 1991, Olff *et al.* 1999) it is hard to transfer the results to previous forests because the ecosystem has changed completely (e.g. it has lost large predators).

10.5.2 Forest improvement: conversion and transition of forest stands

Forests in central Europe cover a wide range of site qualities and stand structures, and were often deeply influenced by former (mis)use of land due to litter raking, exploitation and more recently by atmospheric depositions (Glatzel 1999, Kenk & Guehne 2001; see also above). Mainly beginning in the 19th century pine (*P. sylvestris*) or Norway spruce (*P. abies*) plantations were established in areas that had originally been dominated by deciduous trees. The present commercial forest ecosystems differ widely from natural forests (original ones as well as those preserved in a near-natural state or those naturally regenerating; Schmidt 1998). Most of the stands show increased susceptibility to damage by snow, ice, wind, drought, insects, fungi and possibly soil degeneration as compared to forests composed of site-adapted species due to a low resilience and a lack of ecological self-regulation.

At present, forest management in many European countries is changing. Changing demands of society and an enhanced level of ecological understanding have given an impetus to a lively discussion and intensified research regarding the improvement of these secondary Norway spruce and Scots pine stands (Spieker & Hansen 2002). Two options for increasing the resilience of stands are given (Fig. 10.4).

1 **Conversion**, a change in tree species composition, of secondary coniferous stands into pure or mixed broad-leaved forests, usually by advanced artificial regeneration (Hasenauer 2003).
2 **Transition**, a change in vertical structure, of pure structured stands in multi-layered stands, for example by moving from an even-aged to an unevenaged forest stand and changing the rotation period (or even to strive for continuous cover management; see Möller 1921).

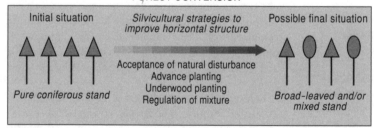

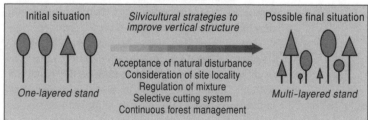

Fig. 10.4 Silvicultural strategies to improve forest stands that are poor in structure.

These silvicultural strategies are the order of the day within forest administration. Besides the ecological aspects (Cannel *et al.* 1992, Kelty *et al.* 1992), mixed stands are widely believed to be potentially more productive than monospecific stands (Bartelink 1999), which may be due to **complementarity** (tree species may differ in their crown morphology, shade tolerance, in height and diameter dynamics, rooting depth and/or phenology) and/or to **facilitation** (positive interactions and feedbacks such as synergism and symbiosis). An improvement will also affect future wood markets, economic results of forestry and different other goods and services that forest ecosystems provide for society.

Despite experience of how to technically convert pure pine stands into mixed stands with broad-leaved trees and of how one can increase structural elements, knowledge about the ecological properties of these mixed stands, for example the ecological interactions between the tree species and their impact on the soil, is generally vague. In this context the soil organic matter in the forest floor and the mineral soil are of particular interest (von Zezschwitz 1985, van Mechelen *et al.* 1997, Puhe & Ulrich 2000). The humus form with its morphological and chemical properties can be viewed as an integrating indicator for assessing the state of soil development in converted forest ecosystems (H. Fischer *et al.* 2002). Due to a decoupling of the carbon cycle in space and time large litter layers may develop in *P. sylvestris* and *P. abies* forests since in these forests the decomposition of plant litter material is limited because of low mineralization or bioturbation intensity. Often the result is typical humus disintegration. In connection with forest conversion the humus form undergoes diagnostic, morphological and chemical changes which are in turn relevant for nutrient cycling and carbon sequestration. The development of humus cover properties in the course of conversion from pine to mixed stands of pine and beech has been studied by comparing four forest stands representing a chronosequence in the northeast German lowland (Buczko *et al.* 2002, H. Fischer *et al.* 2002). The thickness of the dominating humus form, mor, decreased within the chronosequence and simultaneously a change was noted from moder-like humus forms towards mull-like moder and oligomull. The soil organic matter content in the humification layer (H horizon) decreased due to bioturbate processes

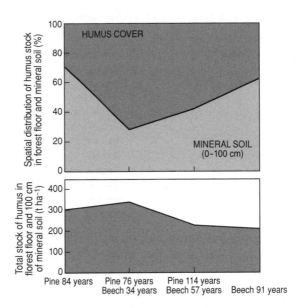

Fig. 10.5 Accumulated humus stock and relative distribution of soil organic matter in forest floor and mineral topsoil (to 100 cm) in a chronosequence of four study sites in north-eastern Germany. See H. Fischer *et al.* (2002).

with an increasing percentage of beech. With respect to the soil carbon stock and its distribution, two developments can be stressed. In the initial phase of forest conversion the total humus stock increased in the younger mixed stand whereas in later stages the soil organic reserves were decreasing continuously (Fig. 10.5). In contrast to the pure stands of pine and beech (initial and final phases of the chronosequence) the mixed stands are characterized by a significant decoupling of the carbon cycle whereas the litter layer acts as the main storage compartment for carbon. In beech forest the humus integration is relatively good and organic matter is not stored mainly in the humus cover and the upper Ae horizon (the horizon in the upper soil layer showing evidence of eluviation; as in Scots pine and mixed stands) but mainly in deeper mineral soil. These horizons are intensively penetrated by roots. The easier the prevailing litter can be decomposed the faster humus integration occurs.

Further studies will have to verify to what extent the results of this forest conversion study can be

Table 10.2 Three main storm events in Europe in the 1990s.

Storm event	Area/country affected	Dead wood (million m³)	Factor of annual cut
28 February–1 March 1990*	Switzerland	4.9	1.1
	Southern Germany	65	2.0
	Czech Republic/Slovakia	11.3	0.6
	Great Britain	6.0	1.5
	Belgium	5.5	1.8
	Austria	4.8	0.3
	Central Europe	100–20	
3 December 1999†	Denmark	3.6	>0.5
26 December 1999*	Western France	114.6	~4
	Switzerland	12.3	2.6 (locally up to 10)
	Baden-Württemberg	23.5	2.4–3 (locally more)
	Bavaria	4.3	0.4
	Austria	0.5	0.04

*Databases: several sources, see A. Fischer *et al.* (2002).
†Databases: Klitgaard (2002) and Brunner (2003, personal communication).

generalized. Forest transition and conversion mark a way back to a higher degree of naturalness, lead to an improvement of soil quality, and also result in a higher degree of stability by reducing the danger of specific pests (insects) and of fires and by anticipating expected global climate change.

10.5.3 Forest-stand regeneration after wind throw

Whereas in the boreal zone wildfire is an important site factor of forest ecosystems, within the temperate zone the most important disturbance impact influencing forest structure are storms. Storm events may occur during winter connected with intense low-pressure systems arriving from the Atlantic Ocean and crossing central Europe, or during summer time as short but strong thunderstorms, restricted to small areas. In central Europe the 1990s were a period with several strong winter storms; as Table 10.2 shows, the damage was great. These storms caused a strong impact on forestry and stimulated new scientific studies on forest dynamics following disturbance (A. Fischer 1998, Mössmer & Fischer 1999, Schönenberger *et al.*

2002). Three main impact options were analysed: (i) management-free stand development, (ii) clearing the wind-throw area, followed by free stand development and (iii) clearing and then establishing a new tree layer, usually by planting.

All these studies show that species composition (plants, animals and fungi) as well as the successional trends are different in cleared and undisturbed wind-throw areas. Plant species with a persistent seed bank are promoted by the clearing procedure (mechanical disturbances of the soil); after soil disturbance buried seeds are stimulated to germinate. Without clearing, soil seed-bank species are restricted to the pit and mound systems around the bases of fallen trees. Seedlings and saplings of trees present before the storm event will grow up immediately after the event (due to improved light conditions), whereas most young trees will be destroyed by clear-cutting. Therefore natural tree regeneration may work much better in uncleared areas. Broad-leaved species may quickly dominate under natural tree regeneration, if they are present in the area (saplings) or in the close vicinity (mature trees). Mycorrhizal fungi need living trees to survive; in uncleared areas young trees are present, and they are carriers of the suitable fungi. Survival rates of fungi are therefore rather high (see in A. Fischer 1998). Mycorrhizae are also fed upon by several small insect species living in the soil, for example the order Protura (Apterygota). The dead wood itself is the major carrier and promoter of biological diversity in unmanaged and managed forests.

Altogether, it becomes clear that wind throw means not only damage but also a chance.

- Windfall in artificial spruce or pine forests may induce a (natural or managed) change back to a higher proportion of deciduous trees (see section 10.2.2).
- Free stand development (without clearing) in a number of forest stands is an option to create new ecological niches and therefore to increase bio-diversity in forests, which is seldom realized in commercial forests up to now. Because each stand is changing its structure and its species composition successively, a certain number of such stands of different ages should be present in a larger forest area to maintain biological diversity.
- Using the natural regeneration potential may save costs (for seeding, planting) and prevent root deformations (causing bad tree growth); besides this, it is a contribution to enlarging the genetic basis of the tree populations involved. Nevertheless, pest outbreak has to be prevented; fallen spruce is critical in this respect (it is a micro-habitat for, e.g., *Ips typographus*, *Heterobasidion annosum*, *Armillaria mellea* agg.), but an increasing share of broad-leaved species will reduce this problem in the future.

Although from the point of view of economic forestry storm events are real catastrophes, on the accepted background of close-to-nature forestry they can be understood as unexpected options to improve the degree of naturalness in forests in the near-term.

10.5.4 Reforestation of former agricultural land

In Europe **reforestation** (restoration of forests on recently non-forested land that was forested in previous times) plays an important role in landscape planning, since abandonment in the 1980s and 1990s was common in many parts of Europe. The EEC

Table 10.3 Areas that have been reforested with the aid of EU regulation 2080/92 (IDF 2001).

Code	Country	Area (ha)
DE	Germany	26,249
DK	Denmark	2,725
ES	Spain	459,395
F	France	45,147
IRE	Ireland	98,258
IT	Italy	64,162
PT	Portugal	205,768
UK	United Kingdom	100,868
BE	Belgium	191
ELL	Greece	16,401
LUX	Luxembourg	12
NE	The Netherlands	2,271
OST	Austria	343
SUO	Finland	19,732
SVE	Sweden	67
Total		1,041,589

regulation 2080/92 from 1992 (EC 1992b) aimed at reducing the area of arable fields, to reduce subsidies for surplus of agricultural products, to contribute towards an improvement in forest resources as well as of countryside management and environmental balance, and to absorb CO_2. In landscapes with high agricultural intensity reforestation diminishes the pollution of seepage water with nitrate, which often is a problem in terms of purification of drinking water (Kreutzer *et al.* 1986, H. Fischer 1998). Reforested natural retention areas can be an important contribution to flood regulation. Up to 1999 about 1 million ha were reforested in Europe, mostly in Spain and Portugal (Table 10.3), and still the calculated reforestation potential is enormous: in Denmark the political reforestation aim is 500,000 ha (1995–2050), in Ireland it is 600,000 ha within 20 years, in Portugal it is about 1.5 million ha within 50 years and in Germany the scenario is 1.2 million ha or more (no time period indicated; Weber 1998).

Reforestation of former agricultural land starts with a soil that is settled by enormous numbers of individuals of many species of micro-organisms and insects; plant species may be present as a soil seed bank or as established individuals, and the site is

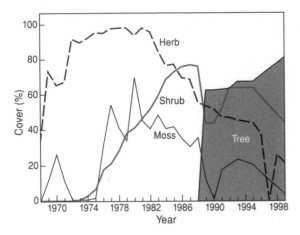

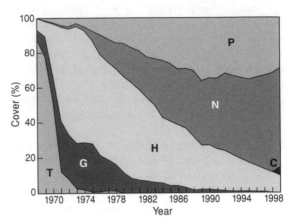

Fig. 10.6 Succession experiment in the New Botanical Garden of Göttingen, Germany. Cover degree of moss, herb, shrub and tree layers 1968–99 (free development). Woody species at first are less than 5 m high (shrub layer); in 1989 some of the growing trees reached a height of more than 5 m: shrub layer decreases and tree layer starts suddenly. The tree-layer curve is shaded. Values for 1994 and 1996–8 are interpolated. Based on Schmidt (1993), with new values for 1993–9 added (Schmidt, unpublished observations).

Fig. 10.7 As for Fig. 10.6: relative importance of cover degree of growth forms in 1968–99 (free development). T, therophytes; G, geophytes; H, hemicryptophytes; C, chamaephytes; N, nanophanerophytes (shrubs); P, phanerophytes (trees). Values for 1994 and 1996–8 are interpolated. Based on Schmidt (1993), with new values for 1993–9 added (Schmidt, unpublished observations).

suitable for tree growth immediately. Two long-term permanent plot studies have been established in Germany for analysing the vegetation processes after abandonment of former agricultural areas, including analysis of the influence of different treatments: (i) an abandonment experiment in Baden-Württemberg (Schiefer 1981, Schreiber 1997) and (ii) a succession experiment in the New Botanical Garden of Göttingen (Schmidt 1993; Figs 10.6 and 10.7). Although the aims of these two studies, as well as the analysed vegetation types, the geographical situation and to some degree the methods of recording, are different, after two decades of unmanaged stand development it is obvious that trees may become established quite early and build up forest stands within a few decades, whereas the successional pathway may be quite different. The species composition and the period involved in the changes depend on the local species pool as well as the vegetation structure at the beginning of the developmental process.

As a consequence, there are two main options for forest restoration.

1 If there is time enough available and no fixed species composition planned for the future forest stand, then doing nothing and waiting is an option. Depending on (i) the presence/absence of mature trees in the close vicinity and (ii) the presence/absence of small individuals of shrubs and trees in the recent vegetation stand, a patch mosaic will establish including a large variety of species and structural types, thus leading to a high degree of biodiversity. The final stage of management-free development will be a relatively closed forest canopy anyway, which then will result in a decline in the species diversity and in the structural diversity of the forest understorey.

2 Usually, however, the highly valued functions of forests, as mentioned in section 10.2.2, require work to start as soon as possible, to adapt both species composition and forest structure to the functions being focused on. Therefore, usually a **guided development** of such forest stands is preferred. Following the concept of close-to-nature

forestry, native trees, in temperate Europe especially deciduous trees (e.g. *F. sylvatica*, *Quercus* div. spp.), are aimed at and should be included in growing-stock objectives to a significant proportion. Tree species such as ash (*F. excelsior*), maple (*A. pseudoplatanus*), cherry tree (*Prunus avium*) and mountain elm (*Ulmus glabra*) may benefit from the generally favourable soil conditions of the abandoned fields and meadows, but this high nutrient status – together with intensive radiation on the ground in the initial phase of reforestation – also induces intensive competition by grasses and herbs (H. Fischer 1999, 2000). Additionally, fencing is a need for a certain period in cases where large populations of wild ungulates (mainly roe and red deer) are present.

10.5.5 Afforestation in post-mining landscapes: starting at point zero

Mining is an important field of European economic activity and often has a long tradition. Devastated areas,

particularly following open-cast mining activities, among which the states with lignite conveyance are the most important, are in need of reclamation or rehabilitation, starting at 'point zero' (Bradshaw 1983, Hüttl & Weber 2001). Most mine industries employ surface techniques which disturb the local landscape fundamentally and have a drastic impact on the nature of the whole region. Existing ecosystems are destroyed completely and the removal of the overburden covering the mineral or coal resource changes the topography and hydrology on a large scale. **Afforestation**, forest establishment on substrate and surface that never before was settled by forests, is often applied (Fig. 10.8). Research into the ecology of afforested mine sites is being conducted in numerous countries (Poland, Czechia, Hungary, Greece, Romania, Bulgaria, Germany and Spain; Bradshaw & Chadwick 1980, Cairns 1991, Häge *et al.* 1996, Strzyszcz 1996, Bradshaw 1997, Hüttl 1998, Bradshaw & Hüttl 2001, Nienhaus & Bayer 2001).

The most important effect of open-cast mining is the dramatic influence on **water balance**. Before the lignite seam can be uncovered, the water must be

Fig. 10.8 Post-mining landscape in Germany. Afforestation with *Quercus petraea*. Photo by H. Fischer.

removed from the seam and the overlying parent material. A long-term pumping action greatly influences both the ground and surface water (Hüttl & Bradshaw 2001). The overburden usually is translocated by the use of conveyor bridges, which mixes different geological materials. While this technique homogenizes the abiotic conditions on a landscape level, it leads to large substrate heterogeneity on a small scale, hence affecting the quality of the top-layer material (Elgersma & Dhillion 2002). The small-scale spatial variability of soil chemistry forms niches for the tree roots and is the key to understanding the successful establishment of forest stands on these – on average – extreme sites (Weber 2000). Concerning mine soils formed from carboniferous overburden, a high acidity of sulphurous substrates caused by chemical weathering and microbial oxidation of the ferric disulphides pyrite and marcasite is usually the main problem (Pugh et al. 1984, Pietsch 1998, Schaaf et al. 2001). This process starts immediately after contact with atmospheric oxygen, it is irreversible (Katzur & Haubold-Rosar 1996) and it causes pH values in the extreme acid range down to 2.1. The sulphate content of post-mining limnic biotopes can be as high as 2000 mg l^{-1} (Grünewald & Nixdorf 1995). In addition, most mine soils are characterized by an initial lack of humus, an insufficient nutrient supply (Schaaf et al. 1998, 1999), periodic and long wetting resistance (Katzur & Haubold-Rosar 1996) and by preferential flow which causes bypassing of water and nutrients in most of the volume of the soil matrix (Gerke et al. 2001, Hangen 2003). Another serious problem for forest plants is the characteristic climate of wide-open land (intensive light radiation, soil-lifting frost, late frost, wind and erosion). Nevertheless, adapted species may grow up and cause spontaneous ecosystem establishment and succession, depending on site quality and diaspore availability (Jochimsen 1991, 1996, Tischew et al. 1995, Wittig 1998).

The first restoration step is a technical rebuilding of the landscape, the creation and levelling of the surface and the moulding of slopes. The subsequent development depends on the dumped substrates. The common sulphurous and carboniferous mine soils remain barren of vegetation for decades if there is no human interference (Felinks et al. 1998, Wiegleb & Felinks 2001). In many cases soil amelioration is needed, which requires detailed knowledge of site conditions, in particular the identification of its limiting soil nutrients (Heinsdorf 1992, 1996). Large amounts of CaO are used to change soil pH value (a cultivable soil layer of at least 100 cm thickness needs 40–50,000 kg of CaO ha^{-1}). Traditionally, inorganic fertilizers are used, whereas in recent years there has been a trend towards the use of organic waste materials (Leirós et al. 1996) and special humus material (Katzur et al. 2001).

The first afforestation experiments on raw mine soils in Europe were carried out and published in Germany by Heuson (1929) and Copien (1942). Even then they promoted mixed stands, and to date there are long-term experiments with many species: pine, oak, lime and larch as well as pioneer trees such as alder, poplar, birch and black locust (Robinia pseudoacacia). The establishment of forest stands by mechanical planting after adequate soil amelioration shows surprisingly good results. After afforestation soil development first leads to the formation of regosols. The results of Rumpel et al. (1999) and Fettweis et al. (2005) indicate that in pine ecosystems on former lignite mine sites internal cycling processes consisting of litter fall, decomposition of litter and nutrient uptake by plants play an important role in nutrient supply, as is known for pine stands on naturally developed soils. In pine stands of about 30 years of age roots were found below the amelioration horizon. Chronosequence studies show that shortly after afforestation a biocoenosis develops spontaneously, including soil fauna (Düker et al. 1999, Düker 2003) and symbiotic micro-organisms such as mycorrhizal fungi (Golldack et al. 2000, Wöllecke 2001) and nitrogen-fixing bacteria (Kolk & Bungart 2000).

As in the case of reforestation of former arable land, abandonment may also be an option in post-mining landscapes. According to the high diversity of microsites a high diversity of species, structures, ecological niches and ecosystems will establish, and in the course of time these ecosystems will change permanently into forest as a final system. There are, however, a lot of social, economic, aesthetic and ecological needs connected with such landscapes and the developing vegetation; fast-growing tree species and alley-cropping plantations as special forms of energy forests have recently come into focus (Bungart et al. 2001). The new forests will be evaluated by how well they are meeting these demands as soon as possible, which may require activities like soil amelioration and tree planting.

10.6 Generalization and concluding remarks

In a century of intense forest loss, reforestation is an important aspect of landscape ecology and landscape management. Human impact has changed site characteristics fundamentally. In many cases soil quality has decreased, for example due to litter removal in central Europe or soil erosion in Mediterranean Europe. The climate has changed during that time, both by natural processes and as a result of human impact

(section 10.4.2). The chemical composition of the atmosphere and, as a consequence, of the soils has also changed drastically (section 10.4.1). Therefore, the new establishing and regenerating forest ecosystems will not be exactly the same as those that stood before human impact. Besides the changing site conditions the migration opportunities for species has changed drastically, too. Many non-woody forest plant species are so-called *K* strategists: they have a relatively low generative reproductive potential but often spread vegetatively; therefore, in an intensively used cultural landscape the chances are small for them to find new potential habitats. On the other hand some introduced species were able to establish self-reproducing populations (e.g. *Prunus serotina*,

---

Box 10.1 The recent situation as well as restoration options for the main European forest types (units are described in section 10.3).

- **Beech forests**: maintain existing extended beech forest areas (Fauna, Flora, Habitat Directive EEC/92/43; EC 1992a). Manage remaining (mixed) beech forests in selection-cutting or group-selection-cutting systems. In pine and spruce plantations on sites with conditions (PNV) for beech forests the portion of beech should be enlarged. Reforestation of such sites in flat land often needs pioneer trees as a first step.
- **Oak and oak–hornbeam forests**: in its natural distribution area in eastern Europe this type is protected, for example in the Bialowieza National Park, Poland. In central Europe it is restricted to places that are too wet or too dry for beech forests. Large areas of former oak–hornbeam forests were developed during the past centuries because of coppicing. Such stands are not according to the PNV and therefore their floristic structure are dependent on permanent human impact.
- **Elder swamp forests**: the remaining limited number of stands of this forest type must be preserved, primarily its water regime. One ecological problem is humus mineralization.
- **Riparian forests**: these are extremely degraded by humans in terms of number (conversion to arable land, infrastructure) and quality (reduced flooding, eutrophication). The first step of restoration needs re-establishment of a flooding regime; secondly

the remains of such forests should be included in restoration projects.
- **Bolder field forests**: these should be left untouched. The stands are naturally very small and isolated, accessibility for forestry is hard, but biodiversity is high; recent quality of many of such stands is still close to primeval.
- **Pine forests**: in central Europe these are very rare (in the vicinity of raised bogs or on very dry sites). Natural stands should be preserved. Their productivity is very low. Boreal forests are managed intensively; close-to-nature stands need protection. Large-scale central European pine forests are usually man-made; a conversion from pure pine to an increasing portion of deciduous trees is proposed.
- **Spruce forests**: according to PNV these are rather rare in central Europe. Such stands need strict protection (e.g. Bavarian Forest National Park, south-eastern Germany). Boreal spruce forests are managed intensively; close-to-nature stands need protection.
- **Mediterranean evergreen sclerophyllous forests**: nowadays these are rather rare because of long-term utilization and site degradation. Fire as well as prevention of sheep and goat grazing are the first steps in a long process of forest restoration. During the 1990s intense reforestation with *Q. suber* and *Q. ilex* (promoted by ECC regulation 2080/92; EC 1992b) took place. Remaining stands should be protected.

Impatiens parviflora, Impatiens glandulifera and *Solidago canadensis*).

In Europe the site potential is good for forest ecosystems nearly everywhere (section 10.1). Despite the changed (and still changing) conditions, the main units of European forest vegetation (on the association level) as described in the past will still exist in the decades to come; therefore, they can be used as reference for forest restoration. Box 10.1 summarizes the recent situation as well as restoration options for these forest types.

The degree of divergence between the recent vegetation on the one hand and the site potential (PNV) on the other hand can be rather different:

- on sterile sediment or bedrock material in large opencast mining areas recent ecosystems are far from the site potential; the soil is poorly developed, and the organisms (plants, animals, micro-organisms) forming the ecosystem are missing;
- on abandoned arable fields there is well developed soil; organisms, although not typical for forests, are present;
- tree plantations belong to forest formation, and a lot of characteristic forest organisms are present, although species composition, structure and processes are significantly different from primeval forests;
- in close-to-nature forests the spectrum of species is similar to that expected in a primeval forest, but some structural features are different (causing differences in species richness and evenness).

In contrast to most of the semi-natural ecosystems in the European cultural landscape – rough meadows, heathlands, wetlands, shrublands and the Mediterranean garrigue – restoration of forests is possible without any input from humans: if there is enough time, a forest ecosystem will establish by itself. However, the newly establishing or regenerating forest stands today have to fulfil several valuable functions (section 10.2.2). Therefore, usually there is no time available for management-free stand development, and we have to accelerate the developmental process. The most important aspects to consider are: the use of native tree species, use of site-adapted species, inclusion of natural processes in management plans and inclusion of remains of former larger forest stands in restored forest landscapes. As has been pointed out (Ammer & Pröbstl 1988, EC 1992b), within the implementation phase of the EU directive on reforestation of agricultural land the reforestation of marginal soils may lead to conflicts with nature conservation, because such marginal sites often harbour the remains of anthropogenic ecosystems of the historical cultural landscape. These considerations have to be taken into account to achieve a general social agreement on the objectives and implementations of forest-restoration projects.

11

Restoration of rivers and floodplains

Jenny Mant and Martin Janes

11.1 Introduction

The rivers and floodplains of Europe are of great importance for plants, fish, birds and mammals (including humans) but the diversity of their habitats has declined particularly over recent years. Floodplains have often been disconnected from their rivers, reducing their flood-storage function and hence the range of low floodplain and marsh areas required to sustain, for example, reed bunting (*Emberiza schoeniclus*), woodpecker (*Alcedo atthis*), European beaver (*Castor fiber*) and otter (*Lutra lutra*). Similarly, aquatic plants such as water plantain (*Luronium natans*) and river-water crowfoot (*Ranunculus fluitans*) have been affected by eutrophication, whereas intervention such as canalization, dredging, draining and vegetation removal have all been instrumental in degrading our river systems. Similarly, fish habitats have been destroyed by these measures and the construction of dams and siltation of spawning grounds has meant that many native species such as salmon (*Salmo salar*) and sea trout (*Salmo trutta*) are now threatened with decline or localized extinction. Today very few of Europe's rivers can be truly defined as natural and this has serious implications for the associated ecosystems. Furthermore, as explained by Postel and Richter (2003), natural river flow is a key element in sustaining a healthy river system, including absorbing pollutants, decomposing wastes, producing fresh water and the redistribution of sediments and habitat replenishment during floods. There is therefore a huge potential resource that could be restored to increase and improve diversity of habitat. It is now accepted that threatened habitats

and species form part of the European Union's (EU's) natural heritage. Threats are often trans-boundary in nature and various EU directives now in place are aimed at implementing measures at the EU level that advocate conservation of sites under pressure from development and associated pollution; see especially the EU Habitats Directive 92/43/EEC (EC 1992a) and the more recent Water Framework Directive 2000/60/EC (EC 2000). One of the objectives of river restoration is to promote activities that initiate or accelerate recovery of degraded ecosystems and hence help to implement some of these Union Directives.

River restoration is a complex subject that affects not only the users of a particular watercourse but also the land and natural ecology within a river catchment. All but a few rivers and floodplains within Europe have been severely degraded over a prolonged period. Thus any restoration cannot simply imply a return to some previous state, such as making an assumption that re-meandering of a river, based on historical evidence, will be sustainable. Instead it should focus on the re-establishment of a self-sustaining system that can allow the form and function of the river to develop as naturally as possible under the present and future climatic regimes and in doing this it should improve the ecological conditions. Ideally, restoration should recognize that floodplains are an integral part of the natural functioning of riverine systems and as such there should be at least intermittent hydrological connectivity between the two. At its most basic, de-culverting (or daylighting, in the USA) equates to a huge increase in the biodiversity value of a watercourse (Fig. 11.1). In Switzerland over the last 130 years,

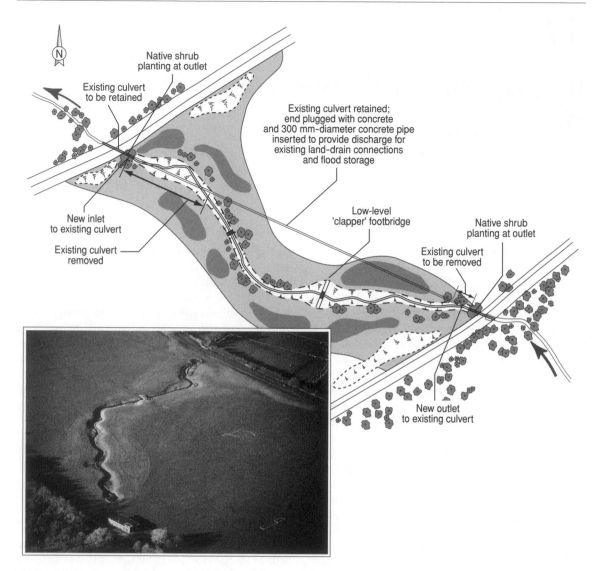

Fig. 11.1 Plan for opening up a culverted channel along the River Ravensbourne, London, UK. The inset shows the new sinuous course (Environment Agency).

for example, approximately 90 km of Zürich's streams have been culverted. New legislation aims to restore over 26 km of surface streams in this city alone (Pinkham 2000).

Ideally, river restoration should encourage a return to natural flow regimes, free passage for fish (i.e. dam and weir decommissioning) and reconnection to the floodplains. Clearly, very few societies are, at present,

in a position to achieve this and hence individual needs and objectives must be acknowledged. This can considerably influence the restoration outcomes of schemes. Luxembourg, for example, has all inland rivers and thus waste water (water quality) is the main problem; Ireland has spent over €15 million on fisheries-rehabilitation measures (see O'Grady 2001; Fig. 11.2); southern Spain's main concern is improving

Fig. 11.2 Example of fisheries enhancement along a tributary stream of Loch Ennell, Northern Ireland.

in-stream flows where intensive water extraction for irrigation has caused severe degradation (Cachón de Mesa 2001). Furthermore, any restoration must be appropriate to the specific landscape and the historical heritage. Although lessons in approach may be learnt from the USA (FISRWG 1998) and Australia (Rutherford *et al.* 2000), where restoration of rivers is high on the agenda, the techniques applied there may be wholly inappropriate for European rivers.

It is essential that the approaches to river restoration be clearly understood and it be appreciated that they do not necessarily provide a quick fix to a degraded habitat. Ecological recovery requires protection and nurturing. Often the success of restoration depends upon the lead discipline and those mainly responsible for the management and funding. What, for example, is the impact of government initiatives as opposed to the non-governmental organizations (NGOs), or public pressure and perception? The Water

Framework Directive has been instigated to address such issues (Chave 2001) but even so each country will inevitably start from a different point depending on the historical use of its rivers, with issues ranging from water quality to inappropriately engineered rivers.

The main purpose of this chapter is to present a broad overview of the state of river restoration and rehabilitation within Europe. This will explore the improved status of in-channel watercourses through, for example, the removal of obstructions and, equally important, examine the connectivity of rivers with respect to their floodplains. Furthermore, it will show how river and floodplain restoration needs to take account not only of wider catchment issues but also the interaction between economics, policy, scientific understanding and public perception. Together these are integral parts of the riverine restoration process. First this chapter provides a brief summary of the types of historical impact that have been instrumental in shaping the riverine landscapes in Europe today. Second, it outlines river types in terms of flow diversity, form and sediment (all of which are essential to understanding how rivers react to change before any restoration is considered) and their potential ecological value. Third, it provides a short discussion about current disturbances and threats. An appreciation of how rivers have been affected by both natural and anthropogenic change both past and present is key to achieving sustainable river management. Yet, despite continued scientific and political discussions about how best to ensure the future 'health' of our rivers, their restoration is still a relatively new phenomenon. Nevertheless there are still good examples of rehabilitation throughout the EU that aim to improve biodiversity within the whole riverine environment, even though concerns about increased flooding and development pressure are always high on any country's political agenda: some examples of this are outlined in this chapter.

11.2 Historical background to European rivers

Wherever rivers are not confined they will change their course as they seek to find an equilibrium state for the specific range of environmental conditions acting upon them. Therefore, changes in climatic conditions mean that rivers and their floodplains are always adjusting to both local changes that affect the annual variability of flood frequency and size of flow events, and global variation that can result in major shifts in river planform pattern and transform their natural geomorphological state. The precise mechanisms of geomorphological change in rivers cannot be explained within the confines of this chapter but many textbooks already cover these issues (see for example Callow & Petts 1992, Knighton 1994, Rosgen 1996, Gregory 1997, Thorne et al. 1997, Wolfert 2001). Nevertheless, the importance of understanding geomorphic processes cannot be overestimated. It can both hold the key to explaining the changes that have occurred in our riverine landscape as a result of anthropogenic intervention and help to predict how specific restoration methods will enhance degraded rivers.

11.2.1 Natural impacts

Although rivers and floodplains have constantly been shaped by changes in climate over time (see Table 11.1) the riverine topography we have inherited today owes many of its features to the impacts of the last major ice age that ended some 10,000 years ago. This formed not only many of the rivers themselves but also the valleys and entire catchments. In northern Europe, the ice pack covered much of the landscape, rounding off mountains and scouring out wide U-shaped valleys, far larger than could have been shaped by their current watercourses. This large-scale process had impacts on both flora and fauna and areas once below the ice now have less genetic diversity than those that survived further to the south, since, in this case, they relied on colonization by plants and animals able to travel large distances. Where the ice failed to advance, new rivers developed and old rivers were displaced to drain the meltwater. The River Thames moved miles further south, discharging into a new estuary where London is now located. The River Skjern valley in Denmark, which drains 11% of Jutland's soil, was gouged out by torrents of meltwater and, even beyond the immediate influence of the ice, deposition of eroded rock and sediment occurred, creating lakes and deltas, thus helping to shape the course of rivers and streams.

The late medieval period, generally known as the medieval warm period, resulted in another significant shift in the riverine regimes of Europe. This period

Table 11.1 Historic climate change and associated channel processes. Adapted from Gregory and Lewin (1987).

Year BP	Climate	Precipitation	Temperature	Vegetation	Hydrology	Channel processes
Pre-15,000	Glacial	Snow	Low	None	Summer flood	Multi-thread
15–10,000	Early part of the late Glacial	Snow	Ameliorating: cold winters	Grass/sedge succeeded by birch woodland	High peak discharge, but decreasing as forest spread	Multi-thread
11–10,000	Late Glacial	Mainly drier	Extreme cold	Herb-rich grassland	Runoff reducing, but high sediment supply	Fluvial deposition, unstable
10,000–4000	Holocene	Rain	Rising	Mixed oak forest with some clearance	Runoff lower than today	Fluvial erosion dominant with single-thread meandering channels
4000–present	Holocene cont.	Rain	Reducing but fluctuating	Deforestation	Seasonally fluctuating discharge	Lowland cut-offs and floodplain accretion of sediment

was represented by hotter drier summers north of the Alpine areas and, although the winters did not differ substantially from today's climatic conditions, the result was more convective rainfall in the summer months resulting in high-intensity but short-duration rainfall. Evidence in England from the middle reaches of the River Trent in the Midlands suggests that as a consequence of this climatic change the character of the rivers shifted from single-thread channels to a braided system, then anastomosing (multi-thread) and back to single thread during this period as the river profiles adjusted to accommodate larger, more unpredictable (flash) flood conditions (see Benito *et al.* 1998 for more details). South of the Alps, the impact on rivers was equally complex. The higher temperatures resulted in higher evaporation, lower annual discharge and groundwater and less snowmelt. Conversely, high sea temperatures increased the amount of convective rainfall. The effects on the rivers and floodplains, therefore, depended on the local antecedent conditions, and the infiltration rate of the substrate. In northern Italy the rivers were unstable and a braided form dominated, while in the south rivers tended towards single-thread systems again where flows were often out of bank and hence deposited large amounts of sand and silt on the floodplain.

Since the medieval warm period, the rivers of Europe have been subjected continually to minor changes in climate. Perhaps the most notable of these was the Little Ice Age that occurred between the 12th and 17th centuries. Documentary evidence suggests that the River Thames, in London, often froze during this period although it is not completely clear if this was related solely to climate change or exacerbated by the hydrodynamic effects of the bridges. Nevertheless, the effects of the Little Ice Age were not restricted to the UK and records across the whole of Europe show a similar pattern, with many lakes and rivers freezing over and evidence of glacier expansion. Furthermore, records of this period imply that this was also a period of dry weather and hence intermittent riverine activity (Benito *et al.* 1998).

Today there is much discussion about future climate change and the impact on biodiversity (see, for example, Gitay *et al.* 2002). By extension, this has implications for the river systems and their future flow regimes. This in turn could have an adverse effect on ecological status and hence any future restoration initiatives should build in some design flexibility to account for changing scenarios and appreciate that rivers are very sensitive to change (see Downs & Thorne 1998).

11.2.2 Human impacts

Effects of climate and local weather, although instrumental in driving the form of Europe's rivers, have gradually been matched by an increase in the degree of human intervention, starting during the Holocene period (within the last 10,000 years) with simple woodland clearance. Although some distinctive phases of human activity that ultimately affect our river systems can be identified since that period, it often remains difficult to disentangle anthropogenic impacts from more natural or non-human determinants. Take for instance the impact of human interference through deforestation compared to the natural decline in forest cover as a result of periods of glacial activity. Immediately it becomes apparent that trying to distinguish the relative impacts of each of these is problematic. Furthermore, especially in northern Europe, many riverine areas are still recovering from the Little Ice Age, while at the same time some have been widened and deepened to drain the land or for navigation purposes. It is important to recognize the additional difficulties associated with a long period of human intervention and constraints when evaluating degraded rivers with a view to restoration. A pristine riverine state to start with is difficult, if not impossible, to identify and in many cases, as expressed by Macklin and Lewin (1997), today's rivers are often 'climatically-controlled but culturally blurred'.

Deforestation in particular has had a major impact on river systems especially during the last 4000 years. Along the Rhine, for example, estimates of forest decline vary but broad figures suggest a 75% reduction; today there is only 150 km² of forest remaining along the Rhine's corridor. Such massive decline in forested areas resulted in an increase in sediment availability. This often deposits on the river bed both directly by means of runoff and indirectly via systems of field and land drains. During the medieval period, rivers throughout much of Europe became harvested as a source of energy to drive mills for a range of activities from cloth- to flour-making industries and hence marked the beginning of a period of major

Fig. 11.3 A river with little or no biodiversity, natural flow dynamics or channel morphology encased in concrete; example from the River Frome, Bristol, UK.

constraint of river courses. The impact of these mills cannot be over-estimated. An entry from the Domesday Book of 1086, for instance, shows that in southern England alone there were 5624 watermills, which equated to one for every 50–60 head of population. Further manipulation of the water courses occurred during the 14th–18th centuries as water meadows became fashionable as a way to increase food crop production by controlling flow on to the floodplains. These subsequently became redundant and fell into decline with the introduction of fertilizers and new intensive farming methods by the late 19th century. Dredging and straightening became the primary objective with a major commitment to improving the drainage of fields during the 1930s. Further arterial drainage schemes were implemented during the 1960s in many countries, partly as a push towards ensuring self-sufficiency of food after the Second World War; in Austria for example, 30,000 km of rivers have been regulated in the last 50 years. At the same time an increase in housing development and associated infrastructure has resulted in further straightening and deepening of rivers, with a disconnection from their associated floodplain areas and water abstraction to support the increased development. Many of the rivers of Europe today have effectively been turned into carriers of floodwaters in an attempt to move water from both rural agricultural land and urban areas as quickly as possible with little or no consideration of either biodiversity or the natural flow dynamics and morphology of river systems (see Brookes 1988, Petts *et al.* 2002 for more details; Fig. 11.3). These changes often result in a far more varied (flashier) flow regime than would naturally occur as water is impounded and prevented from reaching its floodplain under high-flow events. The negative impacts of these historical activities as outlined in Table 11.2 are now beginning to be addressed both in terms of economic viability and potential ecological gain. This has created the opportunity for river restoration to be taken more seriously as a viable alternative to present management practices as there is increasing acknowledgment that it is not always possible or acceptable to channelize and/or regulate rivers (see Petts 1984).

11.3 Characteristics of European rivers

Natural river systems are dynamic bodies that continuously change as a result of their inherent physical conditions such as slope, bedrock geology and the complexity of the drainage network. Yet, as previously discussed, their characteristics are also influenced by

Table 11.2 The main human activities that have influenced river systems.

Activity	Reason	Impact
Land drainage	Agricultural	Loss of floodplains and wetlands; increased sediment supply to the river; increased flood peak through decreased storage.
Flood protection	Urban/economic	Isolation/loss of floodplains; loss of biodiversity.
Reservoirs/dams	Water supply/ hydroelectric power	Ecological deterioration downstream; increase in contamination; disruption to the transfer of sediment.
Weirs	Mill systems	Prevents fish passage to upstream reaches; reduced access to spawning grounds.
Channelization	Erosion prevention; flood control; drainage of surrounding land; navigation; infrastructure	Disrupts the physical equilibrium of the watercourse; impact on riparian vegetation and water temperature; increase in flow velocity and hence reduces habitat and biodiversity.
Dredging	Gravel/sand extraction; increased capacity	Over-widening and -deepening of the river; instability and bank collapse; removal of natural bed material.
Water abstraction	Drinking water and agriculture	Lowering of river-water levels and the floodplain water table.
Urban expansion	Increased populations and economic growth	Increase in hard surfaces, greater/faster runoff; loss of riparian corridor and floodplains; poor water quality; sewage; spillages.

external factors including climatic conditions and human activities such as afforestation, deforestation, urbanization, land drainage, pollutant discharges and flow regulation; it is the combination of the natural river processes that occur over time (Schumm & Litchy 1965, Brunsden & Thornes 1979, Schumm 1979) and the external influences that ultimately determine the form of a watercourse.

From a geological point of view the continent of Europe is relatively young and this has resulted in river catchments that tend to be numerous, but small, compared with much of the rest of the world. Only about 70 rivers in Europe have catchment areas exceeding 10,000 km². Of these the largest is the Volga at 1,360,000 km² (compared with 3,349,000 km² for the Nile) and this, together with the Danube and the Dnepr, drains one quarter of continental Europe. Furthermore, the rivers of Europe account for only 7% of the world's rivers with a total annual discharge of about 3100 km³ (8% of the world's discharge) of fresh water into the sea each year (Kristensen & Hansen 1994); although a relatively small amount of the total discharge the overall impact is not insignificant. The temperate humid climate together with a high percentage of highly erodible limestone means that the amount of dissolved solids in European rivers is considerable. This is exacerbated by a heavily populated continent and associated agriculture resulting in an intense concentration of minerals within the rivers.

The concentration of pollutants varies between European countries. In those where coastlines are dominant (e.g. UK, Norway, Sweden, Denmark, Italy and Greece) there is tendency towards small catchments and short rivers. Thus population tends to congregate towards the coastline and hence waste water is often discharged into the coastal areas rather than into the river systems. In these instances water quality of the rivers upstream is often relatively good and therefore restoration efforts are able to concentrate on ecological habitat initiatives. Conversely, combating river pollution tends to be the main rehabilitation driver in countries with no coastline (e.g. Switzerland) where waste water has, historically, been pumped directly into them.

11.3.1 River flow

River flow varies during the year and this is determined by the seasonal variation in weather conditions together with the nature of the catchment, and the land management. This not only affects flow patterns on a local scale but also is instrumental in affecting them at the Europe-wide scale. Precipitation, for example, is highest in the west of Europe and lowest in those countries to the east of the continent, whereas there is an increase in evaporation in the southern and eastern extremes. Furthermore, the range of variation in runoff is considerable and whereas it may exceed 3000 mm in the Alps, in parts of Spain it can be as low as 25 mm per annum. Localized seasonal differences are also critical to the development of rivers and, while in some areas snowmelt is the driving force (e.g. 25% of the annual discharge of the Torn Alva in Sweden occurs during 1 month during summer), conversely in southern Europe where rivers are fed by rainfall the main flow regime often equates to the cyclonic weather patterns in the autumn and spring with low or no flow occurring in the summer.

11.3.2 Inter-connectivity of rivers with wetlands and floodplains

No river system should be considered in isolation from its floodplain and yet throughout Europe there have, historically, been demands on these areas and as a result many are now left with a legacy of insensitive land management and flood-defence strategies often based on unnecessarily over-engineered solutions. The result of this is disconnected floodplains which, in a natural riverine environment, would not only have been of much higher ecological value but would also have increased the storage capacity of the rivers in times of high flow, thus reducing the need to mitigate against flood impacts downstream. A recent EU-funded trans-national partnership project entitled *The Wise Use of Floodplains* (Zöckler 2000) looked at how sensitive management could contribute to sustainable solutions to water resources at the scale of river basins and catchments. The project was primarily a response to the problems associated with the historic unwise use of floodplains which has resulted in catastrophic flood damage of property, danger to inhabitants and the loss of floodplain wetlands. In addition, these problems have been exacerbated by agricultural subsidies which have not encouraged best-practice (environmentally friendly) farming methods, but rather the eutrophication and heavy modification of river systems.

11.3.3 Ecological value of rivers and floodplains

Rivers and floodplains naturally support a wide variety of flora and fauna and are an essential component in creating pathways or corridors between other habitats such as woodlands, and like any ecosystem they contain characteristic communities and functions. While many species rely on specific hydrological and sedimentological conditions, for others it is the connectivity between the floodplain and the river that is crucial to their survival. For example, fish show highest diversity within the main channel, yet dragonfly and aquatic-plant species richness increases in stagnant backwaters. White bream (*Abramis bjoerkna*) need both the main flowing river channel and backwaters to complete their life cycle, whereas amphibians may migrate to the floodplain areas in spring to deposit eggs. Others play an integral part in ensuring the sustainability of the various habitats. Beavers (*C. fiber*) for one, although in decline throughout much of Europe, can greatly enhance the habitat where they live. Water accumulates behind their dams, creating beaver ponds, which provide extra food and refuge pools for fish and create additional habitat for other aquatic wildlife and plants. Equally, their foraging and feeding behaviour protects valuable wetlands by keeping them free of scrub and provides additional deadwood for many species of invertebrates.

Clearly many species rely on this range of river and floodplain environments which consist of vastly different habitats and therefore can be occupied by very different species assemblages. Upland bolder-strewn streams, for example, can provide spawning grounds for salmon (*S. salar*) and sea trout (*S. trutta*), which in turn are dependent on the food supply provided by aquatic and bankside invertebrates. In the siltier slow-flowing reaches spined loach (*Cobitis taenia*) and lamprey (*Lampetra fluviatilis*) bury into the substrate. In addition to the river, many backwaters, cut-off channels (natural and man-made) and pools can provide permanent and semi-permanent wet habitat for a range of wetland and sheltering riverine species, for example marsh warbler (*Acrocephalus palustris*), black bog ant (*Formica candida*), southern damselfly (*Coenagrion mercuriale*), water vole (*Arvicola terrestris*), marsh fritillary (*Eurodryas aurinia*) and common frog (*Rana temporaria*), to name but a few, whereas the otter (*L. lutra*), for one, travels long distances between rivers, lakes and wetlands to hunt and rest. The diversity of habitats along the river corridor has gradually declined in response to increased urbanization and the decrease in forest floodplains. EU Commission figures suggest that, among others, 45% of butterfly, 38% of bird, 24% of flora and 5% of mollusc species are considered as threatened by extinction (Halaham 2000). Many of Europe's rivers now have little natural floodplain habitat with restricted riverine corridors and a thin strand of open water is often all that remains to connect this network. As these habitats have reduced, the Biodiversity Action Plans have tried to redress some of the issues of habitat destruction and these plans have been introduced across Europe with the aim of rectifying some of this environmental damage. Within these plans specific species associated with rivers and streams have been identified.

11.4 Current disturbances and threats

Despite a growing awareness in the scientific and political worlds that natural riverine environments are a major asset to ecological diversity and that reconnection of the rivers to their floodplains is one way to alleviate current uneconomically viable flooding problems, there continue to be disturbances and threats to river ecosystems that must be acknowledged.

11.4.1 Future climatic scenarios

Climate scenarios vary considerably and there is still much discussion even within the Intergovernmental Panel of Climate Change (IPCC) about both the precise rate and extent of future variation in temperature and precipitation. Nevertheless, there is general consensus that there is likely to be an increase in global warming. Some figures suggest that an average 4°C increase in temperature throughout the world by 2080 could result in an estimated average increase in river discharge of 20% worldwide. Furthermore, it is

suggested that the distribution of rainfall may change and stormier conditions will prevail with associated flash floods resulting in new flow regimes altering the meaning of a one-in-50-year event. Since 1988, an increase in the frequency and magnitude of flooding in many parts of Scotland, for example, has led to increases in the assessed risk of flooding. Marsh *et al.* (2000) showed that the risk estimate for a flood the size of that which occurred in Perth, Scotland, in 1993 increased from one in 2000 years in 1988 to about one in 100 years in 1994. Subsequent work examining the effects of climate change in Scotland (Werritty *et al.* 2002) indicated that, by the 2080s, floods presently estimated to have return periods of 50 years may occur with return periods of as low as 17 years. Such research implies that previous embankments and walls designed to a high level of protection are now woefully inadequate for major flooding events.

11.4.2 Urbanization and development

In addition to these predictions of future climate change, urban expansion continues and whereas historically the floodplain was a wet area avoided by industry and homeowners, with the development of new, improved flood-protection techniques some of this reticence has long been forgotten. Yet, a flood-protection scheme is only as good as its design specification and many of those implemented to withstand a one-in-50- or even a one-in-100-year scenario are not a guarantee that these floods will be kept at bay in the future. These schemes are therefore unlikely to be sustainable economically in the longer term. Furthermore it is unrealistic to expect to be able to design flood defences to cope with an estimate of the probable maximum flood either on the grounds of finance or health and safety. There is now a great deal of conflict between the pressure on local governments to deliver increased housing and infrastructure, and the acceptance that empty floodplains that are allowed to flood can be an economic asset as well as being of benefit to biodiversity. In the UK the government document PPG25 (2001) states that planners should avoid inappropriate development of the floodplains.

In many countries though, the scientific expertise to prove accurately that new developments will not increase flooding in vulnerable highly urbanized areas is still not available. In continental Europe the pressure not to change old flood-alleviation ways is exacerbated by the trans-boundary nature of rivers. Those countries where the precautionary principle of highly engineered rivers and flood-defence systems is widely applied are often at the 'end of pipe'. The Netherlands is just one example where the government has little control over what they receive from the Rhine and Meuse and until a more uniform approach to the management of the rivers is adopted such countries will be cautious about altering existing approaches. There is hence a need for catchment-wide planning to increase catchment flood storage especially to benefit downstream river reaches.

11.4.3 Economic justification of restoration

Without accurate tangible costs to deter floodplain development and instigate river restoration, much restoration work is undertaken piecemeal and on an opportunistic basis. Directly comparing the cost of a road-improvement scheme and the value of floodplain woodland is still, for example, very difficult. How much value for example can be assigned to an otter (*L. lutra*), a water vole (*A. terrestris*), a reed bunting (*E.*

schoeniclus) or a southern damselfly (*C. mercuriale*)? Many current cost-benefit analysis systems cannot adequately integrate biodiversity, aesthetics, public enjoyment and natural-heritage issues. More difficult still is proposing an economic justification for the restoration of a previously degraded river system, on the basis of its naturalness. Fortunately the Habitats Directive 92/43/EEC (EC 1992a) has required Member States to protect rare habitats and improve them to 'favourable condition' habitats where rare species live. Although at present it only applies for the designated species, it has provided a means of protection, and justification of expenditure on restoration, as Member States have accepted the legislation. It remains therefore early days in terms of concerted river-restoration efforts but nonetheless there are still good examples of rehabilitation and restoration throughout Europe from which lessons can be learnt for the future.

11.5 River and floodplain restoration

There is no simple solution to restoring rivers and their floodplains. Each project needs to be considered on its own merit. In some cases a previously straightened reach may be restored to a meandering form (see Fig. 11.4), bringing with it the added benefits of

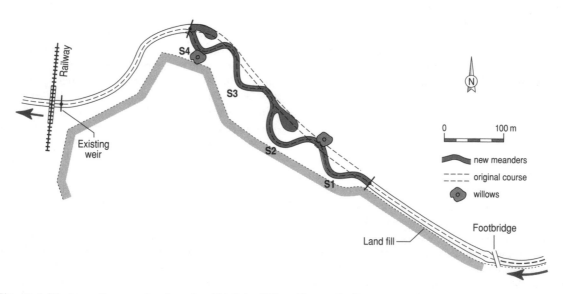

Fig. 11.4 Diagram of a meandered section (S1–S4) of River Skerne, Darlington, County Durham, UK.

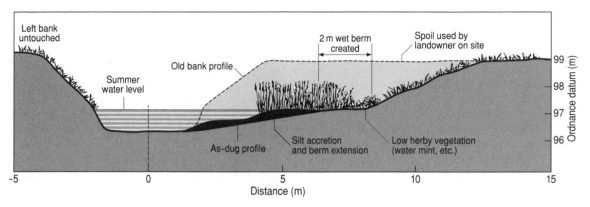

Fig. 11.5 Creation of on-line bay by re-profiling an old bank on the River Tall, Moy, County Armagh, Northern Ireland. M.O.D., ordnance datum (in metres).

increased biodiversity, but this is by no means always the most appropriate method. The recovery of an eco-system may for example be assisted by less-radical restoration principles such as re-profiling river banks and creating wet berms as shown in Fig. 11.5. The fol-lowing section provides a few examples of the types of issue impacting rivers and floodplains and also out-lines those factors that are driving river restoration in various countries across Europe. The list is by no means comprehensive and many more examples can be found in a variety of textbooks (including Ward *et al.* 1993, Middelkoop & Van Haselen 1999, Klijn & Dijkman 2001, Nijland & Cals 2001) and the River Restoration Centre's *Update for the Manual of River Restoration Techniques* (RRC 2002); see below for further details.

11.5.1 UK

In the UK, rivers have been substantially altered over the past few centuries, with at least 80% of low-land river reaches having had at least part of their channel modified (Raven *et al.* 1998). Degradation has been particularly severe in the last 50 years through intensive management. Programmes of draining, dredging and straightening have touched almost all rivers. Since 1990, various groups and government agencies have been working to change the perception of river managers and to promote a more sustainable approach to river and floodplain use and river man-agement, rather than purely flood defence.

In the 1980s conservation and river enhancement was restricted to voluntary bodies and were seen as add-on works to the statutory duties of flood defence, fisheries and water quality. Many rivers were still seen as efficient conduits for the evacuation of troublesome water. In the early 1990s concern for the UK's rivers led to the formation of the River Restoration Project (RRP), and the design and implementation of two RRP river-restoration demonstration projects on the Rivers Cole and Skerne. This work, funded by an EU LIFE grant and in conjunction with works in Denmark (e.g. the River Brede), acted as a catalyst for change, sup-porting a rising number of restoration and enhance-ment projects in the last 10 years. The UK now has an established River Restoration Centre (RRC) whose single aim is to provide information and advice on river management for restoration and enhancement. The Centre has an inventory of over 900 river-restoration projects and has produced a manual detailing 47 river-restoration techniques from 17 UK projects (RRC 1999, 2002, CIRF 2001).

The River Skerne restoration project

The urban River Skerne in Darlington, County Durham, had been progressively straightened and re-aligned during a period of industrial and urban housing development over a period of some 200 years. A 2-km reach was chosen for rehabilitation. The works aimed to re-create a more naturally functioning water-course which would become an attractive landscape for the local community and a local amenity area (see Fig. 11.6). The work involved restoring the once meandering planform, in-channel deflectors to re-create sinuosity and flow diversity, bank re-profiling, channel narrowing, spoil disposal, landscaping, surface-water outfall rationalization, backwater creation, soft-revetment engineering, increasing floodplain storage, extensive marginal planting of native species and the creation of wetland scrapes. This project was funded by the EU, statutory government environmental agencies and the landowner (the local council), with an emphasis on engaging the local public so that river ownership could be assured. Many constraints had to be overcome but the final project has achieved its aims of restoring a more-natural river system which has a greater self-regulatory capacity with benefits for flood management, wildlife, amenity and water quality. The essential urban compromise led to a watercourse with implicit amenity and landscape appeal; 'a piece of countryside in a town'.

11.5.2 Denmark

It is estimated that 97% of Denmark's 30,000 km of natural rivers have been altered in their physical form (Brookes 1984). In addition another 30,000 km of man-made watercourses exist designed to drain wetlands and floodplains and to increase agricultural production. Much of the work to drain Denmark was undertaken in the last century, together with fish farming and industrial and urban expansion. Channelization was also frequently accompanied by mills and other structures, effectively denying free passage to fish and macro-invertebrates. Through legislation and changes to the administration of Danish watercourses, management improvements and river restoration have been implemented since the mid 1980s. Twenty-four such schemes are detailed in Hansen (1996). Many of these consist of repair to the physical form after water quality standards have been achieved. In 1996, arising from the above joint Danish/UK collaboration, Denmark established the basis for the European Centre for River Restoration.

The Skjern River restoration project

Between 1962 and 1968 the River Skjern system, the largest river in Denmark, was channelized and

Fig. 11.6 (a) River Skerne, Darlington, County Durham, UK, before restoration. (b) Log toe revetment during construction on the River Skerne. (c) River Skerne after restoration. Courtesy of Northumbrian Water, UK.

dredged, with approximately 4000 ha of floodplain, wetlands, reed bed, meadows and marshlands drained to allow agricultural production on its fertile soils. Over time a number of issues emerged. The peat soils were desiccating, lowering the entire farmed area. Continuous and expensive updating of the pump systems was required to drain the low-lying ditch systems into the river. Instead of wetland meadows acting as a sediment sink (deposition zone), farmlands were now a major source of sediment into the river. This was evident from the severe water-quality pollution and major deposition occurring in Ringkøbing Fjord. This was fast becoming an excellent example of unsustainable river management. In 1998 the Danish Parliament, with a huge majority, passed a Public Works Act to restore the lower reaches of the Skjern system, recreating over 2200 ha of typical floodplain and the sinuous course of the once multi-threaded river (Fig. 11.7). The project was budgeted at approx €31.9 million (NFNA 1999). The project aims to restore the natural functioning of the system and once again provide a rich variety of habitats for some of Europe's endangered species, such as the bittern, otter, black tern and corncrake.

11.5.3 The Netherlands

The Netherlands has a long history of managing its rivers for settlement and farming. It also has a similarly long history of protecting itself from the results of the river-management practices of other countries. At present the main land use of much of the floodplains in the Netherlands is agriculture. As a result the floodplain area available to store and attenuate major floods is limited compared to the large flat wetland area which would naturally have been available prior to intensive farming. For this reason, floodplain restoration is now of high priority, encouraging a landscape that has more natural variation, including wetland hollows and floodplain woodland, and can support an increase in biodiversity. From a species perspective, the enhancement of stagnant water bodies, which were created as a result of previous gravel and sand extraction, and the reinstatement of secondary channels, are also very beneficial, acting as refuge areas for fish and macro-invertebrates. The long history of flood defences, multiple raised embankments and

Fig. 11.7 (a) Before and (b) after restoration photos of the River Skjern, Denmark (NFNA 1999).

pollution from upstream has meant that today those floodplains that have been frequently inundated are covered with a layer of clay or sandy clay. The original diverse floodplain topography, soil type and vegetation have often disappeared. The solution being adopted today is both giving room to the river and lowering the floodplain (RIZA 1996, Smits et al. 2001). Any sustainable flood-protection measures will necessitate imposing limits on the amount of floodplain land that may be used for purposes that are not directly river related and the most sustainable flood-defence measures are likely to be those that are tied in with the natural processes of rivers. In addition, lowering the floodplain by removing the clay deposits on the floodplain down to the original profile (economically viable through its use in brick manufacturing) provides an increase in habitat diversity as well as renewed storage capacity (Wolters *et al.* 2001).

11.5.4 Italy

At the end of the last century a growing need for energy supply drove the implementation of hydroelectric industrial systems and storage devices throughout the Alp and Apennine regions of Italy culminating in 495 plants with a total capacity of 14,312 MWh, mostly located about 2000 m above sea level. Furthermore, the combination of extensive mountain ranges, urbanization and development along the valley floor and floodplains meant that natural river dynamics were considered to be a hydraulic risk within river corridors. As a consequence many rivers suffered the same fate as much of the rest of Europe, resulting in concrete channels, engineered infrastructure, extraction of gravel and sand, intensive agricultural use of soil and a loss of biodiversity for all but a few rivers. By the end of the 1980s, the Italian Government produced important formative actions concerning the water-management policy that led to the introduction of River Catchment Authorities that no longer followed administrative borders and of ATO (Integrated Management of Water Resources) that follows flow regimes from source to discharge. In this context the Italian Centre for River Restoration (CIRF) was established in 1999 as a non-profit-making association that aims to promote an innovative approach to water and territory management through pilot projects (Beppe Baldo, Director CIRF, personal communication).

Zero River Project

The Zero River Project is one such scheme aimed at developing and implementing a catchment strategy to protect key freshwater and estuarine habitats associated with the Venice Lagoon. The project has involved meeting nutrient (nitrogen and phosphorus) reduction targets through zoning the risk of the catchment and creating buffer zones for pollution entering the rivers within the catchment. These buffer habitats include riverine lakes, large reeded bank margins, floodplain lakes and large areas of riparian woodland. This project is now complete and has resulted in the creation of 16 km of restored river chan-

Fig. 11.8 River Zero, Italy, post-restoration. Courtesy of CIRF, Italy.

nels, 30 ha of riverine and floodplain lakes, 10 ha of reed beds and about 160 ha of riparian woodland of a previously concrete channel (see Fig. 11.8).

11.5.5 Large European river systems

Some central European rivers have the added complication of flowing through a large number of countries, all with a vested interest in the river. Some may see (or may previously have seen) these rivers as conduits for waste and floods and others may see them as the bringers of pollution and floods. Often, interpreting national policies on the sustainable management of rivers is a difficult-enough task, without having to agree multi-nationally. However, some good examples do exist that take into account the requirements of local inhabitants, improve biodiversity and ensure the economic needs relating to flood defence are met. In this context good schemes are those that aim to create a sustainable river and floodplain solution which functions as naturally as possible within the constraints of today's increased pressures on the surrounding land.

The Rhine

Within northern mainland Europe, the Rhine is one of the most important rivers and, while it originates in Switzerland, it flows through France, Germany and the Netherlands to the North Sea. It has been the foundation of settlements and commerce for many centuries and hence human impact has been considerable. One of the earliest recorded impacts was in 1449 when over-fishing and pollution led to declines in the fish population. Even at this point the Strasbourg Regulations were adopted to improve the Rhine, yet further deterioration was still to come. Between 1817 and 1874 considerable straightening occurred due to major engineering works in an attempt to improve navigation, reduce flooding and recover alluvial areas for farming. By 1950 Atlantic salmon in the Rhine had disappeared as habitats diminished and physical barriers increased. Since 1952 the Rhine Commission has worked towards improving water quality, river ecology and pollution emissions. Many small and partnership programmes of works have and are being undertaken including flood retention in Germany and embankment removal in the Netherlands (Fig. 11.9). Restoration strategies and the status of ecological rehabilitation of the lowland basin of the river Rhine have been reviewed by Buijse *et al.* (2002) and Nienhuis *et al.* (2002).

The Danube delta restoration project

The Danube is the second largest river in Europe, covering around 800,000 ha and flowing through 11 countries. The river rises in Germany's Black Forest, eventually discharging into the Black Sea delta area in Romania after 2800 km. The delta itself is the second largest wetland in Europe and was declared a Biosphere Reserve in 1990 by the United Nations Educational, Scientific, and Cultural Organization. Over 60 species of fish and 300 species of birds use the delta wetland (Zöckler 2000). Successive programmes of so-called improvements have led to deterioration in the wetlands. These were initially for navigation purposes, but problems were exacerbated through the construction of dams and channelization for agricultural production during the 1960s and 1980s.

With the opening up of central Europe, two major initiatives have been undertaken to look at the restoration of the hydrological system underpinning the delta wetlands. Firstly the World Wildlife Fund's Green Danube Initiative consists of five restoration

Fig. 11.9 Floodbank removal on the River IJssel, The Netherlands. Courtesy of Ute Menke, ECRR.

projects located in the headwaters, middle and delta reaches (Germany, Hungary, Bulgaria and Romania). The second forms a part of the Danube Carpathian Programme and involves the countries of the lower Danube. Initiatives such as these and the work of the Rhine Commission on rivers with vast catchment areas and physical and national barriers illustrate the importance placed by government and environmental NGOs on the concept of river and floodplain restoration. Furthermore, the Danube Delta Biosphere Reserve Authority (DDBRA) and the Danube Delta National Institute for Research and Development (DDNI) have jointly commenced a polder restoration programme. In this project abandoned or inefficiently managed fishing and agricultural polders were identified for potential restoration and in 1994 the first polder, Babina (2100 ha), was restored by re-wetting. The programme did not stand still, with polders re-wetted at Cernovca, Popina and Fortuna in 1996, 2000 and 2001 respectively.

11.6 Perspectives

Although the projects covered here are forward-thinking and have resulted in examples of best-practice river restoration, not all European countries are at present in a position to implement such ideas. Nonetheless, many are already accepting that natural river systems are part of our future inheritance. In central and eastern Europe water quality remains the main problem but there is little or no money for improvement. In the Russian Federation, for example, 60% of existing sewage-treatment works are overloaded and 40% need repair. This has serious implications for contaminated land, and most fish such as sea trout (*S. trutta*), salmon (*S. salar*) and sturgeon (*Acipenser sturio*) are virtually extinct. Israel has experienced similar problems with sewage outfall, a problem that it shares with neighbouring Palestine. Despite serious political difficulties a collaborative initiative has begun that addresses the pollution issue, recognizing that improvements will be of benefit for all. This has resulted in the Alexander River Restoration Project (Amos Brandeis, Israel, personal communication).

The acceptance that river restoration is important remains a divisive issue among EU member states. Spain has a Restoration Centre, but the country is still grappling with over-abstraction issues for irrigation and increased urbanization, especially in southern areas. This has gone hand in hand with continued building of major dams and an increase in pollutants, both of which have been detrimental to the ecological status of many Spanish rivers. In comparison, the Netherlands already have a structure of government policies and subsidies in place, aimed at putting river restoration high on the agenda. Many river-restoration centres throughout Europe are now promoting the importance of rivers and floodplains to support ecological diversity and sustainable options for flood management. To date, the UK, Italy, Denmark, the Netherlands, Spain, Norway, Romania and Russia have centres, with an overarching European Centre for River Restoration. Furthermore, the World Wildlife Fund (WWF) project Living Rivers for Europe is also championing the idea of restoring rivers and wetlands across Europe. This project is working to protect and restore over 65 partnership river projects within 25 European countries.

In the UK there is growing acceptance that any flood-control measure should ensure that there is no net loss of biodiversity and that wherever possible environmental enhancement should be promoted; to this end the government (DEFRA 2002) has published guidance on the environmental appraisal of flood defence to help ensure that the implications for flood prevention and management are fully considered in decision making. Such schemes are in their infancy yet as many existing flood-defence schemes will soon no longer be viable to maintain on economic grounds alone, more ecologically friendly options are likely to be considered more seriously as sustainable, long-term and cost-effective. Developments in urban landscape planning and design, inclusion of sustainable urban drainage systems (SUDS) in national and local authority planning guidance and better integration between engineers, ecologists, landscape architects, etc. to solve old problems, all add to a wide knowledge base readily available through Europe-wide collaborations.

Inevitably, any new ideas that are put into practice will be subject to a degree of risk and uncertainty. Whereas it is possible to mitigate against this through rigorous scientific analysis, the very nature of a dynamic living system means that exact outcomes cannot always be predicted and a certain amount of flexibility is needed to promote future river restoration and rehabilitation. By ensuring that new schemes are fully appraised both pre- and post-project in terms of their impact on the river geomorphology, ecology and public perception, those charged with restoring rivers today can learn from each completed scheme. To put this into perspective it is essential that new projects are monitored against their initial objectives, be this habitat enhancement or increased channel morphological diversity. Only then can we be certain of the extent to which any scheme can be classified as either a success or a failure and allow scientists to ascertain what types of scheme are best suited to different environments and for various requirements. We can then build on these newly acquired skills to promote a future of sustainable riverine landscape that promotes naturalness at its heart.

Acknowledgements

The River Restoration Centre is very grateful for the data and information sent by Beppe Baldo and his team at CIRF (the Italian River Restoration Centre), Geraldene Wharton's (Queen Mary, University of London) constructive comments on the text and information supplied by Andrew Black and David Gilvear about increased frequency and magnitude of flooding in Scotland.

12

Restoration of freshwater lakes

Ramesh D. Gulati and Ellen van Donk

12.1 Introduction

The development of modern society, especially the human population explosion and intense industrial urban developments, in the last four decades has caused inland surface waters to become heavily enriched by agricultural fertilizers and toxic substances (Forsberg 1987). The man-made alterations to freshwater aquatic ecosystems worldwide – lakes, reservoirs, wetlands and rivers – have been both severe and destructive. Despite the recent attempts to restore ecosystems under stress, freshwaters continue to be perhaps the most vulnerable of habitats. In the Western world, lakes and reservoirs are recreation attractions (for water sports such as swimming, boating and angling). In addition, they are major sources of water for drinking purposes, irrigation, industry, transport and floodwater storage. While the lakes act as sinks for many of the products of human activity in their catchments, rivers drain human and animal wastes and other wastewater effluents into the sea. Increasing demand for fresh water by humans has led to the creation of storage reservoirs in the floodplains of many river systems (Moss 1998). Management strategies for these aquatic ecosystems, including wetlands, often ignore their regional watershed context, hydrology and economic relationships.

For sustained and long-term use, many aquatic ecosystems need not only to be **protected** by abatement of pollution but also **restored** or **rehabilitated**. Several studies are under way in the USA and Europe to facilitate the return of the disturbed ecosystems to conditions prevailing prior to the disturbance (see examples in Gulati & van Donk 2002, NRC 1992, Cooke *et al.* 1993).

The goal of ecosystem restoration is to emulate a natural and self-regulating system that is integrated within its ecological landscape (Berger 1990). In practice, lake restoration is considered synonymous with improvements in water quality defined in terms of clarity, oxygen conditions and the amount of algae, to improve lake conditions designated for human use: recreation, fishing and water supply. Most restoration projects aim at improving the important ecological attributes of lakes, rather than at the return of lakes to a pristine condition. Such attempts are focused primarily on eradicating the undesirable consequences of the man-induced disturbances. Lake-restoration work in the USA and western and northern Europe started in the early 1970s (e.g. Bjork 1972). By 1975 the US Environmental Protection Agency initiated the Clean Lakes Program by amending the Federal Pollution Control Act. Subsequently, more federal funds were provided to clean more than 300 lakes in 47 federal states. The restoration work on Lake Trummen in Sweden that involved sediment removal and fish manipulation is a classic example of lake restoration in Europe (Bengtsson *et al.* 1975, Andersson 1988). Also, restoration of acidified lakes by liming, as in the north-eastern USA, became a relatively common practice in Scandinavia. Restoration is steadily becoming an essential part of national and international efforts to improve both water quality and the ecology of freshwater ecosystems (NRC 1992), especially in western Europe, the USA and Canada (Cooke *et al.* 1993). Most of the national action plans to restore the many threatened rivers, lakes and wetlands were developed in the 1980s. Their execution not only involved cooperation among scientists and engineers, but also called for positive feedback from legislators, environmental protection agencies and industry. In addition, feedback from citizen groups seemed to have been essential for propagating public awareness. During the last three decades some 25 lake-restoration techniques have been developed and tested for their overall effectiveness, type and intensity of recurring problems, cost per unit area, required frequency of employment and range of applicability, etc. Since the early 1980s a biological means of lake restoration, so-called **lake biomanipulation**, has become extremely popular, both in Europe (Benndorf 1987; see references in Gulati *et al.* 1990; see also the review by Gulati & van Donk 2002) and North America (Shapiro *et al.* 1975, Lynch & Shapiro 1981, Shapiro & Wright 1984). Lake restoration is now among the major environmental focal points relating to water management in general (Gulati & van Donk 2002; see also Nienhuis & Gulati 2002).

We present a state-of-the-art résumé of the methods of aquatic ecosystem restoration in progress in Europe and the USA: especially in north-west Europe. We assess the types of disturbance, identify the problems and main symptoms, discuss briefly the lake-restoration techniques in use and draw some conclusions.

12.2 Ecosystem disturbances

Several studies have dealt with restoration or rehabilitation of aquatic ecosystems during the last three decades (Cooke *et al.* 1993; see references in Gulati & van Donk 2002). It is now well known that increases in nutrient input, organic matter, silt and contaminants to lakes and reservoirs cause a deterioration of water quality. The stressed water bodies manifest themselves with an increased growth of algae, water plants or both, causing reduced water transparency and even a marked decrease in water volume due to the accumulation of organic matter including detritus. The most obvious, persistent and widespread water-quality problems related to human use of lakes and reservoirs are **eutrophication** and **acidification**. These two issues have received worldwide attention since the 1970s. In this chapter, while recognizing that there are many other disturbance and stress factors, we focus on the main causes and symptoms of these two major disturbances.

12.2.1 *Eutrophication*

Eutrophication is a natural ageing process of lakes that causes a steady increase in biological production due to a gradual accumulation of nutrients and a slow decrease in lake depth. In classical terms, eutrophication is the enrichment of water by inorganic plant nutrients, especially N and P. The increasing eutrophication in the early 1970s of the Great Lakes and other lakes and reservoirs in the United States, Canada and western Europe (Vollenweider 1968, Schindler 1974, Vollenweider & Kerekes 1981) raised widespread public concern. General recognition of P enrichment and the related eutrophication problems led to large-scale research funding by the state agencies. Whole-lake experiments in eutrophication (e.g. Schindler 1974) and development of eutrophication models provided insight into the quantitative relationships between nutrient loading rates and algal biomass and production (Vollenweider 1987). The eutrophication definition has been expanded to include the loading with silt and dissolved and particulate organic matter (Cooke *et al.* 1993). Human activities typically alter the hydrology and increase the nutrient loads into lakes, thereby accelerating the eutrophication processes. The main causal factors for eutrophication are direct, point discharge of human and animal wastes and non-point, agricultural runoff from the catchment into these lakes. Consequently, the rates of algal production and nutrient accumulation in the lake sediments rise, leading to an increase in loading from in-lake processes (Rast & Thornton 1996), so that the effects of eutrophication persist and build up over time.

In temperate regions the spring-time increases in light and water temperature in eutrophic lakes generally result in enhanced algal production, dominated by filamentous cyanobacteria, formerly called blue-green algae. Consequently, the turbidity of the water rises and the underwater light climate deteriorates. Many genera of filamentous cyanobacteria (e.g. *Oscillatoria*, *Anabaena* and *Aphanizomenon*) and the colony-forming cyanobacterium *Microcystis aeruginosa*, a cosmopolitan species, dominate the phytoplankton of eutrophic lakes. These cyanobacteria reach bloom conditions that may persist during the growing season. The ability of cyanobacteria to grow better than other algae over a wide range of nutrient levels as well as fix the atmospheric nitrogen, together with their relatively poor edibility by zooplankton, allows them to attain both high densities and biomass. They can thus persist and dominate the phytoplankton of many shallow lakes throughout the growth period, and even overwinter. Lastly, several species of cyanobacteria can produce potent toxins, which are a matter of great concern from the viewpoint of human and animal health (Codd 2000, Chorus 2001).

Rast and Holland (1988) provide a scheme with a sequence of decisions to be made in the development and implementation of eutrophication measures in lakes and reservoirs (Fig. 12.1). The scheme takes into account the different concerns and cost-effectiveness in the selection of feasible remedial measures. A practical framework for the management of lakes and reservoirs is to (i) assess the available information, (ii) identify eutrophication problems for establishing management strategies and control measures, (iii) analyse costs and expected benefits of alternative management strategies and the adequacy of institutional and regulatory frameworks for implementation of a given strategy and (iv) select a control strategy and publicity measures and evolve mechanisms to minimize the recurrence of eutrophication problems..

12.2.2 Acidification

Acidification is only second to eutrophication among the most widespread anthropogenic changes in lakes, reservoirs and streams. The regions most vulnerable to acidification, where critical loads have been exceeded, are Scandinavia, North America and the UK. In the first two regions, acidification has caused great damage to fish populations in lakes (e.g. Henriksen *et al.* 1989). Schindler (1988) documented the effects of acid rains on freshwater ecosystems in a paper that attracted a lot of attention. Precipitation in parts of Europe and North America has a pH well below 4.7 (so-called acid rain), compared with a pH of *c.*5.6 for pure rain water. The elevated acidity in the rainwater is due mainly to polluted air masses containing S and N compounds (SO_2 and NO_x), released from the burning of fossil fuels, transported by winds thousands of kilometres away from the place of their origin – across the national boundaries – and transformed by photo-oxidation to sulphuric acid (*c.*70%) and nitric

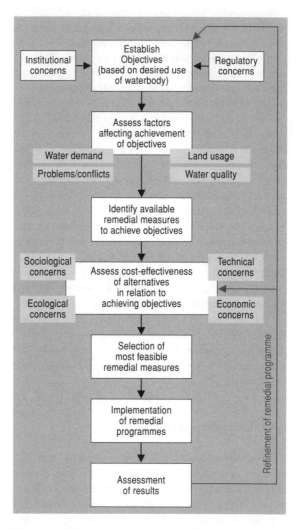

Fig. 12.1 Schematic representation of the sequence of steps involved in the decision-making process for the selection and implementation of control programmes in lakes. Note: the desired use of the waterbody, water demand and socio-ecological concerns will determine the choice of most feasible remedial measures. From Rast and Holland (1988); in Mason (1996). Reproduced by permission of Ambio.

acid (*c.*30%) before they eventually fall as acid rain. These emissions are linked via the runoff from terrestrial ecosystems and via direct transport and deposition to aquatic ecosystems (Fig. 12.2; see also Steinberg & Wright 1994). In addition, in areas with

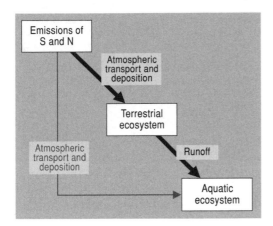

Fig. 12.2 The steps linking emissions of SO_2 and NO_x and deposition, both direct and indirect, to aquatic ecosystems. Note that the catchment area (terrestrial ecosystem) and the runoff from here into lakes is a more important cause of lake acidification than direct, atmospheric deposition alone. From Steinberg and Wright (1994). Reproduced by permission of John Wiley and Sons Ltd.

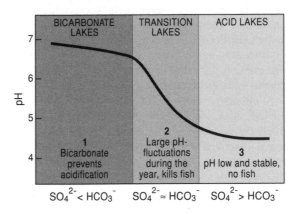

Fig. 12.3 The main steps in the acidification process of lakes. The process is related to the decrease in pH caused by a lowering of the buffering capacity of water due to a shift in the dominant anions, from bicarbonates (HCO_3^- ions) to sulphates (SO_4^{2-}). From Mason (1996). Reproduced by permission of Pearson Education, Inc.

intensive agriculture and animal husbandry, nitrate that is formed from the large amounts of ammonium emitted ($NH_4^+ \rightarrow NO_3^- + 2H^+$) from the farms acidifies the watercourses. In addition, groundwater and forest soils are affected via the runoff in the catchment. Limestone in a drainage basin may help prevent acidification considerably; regions with a calcareous geology are not sensitive to acidification (Henriksen et al. 1989). Figure 12.3 illustrates the pH decreases that occur during the acidification of lakes and the role of bicarbonate ions in buffering of lake water to prevent acidification before sulphate concentrations increase further.

It is not clear if the marked biotic changes in communities of lake organisms caused by acidification are direct, physiological effects of pH decrease (tolerance) or are indirect effects of changes in biotic interactions. For example, community structure could be altered by shifts in the competitive relationships of the algae or by the disappearance of keystone species (Eriksson et al. 1980). Altered solubility and speciation of many metals due to a decrease in pH can cause important biological effects. Aluminium, iron, copper, zinc, nickel, lead and cadmium become more soluble in water

when acidified, but mercury and vanadium become less soluble. Many of the adverse effects on organisms are attributed to the increased solubility of aluminium and its shift to the toxic Al^{3+} form. Increased mobilization of Al^{3+} ions in lakes also causes precipitation of P and humic substances and such acid lakes tend to become oligotrophic and thus more transparent. Reduced rates of organic-matter decomposition and mineralization and O_2 consumption by micro-organisms lead to decreased availability of nutrients such as PO_4-P so that phytoplankton production decreases. In contrast, the development of algal mats at the lake bottom may increase due to improved light climate.

Increased solubility of metals at lower pH will impose physiological stresses on zooplankton: both H^+ and Al^{3+} ions interfere with the sodium balance of most crustacean zooplankton; for example, larger species of Daphnia and calanoid copepods disappear below pH 6.0, whereas Bosmina longispina still occurs at pH values of <4.1 (Brett 1989, Steinberg & Wright 1994). A survey of c.1500 Norwegian lakes showed that snails and bivalves, with calcareous shells, largely disappeared below pH 6. The crustaceans Lepidurus arcticus and Gammarus lacustris, important food items for fish, are sensitive to acidity and their decreases adversely influence species richness and the structure of the macro-invertebrate community in streams in Wales (Wade *et al.* 1989). In invertebrates in acid waters, especially crustaceans and gastropods, the transport of Na^+, Cl^- and K^+ ions is upset and Na^+ in body fluids decreases. In fish stressed by acidity, there may be a decline in body Na^+ and Cl^- contents.

Aluminium is toxic to fish in the pH range 5.0–5.5 and Al^{3+} ions interfere with the regulation by calcium of gill permeability, with enhanced loss of sodium. Five major functions that are adversely affected are: ion regulation, osmoregulation, acid/base balance, nitrogen excretion and respiration (Brakke *et al.* 1994). The loss of Na^+ and a decrease in Cl^- ions in the blood plasma causes the body cells to swell and extracellular fluids to become more concentrated. Reports of the death of Atlantic salmon (*Salmo salar*) in Norway in the 1900s and of brown trout (*Salmo trutta*) in mountain lakes in Norway in the 1920s and 1930s were all attributed to an increase in acidity. The numbers of such lakes doubled by 1986 (Henriksen *et al.* 1989). In Finland, roach (*Rutilus rutilus*) were reported to be the most sensitive species, disappearing from many waters in the 1980s. Even the relatively hardy species whitefish (*Coregonus peled*) and perch (*Perca fluviatilis*) exhibited reproductive damage. There are also reports of similar effects on fish populations in several eastern provinces of Canada. In conclusion, periodic mortality of fish during the early stages of development and growth due to acid episodes causes the populations to decrease and disappear.

12.3 Techniques of lake restoration

Lake and reservoir management and restoration technologies developed rapidly during the 1980s in the USA, Canada and Europe (Cooke *et al.* 1993), especially in the Netherlands, Denmark, Germany and the UK, prompted by research into the nature of the problems faced. The new developments have enabled discernible changes in the perspectives of and approaches to lake restoration. An important basis of the restoration and management measures to be applied depends on users' interests; that is, the economic and recreative utility of the waterbody. Strategic principles of lake restoration combine the social aspects and available technology. They include the nature of water use, problems and public awareness on the one hand and

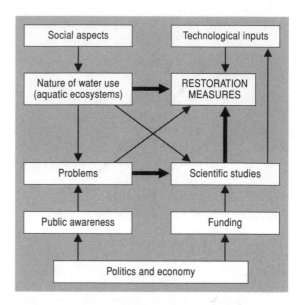

Fig. 12.4 Strategic principles of lake restoration. The choice of restoration measures will depend both on direct and indirect factors related to nature of water use, problems relating to water quality and scientific studies (thick arrows). From Gulati (1989); see also Vollenweider (1987).

funding, scientific knowledge and restoration measures on the other (Fig. 12.4; see Vollenweider 1987, Gulati 1989). Most lake-restoration methods are directed at reducing external P inputs into lakes. The restoration techniques invariably need to be applied simultaneously to ensure some success (see for example in Ryding & Rast 1989, Cooke *et al.* 1993). The choice of lake-restoration or -recovery measures has to be considered in the context of different human influences and the limnological characteristics of the lake to be restored. For restoring eutrophic lakes we can divide the measures into two main types: **external** and **in-lake** control measures (Fig. 12.5; Ryding 1981, Gulati 1989). The external control measures start with the diversion of sewage and wastewater inputs and the prevention of external nutrient-rich inputs into the waterbody to be restored. The **in-lake** restoration measures involve decreasing internal P loading by various physico-chemical control measures (see below), by the so-called biomanipulation of the lake's foodweb structure and functioning, or by using both sets of

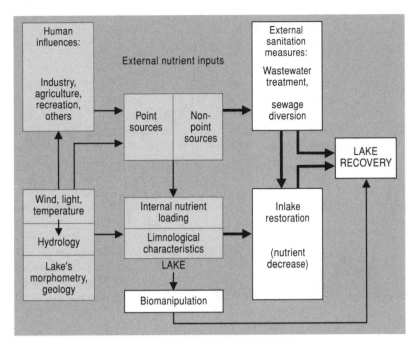

Fig. 12.5 Schematic representation of factors causing an increase in nutrient loading via external inputs, which also result in an increase of internal loading, and restoration measures (white boxes) leading to lake recovery. From Gulati (1989); see also Ryding (1981).

measures. In acid-water lakes, prevention of atmospheric emissions into the lake's watershed will be considered as an external control measure and neutralizing the acidity in the lake by addition of lime as an internal control measure (see below).

We will discuss different restoration measures based on well-documented studies and after identifying and assessing the disturbance effects.

12.3.1 Reduction of external nutrient loads

Ryding and Rast (1989) have assessed in detail the methods to control external sources of nutrients. Several workers (see examples in Sas *et al.* 1989) have reviewed the response of lakes to reductions in external P loads. Both diversion of nutrient-rich inflows and reduction of nutrients in the inflows are principal measures to reduce nutrient inputs into lakes (Cooke *et al.* 1993). During the 1970–80s in western Europe and North America tertiary treatment of sewage and

animal wastes (> 95% P removal) was the main P-control measure, before discharging this treated water into watercourses and lakes. P-containing laundry detergents were banned in most western countries during 1975–85 and replaced with P-free detergents. The measures led to up to an approximately 50% reduction of the allochthonous P inputs into some lakes in the USA.

Edmondson and co-workers (Edmondson & Lehman 1981, Edmondson 1991, 1994) successfully used diversion of nutrient-rich wastes to reduce nutrient input into Lake Washington, Seattle, USA (Cooke *et al.* 1993). In Barton Broad, Norfolk, UK, about 90% of the annual P load in the effluent from sewage-treatment works was reduced by adding ferric sulphate. However, internal P loading (130 mg m^{-2} day^{-1}), which was 10-fold higher than the external load (Phillips *et al.* 1994), resisted the improvement in water quality. In Esthwaite Water in the English Lake District, the removal of P from the sewage effluent, which contributed to 47–67% of the total P loading,

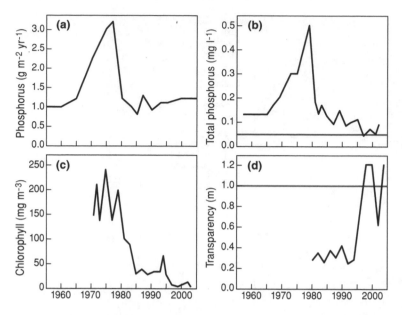

Fig. 12.6 Long-term monitoring of the effects of various P-reduction measures in the shallow Dutch Lake Veluwe almost immediately after the lake's formation. Note the marked increase in Secchi-disc transparency in the mid-1990s. The fluctuations in transparency, especially the decrease during 2001, are caused by wet winter periods. Courtesy of M.L. Meijer *et al.*, unpublished results.

led to no marked changes in water quality, nor reduction in the cyanobacterial dominance (Heaney *et al.* 1992).

In the Netherlands, P loading and total P concentration in Lake Veluwe (one of the Border Lakes formed from Lake IJssel during a land-reclamation programme) have been monitored since the lake's formation in 1957. The different P-reduction measures in the course of years have led to marked decreases in the P and chlorophyll concentrations, causing marked increase in water transparency (Fig. 12.6). In another more elaborate study, P in the inflow water to the hypertrophic Loosdrecht Lakes, in the Netherlands, was substantially reduced in 1984 by reducing P loading via the inlet water by different control measures, and its effects on different parameters including P dynamics in these lakes were monitored (see papers in van Liere & Gulati 1992). The water quality did not improve during the first decade or so, which could be attributed to increased P retention in the sediments, retarding the lakes' response. Similarly, Jeppesen *et al.* (1991) observed no response in Danish lakes to a 70% or greater reduction of P in the inflows. Cullen and Forsberg (1988) distinguished three types of response in chlorophyll and algal biomass based on data from 43 lakes:

1 in 15 lakes there was an improvement in water quality from hypertrophic to eutrophic to mesotrophic;
2 in nine lakes there was insufficient reduction of algal biomass to change the trophic status discernibly;
3 in 19 lakes there was little or no reduction algal biomass, but nuisance species may have been reduced.

Phosphorus from the external inputs may also be removed by passing P-rich effluents into treatment ponds, where much of the element is adsorbed on to sediments and settles. Also, P released due to lysis and death of algal blooms will be taken up by the sediment. Regular harvesting of aquatic vegetation (e.g. duckweeds, *Lemna*; water fern, *Azolla*; and other floating forms) can remove P (e.g. Viessman & Hammer 1993). For a reduction in non-point nutrient loading, agricultural farming practices in a lake's catchment area need to be altered radically by prevention of erosion, rationalizing the frequency and timing of fertilizer applications, minimizing P use and reducing the amount of fertilizer P imported to the watershed. In addition, more pasture land needs to be created and buffer strips should be planted between farmland and watercourses, as recommended for protecting the deteriorating Slapton Ley, a nature reserve in south-west England (Wilson *et al.* 1993).

The lack of a clear response to different external P reductions does not, however, preclude an improvement in water quality (Cooke *et al.* 1993). Most studies apparently did not monitor changes in lakes long enough for the P levels to record a decrease below a certain threshold level. A lake may take longer than a decade to respond, as observed during a long-term study in Loosdrecht Lakes (see papers in van Liere & Gulati 1992).

12.3.2 In-lake measures of P-reduction

External P-load reductions do not guarantee improvements in water quality of lakes in the short term (van Liere & Gulati 1992), primarily because of accumulated P in lake sediments. Apparently, recycling of this P from the sediments becomes more important if the P in inflows is reduced. Therefore, high P concentrations, often observed in lakes during summer (Hansen *et al.* 2003), and algal photosynthesis and production can go on unabated as they did before the external control measures were taken. Thus, increased internal loading may keep lakes in a eutrophic state for years after the reduction of external loading (Sas *et al.* 1989). We briefly discuss below a few restoration methods (Cooke *et al.* 1993) used for reducing internal loading.

Mixing of the water column

Artificial mixing involves aerating or oxygenating lakes using pumps and jets and bubbled air from perforated pipes at the lake bottom. The technique has been used in the states of Minnesota and Wisconsin in the USA to prevent fish dying during ice cover (Cooke *et al.* 1993). Aeration oxidizes substances in the water column, with an increased complexation of Fe and Mn in the sediments, and reduction in internal P loading due to a release from sediment. In addition, algal biomass is reduced due to expansion of the mixed layers and increased light limitation. However, nutrient availability for phytoplankton in the photic zone may even increase by circulation, causing an increase in phytoplankton biomass.

Although literature on the effects of artificial aeration on water quality is scarce, the mixing seems to prevent cyanobacteria from exploiting the optimal light conditions in the upper mixed layers (see Visser *et al.* 1996), leading to shifts in phytoplankton dominance by non-cyanobacterial forms. It may be concluded that mixing or circulation, unlike nutrient reduction, produces instantaneous improvements in water quality, without decreasing the nutrient concentrations or loading. Moreover, it is more effective for deeper lakes and long-lasting, positive effects of artificial circulation are virtually absent, implying the need for its repeated application.

Preventing the internal P loading from sediments

Phosphorus cycling between the sediments and water is a complex and a relatively poorly understood phenomenon. The release of P from sediment depends on both redox and pH and involves bacterial decomposition of organic matter, including algal blooms (Brunberg & Boström 1992). Oxygen of organic matter will enhance greater binding of P in the sediment complexes. The sediment phosphates can thus be inactivated by 'sealing' or stripping, using the salts of Ca, Fe or Al, to precipitate both inorganic P and particulate P, which then sediment as a floc (see Cooke *et al.* 1993). The process removes up to 90–95% of P. However, if not bound firmly in the sediment this P will contribute to a later increase in internal loading. In the Netherlands, Boers *et al.* (1992) applied $FeCl_3$ solution (100 g of Fe^{3+} m^{-2} d^{-1}) for P stripping. The $FeCl_3$ solution was diluted 100-fold with lake water and mixed with surface sediments using a water jet. High external loading of P, short residence time of water and loss of binding capacity of $FeCl_3$ nullified the positive effects. Al immobilizes P more efficiently than Fe (Hansen *et al.* 2003): it has a higher sorptive capacity (for details see also Lewandowski *et al.* 2003). However, Al is potentially toxic and its use for removal of P in public water supplies should be avoided.

Sediment removal by dredging

A drastic but more expensive technique to reduce internal nutrient loading is to remove the P-rich sediment. Dredging simultaneously eliminates the toxic and hazardous compounds and rooted aquatic plants (Peterson 1981), and has the advantage that it does not introduce alien substances into the waterbody.

However, the extracted sediments have to be disposed of somewhere else. The sediment removal of Lake Trummen in Sweden is perhaps among the earliest well-documented work of its kind (Bengtsson *et al.* 1975). Generally, long-term sediment removal results in good reductions of nuisance algae and aquatic weeds (Olem & Flock 1990). However, removal of the often loose, upper sediment layer and its disposal elsewhere make this technique more expensive than P immobilization within the sediments. Case studies from the UK (Moss *et al.* 1986, Moss 2001) and the Netherlands (Van der Does *et al.* 1992) are examples of only transient successes: 'The problem is that even if the sediments are dredged down to the layers that were laid down even when P was low, there is, after a pause of a few months, renewed release' (Moss 2001).

Hydrological management: flushing and dilution with nutrient-poor water

Hydrological management involves replenishing the lake with water from an extraneous source or from another lake with lower nutrient levels but preferably rich in Ca^{2+} and HCO_3^-. Dilution as a restoration tool, therefore, implies necessarily reducing the nutrient levels in in-lake water to limiting concentrations (Cooke *et al.* 1993). The success of these measures depends greatly on the sustained availability of good-quality water for flushing and the timing of the flushing: the winter period is the best since both the dilution water and lake to be diluted have less suspended material. Dilution by flushing has generally been successfully employed in lakes in the US and Europe to improve water quality (Cooke *et al.* 1993).

In the Netherlands, Lake Veluwe and Lake Dronten have been diluted since 1979 with relatively P-poor water, with quite positive results (Hosper 1984, Hosper & Meyer 1986). Hosper (1998) considered 'washout' by winter flushing to be a powerful management tool, provided the flushing water is available in sufficient quantity and is of good quality. He suggested the winter period (November–February) to be most effective for wiping out the blooms of filamentous cyanobacteria (*Oscillatoria* spp.), assuming growth rates of virtually zero in winter. A flushing rate of > 0.75 lake volumes $month^{-1}$ for the 4 months can remove >95% of the algal bloom. However, because good-quality dilution water is scarce and

the logistics of its transport to lakes are generally inadequate and expensive, the technique has not gained popularity in the Netherlands.

12.3.3 Restoration of acidified systems

Reducing emissions

Control actions in Europe, the USA and Canada, starting around the mid-1970s, when these emissions peaked, led to a perceptible decrease in annual emissions to the atmosphere of SO_2 and NO_x by the mid-1980s (Mason 1996), although NO_x emissions have tended to increase, for example in the UK (Fig. 12.7). Thanks to the national and international clean-air acts almost all countries committed themselves by 1983 to reducing sulphur emissions by 30% within a decade, and many European countries agreed to a reduction in emissions of 70–80% by 2010 compared with 1980 (UN 1994). However, acidification will continue to be a problem for many decades in large areas (Henriksen & Hindar 1993, Brodin 1995a, 1995b). In the acidified areas of southern Norway, deposition of sulphur compounds has decreased by about 40% since 1980 but the decrease of nitrogen compounds has not been

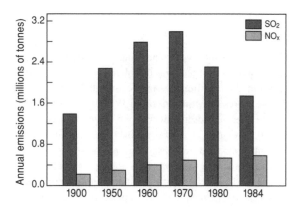

Fig. 12.7 Annual emissions of sulphur and nitrogen in the UK over more than eight decades starting from the beginning of the 20th century. Note the decline in SO_2 emissions starting in the 1970s but a trend of continuing increase in NO_x emissions. After Mason (1996). Reproduced by permission of Pearson Education, Inc.

great. Therefore, acidification due to N compounds is now a more important issue.

Liming: methods, strategies and ecological effects

Liming is by far the most common in-lake technique for restoring acidified lakes, for example in Scandinavian countries, the UK (Scotland and Wales) and eastern Canada (Wright 1985, Schindler 1997). Limestone is the most commonly used compound. In most treatments it includes calcite ($CaCO_3$) and the powdered dolomite that contains a relatively high proportion of magnesium carbonate ($CaMg(CO_3)_2$). Lime dissolves slowly depending on the grain size so that it produces longer-lasting buffering effects. In the field it seldom raises pH above 7.0, even on excess addition. Other forms of lime used for de-acidification of lake water are quicklime (CaO) and slaked lime ($Ca(OH)_2$), in addition to alum ($(Al_2SO_4)_3 \cdot 14H_2O$; Dickson & Brodin 1995). The following equation describes the dissolution of limestone ($CaCO_3$) in acid aqueous solution.

$$CaCO_3 + H^+ \rightarrow Ca^{2+} + HCO_3^- \qquad (pH<6)$$
$$CaCO_3 + H_2O \rightarrow Ca^{2+} + HCO_3^- + OH^- \qquad (pH>6)$$

Liming is a temporary remedial measure in anticipation of more enduring reductions of acidifying compounds. Moreover, continued inputs of acidic water from the drainage basin during the liming process can nullify the effects of liming.

Both Sweden and Norway have chosen large-scale liming as a national strategy for preserving species threatened by acidification (Henrikson & Brodin 1995, Svensson et al. 1995). In Sweden, between 7500 and 11,000 km of streams are limed repeatedly every year (Svensson et al. 1995) to raise the pH above 6.0 for the natural fauna and flora to survive. Between 80 and 90% of the acidified surface waters have thus been restored. Lime treatment of River Tovdalselva (1885 km²) is perhaps the largest integrated liming project in the world (Hindar et al. 1998). Watershed liming, as attempted in Germany and the USA, is expected to have more long-lasting beneficial effects but is a costly option. In south-west Scotland successful liming of Loch Fleet, Galloway, facilitated the introduction of a self-sustaining trout population (Dalziel et al. 1994).

In the Netherlands, Lamers et al. (2002) and Roelofs et al. (2002) have reviewed works on the restoration of fens and macrophytic vegetation in acidified and eutrophic wetlands. The desiccation of the fens by drainage (infiltration), rather than direct acid emissions, has caused a decrease in the acid-neutralizing capacity of the fens. Vegetation in some shallow soft-water bodies is strongly endangered due to atmospheric deposits of SO_4^{2-} and NH_4^+ estimated at 44–50 and 84–103 mmol m^{-2} yr^{-1}, respectively. During dry summers, the pH values range between 4.1 and 5.4 but decrease to c.3.7 due to oxidation of sulphur compounds in the sediments. In a restoration study (Roelofs et al. 2002) small amounts of alkaline, nutrient-poor ground water extracted from deeper aquifers was used to raise alkalinity of small acidified, soft-water lakes after the top sediment layer had been removed by dredging. A rapid decline of NH_4^+ and CO_2 led to an increase in pH from 4 to 7. In contrast to fens and bogs, the larger Dutch lakes are highly buffered due to naturally high inherent concentrations of calcium and bicarbonates, and therefore not affected by atmospheric emissions (Gulati & van Donk 2002).

A critical analysis of the effects of liming of freshwaters is lacking. The effects were examined in plankton, benthic fauna and fish in Sweden (Henrikson & Brodin 1995). Sensitive species of fish were reported to re-colonize and increase in densities after liming but fish deaths were reported if aluminium levels remained high (Leivestad et al. 1987, cited by Mason 1996). Also, whole-catchment liming is reported to cause terrestrial vegetation to die, especially Sphagnum mosses (e.g. Hindar et al. 1998). The growth of Sphagnum and Juncus bulbosus, which are especially stimulated by elevated levels of both CO_2 and NH_4^+ (Roelofs et al. 1994, Lucassen et al. 1999), should become limited on liming and sediment removal. Liming leads to an increase in acid-neutralizing capacity of water and stimulates decomposition of the accumulated organic matter, and thus enables growth of macrophytes (Myriophyllum spicatum and Elodea canadensis) using bicarbonates as a carbon source. For more sustainable results, the liming of a catchment, or watershed, rather than a waterbody itself, might prove more effective (Mason 1996). Nevertheless, the best way to prevent lake acidification is to control the source of acidification (Schindler 1997).

12.3.4 Lake biomanipulation

Lake biomanipulation or foodweb manipulation has become a routine technique for improving water quality of lakes and reservoirs (Kasprzak *et al.* 2002). The technique has an ecological basis (Reynolds 1994) and therefore complements the methods involving nutrient reduction for lake restoration. If applied in conjunction, biomanipulation and nutrient reduction measures can speed up the process of lake rehabilitation. Conceptual works culminating in two important hypotheses have greatly enhanced our insights into foodweb relationships: (i) the **size–efficiency hypothesis** (Brooks & Dodson 1965, Hall *et al.* 1976) and (ii) the **cascading trophic interactions hypothesis** (Carpenter *et al.* 1985). The hypotheses together elucidate ecosystem functioning and structure, and inter-trophic feedback effects. The size-efficiency hypothesis enhances our insight into the fish–zooplankton dynamics, in that fish will tend to prey most heavily on the largest zooplankton taxon, *Daphnia*. According to the trophic cascade hypothesis

the inter-trophic effects in the food chain cascade down the food chain:

Piscivorous fish

↓

planktivorous fish

↓

zooplankton

↓

algae

In short, the lake foodweb is influenced by nutrient inputs (bottom-up control) on one hand and zooplankton grazers and predatory fish (top-down) on the other (Fig. 12.8). Intensifying predation by carnivorous or piscivorous fish will trigger changes that will lead to a decrease in the biomass of planktivorous fish and an increase in that of zooplankton, especially that of larger-bodied *Daphnia* spp. These changes will culminate in a reduction of phytoplankton biomass and lead to improvements in water clarity and the promotion of diverse biological communities (Perrow *et al.* 1997, Perrow & Davy 2002a, 2002b). Among

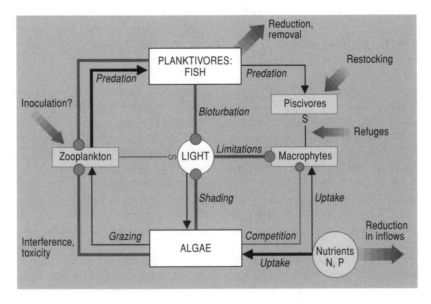

Fig. 12.8 A simplified diagram of top-down and bottom-up control measures in the foodweb of shallow lakes. Nutrient reductions in the inflows and major biomanipulation measures are indicated with arrows. Among the latter, fish-stock manipulation (reduction of planktivore standing stock and restocking with pike or pike and perch) are the main steps. Refuges for young fish and zooplankton against predation by fish are created by fixing stacks of willow twigs to the lake bottom. From Gulati and van Donk (2002).

the resultant changes in shallow lakes the development of macrophytes is most striking; the macrophytes compete for nutrients with phytoplankton and limit the latter's growth, so that the clear-water phase often observed in many lakes during spring or early summer is a culmination of the trophic cascades.

Lake biomanipulation as a restoration technique, started in the 1980s, is now well documented in the literature on lake restoration (Gulati *et al.* 1990, Lammens *et al.* 1990, Kufel *et al.* 1997, Harper *et al.* 1999, Walz & Nixdorf 1999, Kasprzak *et al.* 2002, Gulati & van Donk 2002). In the Netherlands, which is leading lake biomanipulation research and its application, more than 20 lakes and ponds ranging in area from 1.5 to 2650 ha and depth from 0.8 to 2.5 m have been biomanipulated. In virtually all these cases the standing stocks of planktivorous fish were reduced drastically. More than a 75% reduction of the fish stock in winter has been found to be critical for generating pronounced effects on water clarity by early spring. The spring peak densities and grazing maxima of *Daphnia* spp. in lakes that generally precede the clear-water state were invariably not prolonged (Gulati 1990a, 1990b) if reductions of nutrients or planktivorous fish, or both, were inadequate. In case of failure, it was difficult to ascertain whether *Daphnia* populations declined due to poor food quality (high cyanobacterial densities), or due to predation by planktivores, or both. The northern pike (*Esox lucius*), which was introduced to control the planktivorous fish, did not establish well over the years in many lakes. Size-selective predation of larger-bodied zooplankton (*Daphnia*) by the planktivores led to a decrease in *Daphnia* grazing on phytoplankton, which thus increased and retarded the improvement in quality.

The biomanipulation theory and its applications have developed concurrently. The hypothesis of **alternative stable states** – a **turbid state** dominated by phytoplankton, and a **clear state** dominated by macrophytes (Moss 1990, 1998, Scheffer *et al.* 1993) – is interesting. The literature evidence to support the existence of these alternative stable states is, however, not overwhelming (Gulati & van Donk 2002). Extreme disturbance may be needed for a lake to shift from a turbid state to clear-water state; repeated and sustained reductions of the planktivore fish stocks may be required to ensure the establishment of macrophytes.

Thus, improvement in the underwater light climate has been used as the main indicator of success of the top-down, cascading effects (Meijer *et al.* 1994a, 1994b, Hosper 1997, Meijer *et al.* 1999, Van den Berg *et al.* 1999, Meijer 2000, Van Nes 2002).

The literature on lake biomanipulation reveals more long-term failures than successes, mainly because bottom-up (i.e. nutrient input) effects on the structure of pelagic foodweb tend to persist even after strong top-down manipulation (McQueen *et al.* 1986). It needs stressing that reduction of nutrients from the catchment is an important prerequisite for success of biomanipulation measures (Benndorf 1987): the P input rate must fall below a certain threshold for grazers to be able to contain phytoplankton biomass. De Melo *et al.* (1992), who reviewed the results of 18 enclosure and 26 whole-lake experiments, cast doubts on the trophic cascade theory of Carpenter and Kitchell (1992, 1993), mainly due to a weakening of the cascading effect or absence of the top-down response at the zooplankton/phytoplankton level in 80% of the cases analysed. On the other hand, Benndorf *et al.* (2002) have attributed the failures of most biomanipulation work in deeper lakes to extremely high P loading, implying some bias in the analysis of De Melo *et al.* (1992), probably because of more deep lakes than shallow ones in the analysis. Drenner and Hambright (2002), however, found no support for the analysis of Benndorf *et al.* (2002). We expect the nutrient dynamics and the efficacy of restoration measures to markedly differ with lake depth (Moss 1998), due to differences in the sediment–water interactions. In addition, shallow lakes are more likely to be colonized by macrophytes and shift to a clear-water state earlier than the deeper lakes.

Here we highlight a number of conditions that should be met before applying biomanipulation as a technique for lake restoration.

Importance of fish in lake restoration

In shallow lakes, the fish are relatively easy to manipulate (Lammens 1999) to produce virtually instantaneous effects (Jeppesen 1998). Fish removal varies from 25 to 100%, but biomanipulation measures seem to be more effective, at about a 75% reduction of the fish community (Hansson *et al.* 1998, Moss 1998, Meijer 2000). However, this percentage appears to be rather

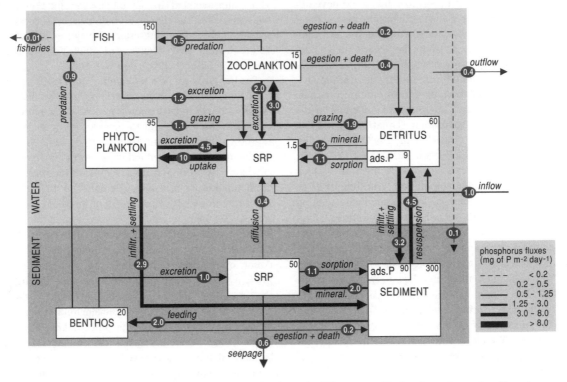

Fig. 12.9 Simulated P-flow scheme of Lake Loosdrecht through different trophic levels, water and sediment (boxes), based on the summer averages of concentrations (mg of P m^{-2}) and flows (mg of P m^{-2} day^{-1}) during 1987. Because fish comprise an important P stockpile, they apparently contribute to P continuously to the P input of the soluble reactive phosphorus (SRP) pool, which is very small but very dynamic and with a very high turnover. ads.P, adsorbed P; infiltr., infiltration; mineral., mineralization. From Janse *et al.* (1992). With kind permission of Springer Science and Business Media.

arbitrary considering the enormous variations of fish stocks in lakes that need to be reduced to < 50 kg ha^{-1} to produce effects. Moreover, continual fish management appears to be indispensable to produce sustained, positive effects on water quality, but this strategy may not be a realistic one considering the cost-benefit aspect. About an 80% reduction in the fish standing stock was needed to achieve a standing crop of c.20 kg of fresh weight ha^{-1} in many Dutch lakes, during a 5-year period of fisheries management (Lammens *et al.* 2002). Such fish reductions in shallower lakes stimulated the *Chara* beds to expand (Hosper 1997) as well as zebra mussel populations to increase. Consequently the light climate in open water improved distinctly. In Frisian Lakes, however, good recruitment and higher growth rates of bream

generally nullified the effects of management measures so that improvements in the light climate were only marginal.

Planktivorous fish generally play a crucial role in P recycling, especially in shallow lakes, and retard the pace of restoration. Because of their high P content per unit body weight and their very high standing crop the planktivores comprise a major P store and contribute to a lake's continuous and steady P regeneration via egestion, mortality and metabolic excretion (Fig. 12.9; Janse *et al.* 1992, see also a review by Lazzaro 1987). Van Liere and Janse (1992) estimated that in Loosdrecht Lakes the P regenerated from fish (excretion, egestion and death) was about 140% of the external loading. Nutrients and algal concentrations are expected to decline upon removal of planktivores

Fig. 12.10 A diagrammatic representation of the mechanisms and factors causing sediment resuspension and turbidity in shallow lakes in relation to macrophytes (submerged plants). After lake restoration the increase in macrophytes plays an important role in reducing sediment resuspension and turbidity and improving underwater light climate. Different feedback mechanisms and their strength are indicated with arrows. (Allelop. subs. = allelopathic substances.) From Gulati and van Donk (2002).

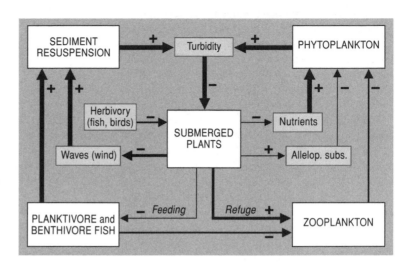

(Horppila *et al.* 1998). The decrease in P input due to a fish-stock reduction of 150 kg of fresh weight ha^{-1} in Lake Wolderwijd, the Netherlands, equalled *c.*60% of the external loading (Meijer *et al.* 1994a, 1994b). Secondly, such a reduction strongly relieves predation on large-bodied zooplankton so that their grazing pressure on phytoplankton and detritus increased, and the light climate improved. By their active foraging for benthic prey in the lake sediments, the planktivores also resuspend sediments, causing an increase in turbidity and a deterioration of the light climate. This sediment resuspension also stimulates aerobic mineralization of P, which may be refixed in Fe complexes if redox potential is high. Model studies on shallow Dutch lakes reveal that fish-induced resuspension causes more than 50% of the turbidity in shallow lakes (Meijer *et al.* 1990).

The reductions in the standing stocks of fish, both planktivores and benthivores, often stimulate recruitment of the YOY (young of the year) fish, thereby considerably cancelling out the positive effects. Both this and the inadequate fish-biomass reductions might explain the failure of biomanipulation measures in many Dutch lakes and elsewhere. That a reduction of planktivore biomass and its maintenance at < 50 kg of fresh weight ha^{-1} will indeed enhance the chances of success needs reaffirming in future studies. However, most failed studies reveal that it is extremely difficult to maintain over long periods a consistently low standing stock of piscivores, such as northern pike,

in biomanipulated lakes. The pike, for reasons not yet fully understood, fail to develop even a moderate population size on introduction into these lakes.

Role of macrophytes

The stabilizing role of macrophytes on lakes after biomanipulation is now well established (van Donk *et al.* 1993, Jeppesen 1998; see papers in Kufel *et al.* 1997; see reviews by Coops 2002 and van Nes 2002). Macrophytes influence various processes in shallow lakes (Fig. 12.10). Because of their huge biomass relative to phytoplankton, macrophytes can accumulate large amounts of N and P from both water and sediments (Barko & James 1998) and thereby reduce the bioavailability of these nutrients for algae (Gulati & van Donk 2002, van Donk & van de Bund 2002). Moreover, macrophytes and the colonizing periphyton act as major nutrient sinks throughout most of the vegetative period. Secondly, they provide refuge for larger-bodied zooplankton and young fish against fish predation (Moss 1990, 1998), and thus promote zooplankton grazing (Timms & Moss 1984). Thirdly, they considerably reduce fish-induced bioturbation, as well as restrict the wind-induced resuspension of the bottom sediments; that is, they increase sedimentation of phytoplankton and detritus. All these features lead to improved underwater light climate in lakes (Barko & James 1998). Consequently, one of the first positive effects in the biomanipulated lakes is the seasonally persistent clear-water patches due to the development of *Chara* meadows. Fourthly, both denitrification in the macrophyte beds (Meijer *et al.* 1994b) and the release of allelopathic substances by macrophytes can adversely affect phytoplankton (Fig. 12.10). The mechanisms for these negative feedbacks are, however, as yet unclear (van Donk & van de Bund 2002).

The nuisance aspects of prolific macrophyte development in restored lakes for recreation, for example in Lake Veluwe (Hosper 1997), need further investigation. Removal of plants by harvesting is a common practice in ditches and canals in Europe. Harvesting in such cases is a sequel to lake restoration rather than a restoration technique in itself: it does, however, imply a net export of nutrients from the lake ecosystems so that its long-term, positive effects on the restored system cannot be disregarded.

Role of zooplankton

Gulati and van Donk (2002) stressed the importance of fundamental research on zooplankton dynamics and zooplankton grazing activities for their application in lake biomanipulation (see Gulati 1990a, 1990b, 1995, Gulati *et al.* 1992). Manipulation of fish communities for lake restoration has the main objective of developing populations of large-bodied *Daphnia* to increase grazing pressure on algae (Moss 1998). In eutrophic Danish lakes, Jeppesen *et al.* (1999) observed that zooplankton grazing on phytoplankton caused clear-water conditions in early summer. This led them to hypothesize that the role of zooplankton grazing for water clarity, especially in macrophyte-rich lakes, may increase with increasing lake trophy. That the marked increase in larger-bodied daphnids and their grazing pressure on phytoplankton following reduction of the planktivorous fish causes clear-water conditions has been reported from several lakes in western Europe (e.g. Timms & Moss 1984, Gulati 1990a, 1990b, Søndergaard *et al.* 1990, Hosper 1997, Meijer 2000). The prospects of such a clear-water phase can be predicted from net losses of lake seston (phytoplankton and detritus) due to grazing, using regressions of animal length versus filtering rate, grazer densities and seston concentration (Gulati 1990a, 1990b, Gulati *et al.* 1992). Since larger-bodied cladocerans (*Daphnia* spp.) are superior competitors for

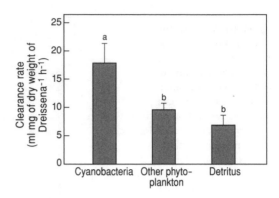

Fig. 12.11 Clearance rates of adult zebra mussel (*Dreissena*) on different food particles from Lake Zwemlust, the Netherlands. The clearance rates on cyanobacteria are significantly higher than those on other phytoplankton and detritus as indicated by the letters a and b above the bars (*P* > 0.05; Tukey's test). From Dionisio Pires *et al.* (2004). Reproduced by permission of Blackwell Publishing.

food compared with rotifers, if fish predation is low it is important that conditions for *Daphnia* growth and development are stimulated to inflict greater mortality on phytoplankton (Gulati 1990b).

Role of the zebra mussel

The zebra mussel, *Dreissena polymorpha*, a bio-invader in many lakes, is a potential tool for lake management. Exploratory works in the Netherlands (Reeders & bij de Vaate 1990, 1992, Noordhuis *et al.* 1992) formed the basis for studies by Dionisio Pires *et al.* (2004). They demonstrated that adult mussels showed higher clearance rates on cyanobacteria than on other phytoplankters and detritus (Fig. 12.11). In Lakes IJssel and Veluwe and other Border Lakes, the areas colonized by zebra mussels are observably clearer than other lake parts (Harry Hosper, personal communication). However, we know little about the sudden disappearance *en masse* of these mussels from lakes. Both lack of suitable substratum and lake eutrophication are plausible causes for the inability of the mussels to establish and build up large populations in these lakes. That the macrophytes form a suitable, natural substratum for the larvae of the zebra mussels (Reeders & bij de Vaate 1990) augurs well for the return of these large grazers after lake biomanipulation.

12.4 Conclusions and perspectives

Although in many European countries freshwater lakes and reservoirs constitute a small fraction of the land area, their importance for human health, recreation and national economies is indisputable. There is obviously an urgent need to further curtail, divert and treat the unwarranted inputs into the lakes and reservoirs of nutrients, organics, silt and contaminants from the watershed. Despite more than two decades of nutrient-reduction measures in runoff waters, many lakes still exhibit an insufficient improvement in water quality. We now understand well that sustainability of the positive effects on water quality is central to the remedial measures. An important principal cause for this hold-up of responses to corrective measures is the large stockpiles in lake sediment of P, which due to its slow release allows the eutrophication symptoms to persist. Secondly, inadequate nutrient reductions in the runoff waters and the virtually unabated and diffuse inputs from the agricultural grounds in the catchment act as a major bottleneck to lake restoration. It is therefore difficult to predict the response of an ecosystem under restoration. Thus, further manipulations of both chemical and biological processes are needed to sustain the positive effects of the corrective measures. However, monitoring of water quality and restoration of in-lake processes alone will be futile exercises if not carried out in a watershed context. This latter involves documenting the entire landscape setting, including habitat type, hydrological regime, soil properties, topography and invasive species, all of which can hamper restoration measures. Moreover, chances of improving a lake are often better if the chosen measures complement each other.

Future lake-restoration plans, for example in the Netherlands, typically envisage near-nature development, emphasizing that a lake is an integral part of a landscape comprising other aquatic, semi-aquatic and terrestrial ecosystems (see papers in Nienhuis & Gulati 2002). Such a measure includes reinforcing the shoreline vegetation of lakes to prevent wind- and wave-induced erosion and improving the propensity of the land–water transition to develop a natural biodiversity. The water authorities have also started to excavate several 20–40 m deep pits within the shallower lake parts to allow wind-induced shifting and burial of the loose, nutrient-rich lake sediments into these pits to retard in-lake nutrient release rates from wind-induced resuspension of the sediments. Moreover, creation of artificial islands to reduce the wind fetch factor and erosion is planned in some lakes. Feasibility plans are also under way to deploy water-level management, encourage the shoreline macro-vegetation and to develop greater natural evolution of the aquatic and semi-aquatic ecosystems. The plans envisage extending the upper and lower limits of the permissible annual water-level fluctuations and exploring the effects, especially of transient water-level draw downs (Coops & Hosper 2002). Near-natural water levels that allow wider fluctuations than the current 'fixed' levels are considered the best option. However, in the light of long-term climate change and its consequences for hydrology and water-management practices, the impact of flooding and recession on the ecosystems as well as water use by humans need to be investigated thoroughly.

Sustainability of the positive effects on water quality is central to remedial measures. The experience gained from the failures and occasional successes of the last two decades should make it possible to develop more-enduring strategies for greater sustainable restoration of our lake ecosystems. Lastly, the long-term aims of water management should be to create and maintain sustainable and healthy aquatic environments that possess optimal properties for their assigned functions.

13

Restoration of intertidal flats and tidal salt marshes

Jan P. Bakker and Theunis Piersma

13.1 Introduction: the historical context

This chapter deals with intertidal flats and the adjacent tidal salt marshes. Tidal flats and salt marshes occur along the edges of shallow seas with soft sediment bottoms where the tidal range is considerable, at least a metre or so (Eisma *et al.* 1998, van de Kam *et al.* 2004). The low-lying **intertidal** areas are largely barren except for the occurrence of *Zostera* (sea- or eelgrass) or *Spartina* meadows and reefs formed by shellfish or tubeworms. Intertidal areas are inundated at least once a day, and make a place for more irregularly inundated areas of salt marsh higher up. In tropical areas, and even some benign temperate areas such as northernmost New Zealand, the upper parts of intertidal areas may be covered by mangrove forests rather than salt marsh. Such mangroves have the tendency also to cover the regularly inundated parts of intertidal soft sediments, thus reducing the extent of mudflats in many tropical areas. No intertidal deposits or salt marshes occur at high latitudes (further north than 70–73°N). Here coastlines are either ice-covered for most of the year or disturbed by moving ice too frequently for soft sediment deposits or vegetation to build up.

Intertidal flats and salt marshes are under complex natural controls. In most parts of the world, and certainly in Europe, intertidal flats and salt marshes experienced human exploitation from the mid-Holocene period onwards. Most of the human exploitation of intertidal flats was relatively unintrusive for a long time, as it consisted of small-scale fishing and

the taking of shellfish by hand. With the advent of motorized power over the past century and the use of large nets and dredges, however, human exploitation patterns of intertidal flats have come to influence the natural processes a great deal. It is not entirely clear whether the same can be said for salt marshes, where, reclamations aside, grazing has been the main human factor. It is quite possible that the grazing by domestic animals has replaced the grazing that took place before human times by large herbivores. Nevertheless, the main external controls for the tidal lands are the sea-level and sediment-supply regimes. Upward sea-level movements and autocompaction – that is, diminishing of the volume of the sediment – combine to provide accomodation space within which marshes build upward. Mineralogenic marshes consist of a vegetated platform dissected typically by extensive networks of blind-ended, branching tidal creeks. The flow-resistant surface vegetation both traps and binds tidally introduced mineral sediment, but also contributes an organic component of indigenous origin to the deposit. When the sea level becomes stable or falls, however, in response to century- or millennium-scale fluctuations, the organic sediment component becomes dominant and mineralogenic marshes are transformed into organogenic ones. Because peat is such a porous and permeable sediment, and there is little or no tidal inundation, organogenic marshes in north-west Europe typically lack surface channels for intertidal drainage (Allen 2000). At present very little peat marsh occurs in Europe (Dijkema 1984). In contrast, the north-east coast of North America features large coastal peat deposits (Niering 1997b). Along the south-east coast of North America the vertical accretion rate of salt marshes is directly related to the accumulation of organic matter, but not to inorganic matter (Turner *et al.* 2000).

Sea-level rise caused a deterioration of the drainage of the hinterland and a subsequent rise of the groundwater table in the adjacent low-lying inland zone parallel to the coast. This zone became marshy, which allowed peat formation on top of the underlying Pleistocene subsurface. As a consequence of increased marine influence, the fresh marsh transformed into an area of tidal salt marshes and intertidal mudflats or brackish lagoons. As a result the basal peat layer was covered by marine sediments, before the area became totally submerged. This transgressive process continued until the mid-Holocene, after which the coastline stabilized more or less at its present position. As a result of the decline in sea-level rise, sedimentary processes became more dominant (Esselink 2000). The stratigraphic sequences that accumulated during the Holocene beneath coastal marshes and high intertidal flats typically present an alternation on a vertical scale of silts (mineralogenic marshes, high intertidal flats) and peats (organogenic highest intertidal/supratidal marshes). Coastal barriers are represented by local accumulation of sand and/or gravel. The silts and peats form vertical alternations which are generally considered to be related to fluctuations of the sea level around the general upward trend (Allen 2000). The coastal landscape was not only affected by natural processes. Gradually it became more and more shaped by human activities, eventually resulting in artificial salt marshes in front of seawalls along large parts of the coast.

The first seawalls were constructed against the increased risk of flooding in the 10th century in the northern Netherlands (Oost & de Boer 1994), and during the 11th century in adjacent Germany (Behre 1995). The entire coastline was protected in the 13th century. The first seawalls were constructed in the salt marshes above the level of mean high tide, and hence a strip of unprotected salt marsh remained generally in front of the seawalls. The remaining salt marsh was agriculturally exploited and was managed for coastal protection. New marshes developed after the construction of the first seawalls, especially in sheltered bays (Oost & de Boer 1994). The new marshes, which originated during the 12th to the 14th centuries, may

have evolved without human intervention. However, human intervention was very likely from at least the 17th century onwards. Several techniques have been applied to promote both vertical accretion and horizontal expansion of salt marshes. At present the majority of mainland salt marshes are artificial, such as those currently found in Germany, the Netherlands (Dijkema 1984) and the UK (Pye 2000). Back-barrier marshes developed in the shelter of dunes, but also as a result of the construction of artificial sand dikes during the 20th century. These marshes always had a natural drainage pattern.

13.2 Patterns and processes

13.2.1 Patterns and processes in intertidal flats

The zonation of intertidal flats is largely a function of the duration of submergence during high tide. Parts of intertidal flats that are only submerged during extreme spring high tides (and are not covered by salt marshes or mangroves) are called supratidal flats. Soft sediment shores in general, and intertidal sand- and mudflats in particular, in contrast to rocky shorelines, are in a state of dynamic flux (e.g. Edelvang 1997). The nature of the sediments is determined by the sediment types available, the nature of the currents, the tides and wind-generated waves, the presence of ecosystem-engineering types of organism (e.g. reef-building oysters, but also infaunal species producing faecal pellets) and human activities such as bottom-fishing and dredging. Good general introductions to the nature and occurrence of intertidal soft sediment habitats can be found in Reise (1985), Raffaelli and Hawkins (1996), Eisma *et al.* (1998) and, especially, Little (2000).

Coarse-grained sediments are mostly found on wave-exposed shores, but may even be found at sheltered places if the currents are forceful enough. Coarser sands can also be found in the surf-zone, forming a beach with a steeper slope than the zone of fine-grained sediments below (e.g. Rogers *et al.* 2003). Fine-grained sediments accumulate in areas with some shelter, with lower currents and less wave action. As testified by the turbid colour of the surface waters above sheltered muddy sediments, even here there is no stasis but a continuous process of sedimentation combined with resuspension. Biofilms of microscopic algae and bacteria (which produce polymeric substances) may trap and bind sediments that render the sedimentary surface more resistant to erosive forces and help to retain particles (Paterson 1997, Austen *et al.* 1999). In both wave-tank (Gleason *et al.* 1979) and flume (Fonseca & Fisher 1986) experiments, sandy bed material was increasingly retained within seagrass (*Zostera* spp.) stands with increasing density of stems. In shallow waters, seagrass meadows not only suppress bed erosion, but also increase the accretion rate relative to other similar unvegetated areas (Ward *et al.* 1984).

Of course, the causal arrow can also point in the other direction. For example, in a study on the effects of storms on the distribution of mussel banks in the Wadden Sea of Schleswig-Holstein, Germany, Nehls and Thiel (1993) concluded that the longest-living mussel banks occurred in areas where there is some degree of shelter, where the banks get some degree of protection from westerly storms. In more general terms, wind and tidal stress factors seem to influence benthic community structures quite strongly (e.g. Warwick & Uncles 1980, Thistle 1981, Emerson 1989). Thus, one of the most interesting phenomena affecting the appearance and biodiversity of intertidal flats is the mutual interaction between abiotic factors and the biota present (Verwey 1952, Bruno & Bertness 2001). The establishment on bare intertidal flats of species that influence the complexity of the habitat (e.g. seagrasses, oysters, mussels or tubeworms) generally generates even greater habitat complexity, more variations in sediment structure and greater biodiversity (Table 13.1). When the complex intertidal structures that provide shelter, nutrition and other favours to various species disappear, for example due to the scouring of the flats by ice or dredging equipment, local biodiversity and the generative processes of this biodiversity are greatly reduced.

13.2.2 Patterns and processes in salt marshes

Elevation and sedimentation

The driving force in salt-marsh development is the tidal amplitude, causing inundation and subsequent sedimentation of silt. The mean spring-tidal range in Europe varies from 12.3 m in estuaries to 1.6 m in the

Table 13.1 List of environmental factors responsible for the high species diversity in physically complex habitats such as oyster and mussel banks, *Zostera* beds and *Sabellaria* reefs, relative to the low species diversity of the bare sand flats that are left after their destruction. Based on Boström and Bonsdorff (1997); after Piersma and Koolhaas (1997).

Factor	Bare sand flat	Complex habitat type
Habitat complexity	Low	High
Shelter	Low	High
Flow velocity	High	Low
Deposition	Reduced	Enhanced
Sediment	Coarse	Fine
Sediment stability	Low	High
Organic content	Low	High
Food availability	Low	High

Wadden Sea (Allen 2000), to nearly zero in the Baltic Sea. The pioneer zone of salt marshes consists of annual plant species, and they do not trap sediment. The perennial grass *Puccinellia maritima* at the low salt marsh catches sediment, whereas erosion can take place of unvegetated soil (Langlois *et al.* 2001). In the *Festuca rubra* zone, higher up the salt marsh with less inundation, the rate of sedimentation is lower than in the *P. maritima* zone (Andresen *et al.* 1990). Sedimentation patterns show spatial variation. Over comparatively wide marshes a landward decrease of sedimentation was found in a natural mainland marsh in Sussex, UK (Reed 1988), and along the Westerschelde, the Netherlands (Temmerman 2003), a back-barrier marsh at Skallingen, Denmark (Bartholdy 1997), and an artificial marsh in the Dollard, the Netherlands (Esselink *et al.* 1998). Superimposed on the large-scale differences, the rate of sedimentation also declines away from creeks and ditches (Fig. 13.1). Moreover, higher rates are found on ungrazed than on heavily grazed artificial marshes in the Dollard (Esselink *et al.* 1998). Dense, tall vegetation positively affects the rate of sedimentation (Leonard *et al.* 1995). An experiment with different cattle-grazing regimes over 18 years revealed a higher position with respect to mean high tide on ungrazed sections than on heavily grazed sections of the artificial Leybucht Marsh, Germany (van Wijnen 1999).

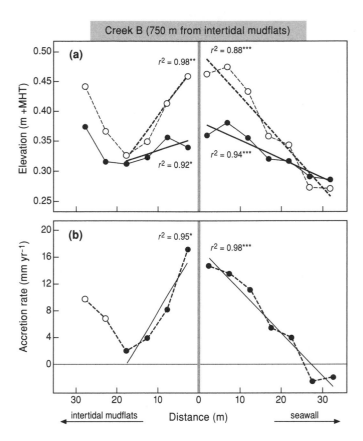

Fig. 13.1 (a) Levee development near a minor creek at a distance of 750 m from intertidal mudflats and (b) annual vertical accretion rate (means ± S.E.M.) at different distances from a main creek as a function of the distance from the intertidal mudflats. In (a): •, 1984; ○, 1991. Points in the left-hand panel at higher elevations were located on the neighbouring levee of the next minor creek. MHT, mean high tide. After Esselink *et al.* (1998). Reproduced by permission of *Journal of Coastal Research*.

Detailed measurements have revealed vertical accretion during winter periods, but shrinkage during the dry summer period (Erchinger *et al.* 1994). The net result of sedimentation and autocompaction should be referred to as net surface elevation change. This change was negatively related to the thickness of the sediment layer in back-barrier salt marshes in the Wadden Sea. Hence, the surface elevation change was negatively related to the age of the salt marshes, ranging from 5 to 100 years (van Wijnen & Bakker 2001). On a century or millennium timescale, the rate of autocompaction also increases with the age of the marsh (Allen 2000).

Zonation and succession

The vertical accretion results in succession from pioneer communities, via low-salt-marsh communities to high-salt-marsh communities. This successional sequence, derived from aerial photographs, is mirrored in the spatial zonation from pioneer to low and high marsh in mainland marshes in the south-west Netherlands (de Leeuw *et al.* 1993). In contrast, long-term permanent plot studies revealed different successional patterns at the low and high marsh of the back-barrier island of Terschelling, the Netherlands, where low-marsh plant communities did not transform into high-marsh communities (Roozen & Westhoff 1985, Leendertse *et al.* 1997). Hence, the zonation does not mirror succession in back-barrier salt marshes, but reflects the underlying geomorphology of the sandy base elevation (de Leeuw *et al.* 1993, Olff *et al.* 1997; see Box 13.1). At many older ungrazed marshes in north-west Europe the tall grass *Elymus athericus* is taking over gradually. With increasing age of the salt marsh *E. athericus* is spreading downwards along the elevational gradient. It is hypothesized that the grass can cope with the salinity stress at lower elevations when more nutrients are available (Olff *et al.* 1997). Salt marshes are nitrogen-limited systems

Box 13.1 Salt-marsh succession and plant–animal interactions

The back-barrier island of Schiermonnikoog, the Nether-lands, extends eastward, thus featuring a chronosequence from east to west. With increasing age, the layer of deposited silt gets thicker. A positive correlation is found between the thickness of the silt layer and both the nitro-gen pool (Olff *et al.* 1997) and the availability of nitro-gen for plants. Hence, the chronosequence represents a productivity gradient featuring low-statured plants in early stages and tall grass in later successional stages, and a decrease in the number of plant species (Bakker *et al.* 2002b). As a result, the forage quantity for natural herbivores such as spring-staging geese and resident hares increases, but the quality, expressed as leaf/stem ratio, decreases (van de Koppel *et al.* 1996). The distribution of the palatable species *Triglochin maritima* is sandwiched between intensive grazing by geese and hares in the younger marsh and increas-ing competition for light in the older marsh (van der Wal *et al.* 2000a, 2000b). Hence, geese and hares are evicted by vegetation succession. A resetting of the suc-cessional clock is only possible when large herbivores such as livestock are introduced.

The North American mid-continent population of lesser snow geese (*Chen caerulescens caerulescens*) increased strongly in recent decades. The sustained increase in population size is thought to be the result of increased dependence of the birds on agricultural food sources and food provided in refugia on wintering grounds and along flyways towards their breeding sites along the coast of Hudson Bay, Canada. The increase in snow geese numbers has resulted in a widespread consumption of the coastal marsh vegeta-tion in breeding colonies. High consumption rates have resulted in loss of vegetation cover, subsequent expos-ure of surface sediments and development of hypersaline soil. Several thousands of hectares of natural coastal marshes have already transformed into 'arctic desert' (Jefferies & Rockwell 2002).

In an artificial salt marsh in the Dollard, the Neth-erlands, drainage works were abandoned in the early 1980s. This resulted in filling the ditches with sediment and subsequent waterlogging of the cattle-grazed salt marsh. Wintering greylag geese (*Anser anser*) main-tained bare soil in inundated depressions by grubbing below-ground parts of *Spartina anglica*. During recent decades the wintering population of barnacle geese (*Branta leucopsis*) increased strongly, but seemed to reach carrying capacity during the past few years. Barnacle geese grazed the *P. maritima* sward that degraded into secondary pioneer vegetation. Experimental exclosure of cattle revealed that the geese can only affect the vegetation when facilitated by the large herbivores, in contrast to the above-mentioned lesser snow geese (Esselink *et al.* 2002).

(Mendelssohn 1979, van Wijnen & Bakker 1999). Nitrogen supply can dictate the competitive relations of marsh plants and hence has important consequences for the abundance and distribution across marsh land-scapes (Levine *et al.* 1998). The lower tidal boundaries of marsh-plant distributions are generally set by phys-ical stress, whereas the upper boundaries of plants are set by competitive exclusion. Hence nitrogen supply may affect the elevation of zonal borders in marshes.

The influence of fresh water discharged by rivers from the hinterland creates a gradient of decreasing salinity away from the sea. The vegetation features *Scirpus maritimus* and especially *Phragmites aus-tralis*, as in estuarine marshes (Esselink 2000), the Baltic Sea in Europe (Jeschke 1987, Dijkema 1990, Puurman & Ratas 1995) and the Atlantic coast in the USA (Bertness 1999).

13.3 Changes in intertidal flats

13.3.1 Human exploitation and roads towards extinction of benthic organisms

Summarizing the extensive and detailed faunistical information from the Sylt area in the northern part of the Schleswig-Holstein Wadden Sea dating back to 1869, Reise (1982) came to the conclusion that whereas bivalves and some other groups of inver-tebrate animals show long-term decreases in species diversity, the smaller polychaete species with short life-spans are doing well (Table 13.2). Small polychaetes can take rapid advantage of environmental disturbances leading to faunal depletions. Reise (1982) attributed the disappearance of 28 common macro-invertebrate

Table 13.2 Long-term changes for the period 1869–1981 of 100 macrobenthic species common at one time or another in the Wadden Sea area close to Sylt, Schleswig-Holstein, Germany. This table gives the numbers of species showing a long-term decrease, no change or an increase. Modified from Reise (1982).

Taxonomic grouping	No. of species		
	Decrease	No change	Increase
Bivalves (Bivalvia)	8	3	1
Other molluscs (Gastropoda and Placophora)	1	7	3
Crustaceans (Crustacea)	6	12	8
Other invertebrates (Porifera, Coelenterata and Echinodermata)	8	7	0
Polychaete worms (Polychaeta)	5	13	18

species to the loss of the many microhabitats provided by the complex physical structures such as oyster beds, tubeworm (*Sabellaria*) reefs, and seagrass (*Zostera marina*) beds. He noticed that of the 30 species showing a long-term increase, half are particularly abundant in mussel beds, a habitat that had only disappeared in the period 1970–80. When these complex intertidal structures (see Table 13.1) disappear, so does the fauna that associates with them. Reise's (1982) assessment can be regarded as a summary of the changes to intertidal flat communities brought about by a century of human use.

Human fishery activities are no doubt very old, but the scale and the intensity of the fisheries has shown dramatic increases over the past century, and especially over the last 30 years. For example, so-called mussel farming has been around in the western Dutch Wadden Sea since the specialists from the province of Zeeland moved in during the early 1960s. This industry not only involves the filling up and dredging out of the artificial subtidal mussel beds, it also entails the bringing together of mussel spat from much larger areas, including mussels from the nearby intertidal zone and from stocks outside the Netherlands (Kamermans & Smaal 2002). Such replacement may take place several times in the beds before mussels are finally transported to market, and each replacement involves bottom dredging.

In response to the development of markets for bait used in sport fishing (angling), techniques to mechanically dredge for lugworms *Arenicola marina* were developed in the Netherlands in the early 1980s.

Given the considerable depth at which lugworms live (30–40 cm), dredging for lugworms is very invasive (leaving gullies 40 cm deep) with considerable consequences for the intertidal biota (Beukema 1995). Over a 4-year dredging period, lugworm densities over 1 km² declined by half. Simultaneously, total biomass of benthic organisms declined even more with an almost complete local disappearance of the large sandgaper (*Mya arenaria*) that initially comprised half of the biomass. The small, short-lived species were quite resistant to the dredging, although the worm *Heteromastus filiformis* showed a clear reduction over the 4 years. Recovery took several years.

Common cockles (*Cerastoderma edule*) have not been a popular food in the Netherlands, but the demands by foreign markets nevertheless made this fishery profitable (albeit on a limited scale) from the early 1900s onwards. With the discovery of new markets, notably in countries in the Mediterranean region, and the development of mechanical harvesting techniques this fishery has, over the past decades, seen a large expansion. From the late 1970s onwards, dredging for cockles became a veritable industry. Ecological studies have shown long-term, near decadal, effects on rates of recruitment of cockles and Balthic tellins (*Macoma balthica*; Fig. 13.2; Piersma et al. 2001). Also in the shorter term there appear to be strong negative ecological effects of mechanical cockle-dredging practices. In 1998–2001 the cockle dredgers concentrated their efforts on the most biodiverse intertidal areas in the western Dutch Wadden Sea, with the declines a year after fishing being

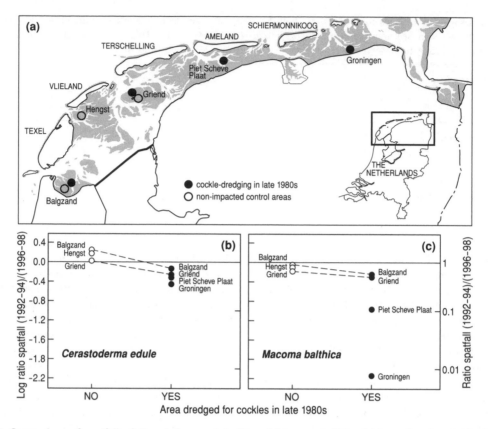

Fig. 13.2 Comparison of spatfall of *Cerastoderma edule* (b) and *Macoma balthica* (c) just after the cockle-dredging events in the 1988–90 period: average spatfall densities in 1992–4 and somewhat later in time (1996–8) on five different intertidal flat areas in the Dutch Wadden Sea (a) that were either exposed or not to cockle-dredging. None of the areas was dredged for cockles after 1990. Two of the areas (Griend and Balgzand) contained a fished and a non-fished part and these data points are connected by dashed lines. For *Cerastoderma* the difference in log ratios of fished and unfished areas was highly significant (separate-variance model, $t = 5.720$, df $= 4.4$, $P = 0.003$). The long-term absence of high cockle densities in areas dredged for cockles appears to be a consequence of reduced recruitment and a genuine effect of the mechanical cockle fishery. For the non-target species *Macoma* a difference in settlement between fished and unfished areas could not be confirmed statistically. The direction of the difference was the same as for *Cerastoderma*, but the power of our test was not large enough to show that the ratios for fished areas were significantly lower than ratios for unfished areas ($t = 2.053$, df $= 3.7$, $P = 0.116$). From Piersma *et al.* (2001).

proportional to the degree to which the benthic species were affected (C. Kraan, A. Dekinga and T. Piersma, unpublished observations). Cockle-dredging thus transforms rich benthic intertidal communities that include the longer-lived bivalve and worm species into much poorer communities that consist of small, short-lived polychaetes and crustaceans.

In the 20th century intense oyster fishery in the Wadden Sea led to the extinction of flat oysters (*Ostrea edulis*; Reise 1982). This process has repeated itself on the other side of the Atlantic, in Chesapeake Bay, the largest estuary of the USA and a region that shows many similarities to the Wadden Sea. In the course of the past 100 years an important shellfish resource, the eastern oyster (*Crassostrea virginica*) has become overharvested (Rothschild *et al.* 1994, Lovejoy 1997). As the oyster fisheries developed, the physical integrity of the oyster banks was damaged by oyster-fishing gear. Subsequently, the oysters lost much of their preferred habitat as a result of sedimentation. After a long-term decline since the early 19th century, eventually, in the late 1980s, the harvests crashed completely. Although short harvesting seasons and a few sanctuaries have now been implemented, with 1% of the previous stocks remaining, little has been left to protect in Chesapeake Bay.

Several species other than flat oysters have disappeared from the Wadden Sea, and in many cases human exploitative activities appear to be involved. For example, the disappearance of the extensive reefs formed along tidal channels by the tube-living polychaete *Sabellaria* are likely to have been due to trawling for shrimp and perhaps other fisheries (Reise 1982). The decline and eventual extinction of the whelk *Buccinum undatum* (a large gastropod) in the Wadden

Sea was kicked off by intense fisheries in the first half of the 20th century, even before any pollution problems occurred (Cadée *et al.* 1995). A small snail, *Rissoa membranacea*, went extinct when its habitat, subtidal seagrass meadows, disappeared from the Wadden Sea (Cadée & Reydon 1998).

In the 1930s a virus pandemic exterminated most of the submerged seagrass in western Europe, including the Wadden Sea, taking with it a rich and biodiverse marine community (de Jonge *et al.* 1997). The extensive seagrass beds of the western Dutch Wadden Sea have not returned, and the greater turbidity and silt load of the waters have been implicated. Although fisheries were unlikely to be responsible for the disappearance of the seagrass beds, they might well help to prevent their re-establishment (de Jonge *et al.* 1997).

In summary, evidence for serious negative effects of trawling, digging and dredging on the sediment characteristics and community structure of intertidal flats and other sea bottoms is now overwhelming (see Table 13.3 and e.g. Hall *et al.* 1991, Dayton *et al.* 1995,

Table 13.3 Some examples of studies documenting consequences for sediment characteristics of perturbations made by humans in search for harvestable marine biological resources.

Biological resource	Area	Type of fishery	Effects on sediments and biota	Authority
Bottom fishes	Offshore north-east USA	Trawling	Sediment resuspension and loss of muds from fished areas	Churchill (1989)
Hard clam (*Mercenaria mercenaria*)	North Carolina, USA	Clam 'kicking'	Sediments are reworked and made more 'dynamic'; no recovery of seagrass beds within 4 years	Peterson *et al.* (1987)
Scallop (*Placopecten magellanicus*)	Gulf of Maine, USA	Dredging	Substratum change from organic silty sand to a sandy gravel, possibly due to disruption of amphipodal tube mats	Langton & Robinson (1990)
Blue mussel (*Mytilus edulis*)	Limfjord, Denmark	Dredging	Sediment resuspension, and increasing likelihood of wind damage to substrates	Riemann & Hoffmann (1991)
Bloodworm (*Glycera dibranchiata*)	Bay of Fundy, eastern Canada	Intertidal manual digging	Contributing to a decrease in sediment stability in the intertidal zone	Shepherd *et al.* (1995)
Date mussel (*Lithophaga lithophaga*)	Mediterranean Italy	Subtidal sledgehammering	Rocky substratum destroyed and recolonization of bare rock prevented by sea urchin grazing	Fanelli *et al.* (1994)

Hall & Harding 1997, Roberts 1997, Watling & Norse 1998, Collie *et al.* 2000, Kaiser *et al.* 2000), although perhaps not so pervasive as when rocky substrates are simply knocked off in the search for an economically valuable shellfish resource (Fanelli *et al.* 1994). We therefore concur with Hall (1994, p. 194), who in his review of physical disturbance and marine benthic communities states that 'there is increasing recognition of the role man plays in physically disturbing marine sediment environments, the most obvious and widespread being commercial fishing'.

13.3.2 Invasions

Invasions of exotic species are now considered a major problem in coastal marine communities, especially the intertidal areas (Carlton 1999, Grosholz 2002). An assessment for the North Sea coasts showed the presence of a minimum of 80 exotic invaders introduced by transoceanic shipping and aquaculture, especially in the 1970s (Reise *et al.* 1999). Most introduced invertebrates were brought in by ships and came from the western side of the Atlantic, whereas the Pacific supplied exotic oysters that brought with them invading algae. Some invasions, such as that of *M. arenaria*, thought to have been brought in from the north-west North American coast by Vikings, are ancient and may actually represent recolonizations after a Pleistocene extinction (Strasser 1999). Others, such as the successful invasion of the Wadden Sea by the American razor clam (*Ensis americanus*) in the late 1970s, are much more recent (Essink 1986, Armonies & Reise 1999).

Although there is as yet little evidence of direct competition between these intertidal benthic invaders and the resident species, the recent establishment of the Pacific oyster (*Crassostrea gigas*) could start an era of serious ecological problems. Between 1964 and 1977 small Pacific oysters from British Columbia, Canada, were deliberately released in north-western Europe (Drinkwaard 1999). By 2000, Pacific oysters had become firmly established and reproductively successful at many sites around the North Sea. In the Dutch Delta and the international Wadden Sea they are now forming extensive oyster beds covering many hectares, sometimes overgrowing or displacing native mussel beds. Unlike the native bivalve species, Pacific

oysters in western Europe suffer little predation from birds and crabs. Pacific oysters may soon compete for space and food with the native biota.

13.3.3 Water-borne influences: pollution and eutrophication

Chemical water pollution comes in many different forms, some of which are known to affect the inhabitants of intertidal flats. A well-known case of the near-disappearance of a sensitive fish-eating bird in the Dutch Wadden Sea is the poisoning of Sandwich terns (*Sterna sandvicensis*), with a drop from 20,000 to fewer than 2000 pairs in the early 1960s. Once this was traced back to releases in the Rhine of the insecticides telodrin and dieldrin (Koeman *et al.* 1967), the releases were stopped and there has been a partial recovery of the Sandwich terns in the ensuing decades (van de Kam *et al.* 2004).

A toxic chemical that continues to damage coastal environments worldwide is the antifouling chemical tributyltin (Page *et al.* 1996). Even at very low concentrations, tributyltin disrupts the gonadal development of marine invertebrates such as the littorinid snail *Littorina littorea* (Bauer *et al.* 1995) and the whelk *B. undatum* (Mensink *et al.* 1996). The International Maritime Organisation passed a resolution prohibiting the use of tributyltin in 2003, but this resolution still awaits full implementation until 2008. Meanwhile, new chemicals continue to be developed at breathtaking speed. Some old ones, and an unknown fraction of the new ones, continue to be a concern in many estuaries around the world. The incredible variety of chemical products means that only a small selection of chemical substances – those that are established to be poisonous – can be monitored in water samples and in bird eggs (Bakker & de Jonge 1998, Becker *et al.* 1998).

Eutrophication, the increase in the amount of organic material entering an ecosystem, is a very different type of pollution (Nixon 1995). Depending on your viewpoint, eutrophication may be considered a good thing or a bad thing for intertidal flat ecosystems. Eutrophication can originate from a single point of discharge, or can come with the tidal streams as increased input of the nutrients nitrate and phosphate. If the production of pelagic and benthic algae is limited by one of these nutrients rather than by light (Colijn & Cadée 2003), an increase in organic input would lead to an increase in primary production (Heip 1995). One might expect that an increase in primary production by bacteria and algae would automatically lead to an increase in the intertidal stocks of benthic invertebrates that feed on these primary producers. However, despite there being some temporal correlations between benthic biomass and nutrient levels (Tubbs 1977, van Impe 1985, Beukema & Cadée 1986, Beukema *et al.* 2002), the evidence for a firm causal relationship between the two is lacking (Essink *et al.* 1998). At the negative side, near the point of discharge of enriched organic material the upper layer of intertidal flats may become anoxic, and in such stinking black-coloured areas benthic species usually do not do well. However, further away from the point of discharge, organic enrichment may be the cause of particularly high stocks of some species. In the Dollard estuary on the border between the Netherlands and Germany, ragworms (*Nereis diversicolor*) were particularly abundant near the point where the effluents were discharged. In years when the discharge was high, ragworm densities were high,

as was the predation pressure by waterbirds feeding on these worms (Essink & Esselink 1998). Typically, however, only some organisms benefit from organic discharges. In the Dollard example decreasing eutrophication was associated with an increase in the densities of the mudshrimp, the amphipod *Corophium volutator*, and the avian predators associated with it.

13.4 Changes in salt marshes

13.4.1 Exploitation

The first colonists in north-west Europe settled on the highest parts of the salt marsh, on levees along watercourses in the 7th century BC. Farmsteads were initially built on the marsh bed. In response to increased risk of flooding during the next ingression, people started to build their dwellings on artificial mounds. The number of mounds along the north coast of the Netherlands suggests intensive exploitation of the salt marshes (Fig. 13.3). When salt marshes extended seaward, new settlements were built on the younger marshes until the entire coastline was

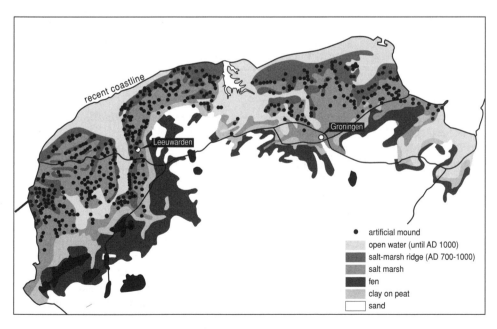

Fig. 13.3 Landscape types and the occurrence of artificial mounds along the north coast of the Netherlands about AD 1000. Rijksdienst voor Oudheidkundig Bodemonderzoek, personal communication.

protected by seawalls in the 13th century. Initially arable crops were grown on the levees, but ditching and the construction of embankments, dated from the 1st century BC to the 2nd century AD, allowed crops to be grown on the salt marshes. The majority of the marshes were exploited for livestock grazing from the early settlements onwards, and haymaking from the 1st–3rd centuries AD onwards. The high frequency of subfossils of *Juncus gerardi* and the low frequency of *Elymus* spp. in artificial mounds led to the conclusion that pristine salt marshes were scarce during most of the occupation period (Esselink 2000). Nowadays salt marshes that have never been grazed by livestock are only found at the eastern point of the Wadden Sea islands. The oldest dates back to 1930 and is found on the island of Terschelling, the Netherlands.

Mature marshes were embanked (see section 13.1). The incentives for embankments have gradually changed during the 20th century from land claims for agriculture to coastal protection. During the early 20th century large-scale accretion works have been started to create artificial salt marshes that were to be reclaimed for agriculture, but some decades later this was no longer economically feasible due to changes in both socio-economic conditions and agricultural policies. From about the 1970s there has been a growing recognition that the remaining salt marshes, though largely developed as a result of human intervention, have an important nature-conservation interest (Esselink 2000). During the 1980s a no-net-loss policy was developed in the Netherlands for the existing mainland marshes. The accretion works protect the existing marshes from erosion. Hence this policy has become the main reason to maintain accretion works in the Netherlands during the 1980s and 1990s. At the same time, increasing areas of artificial salt marshes were designated as nature reserves in Denmark, Germany and the Netherlands, and were included in national parks. Livestock grazing decreased, especially in Germany. Nowadays about 40% of the salt marshes of the Wadden Sea support livestock grazing. The cessation of the creation of artificial marshes, including zonation, resulted in a loss of pioneer communities. The existing marshes undergo a process of maturation, and pioneer and young marshes become lost as a result. This process is still enhanced when livestock grazing on the marshes ceases.

North American salt marshes have a different history. Along the east coast, tidal marshes formed within the last 3–4000 years as sea-level rise slowed to about 1 mm yr^{-1}, favouring the establishment of the initial colonizer *Spartina alterniflora*, a 1–2 m tall grass (Niering 1997b). The accumulation of organic matter controls the accumulation of inorganic matter, not the reverse. Below-ground plant material is very important in maintaining salt marshes once they are established. During the Colonial period (17–19th centuries) salt marshes were mown, grazed, ditched and embanked in order to make them more suitable for agricultural exploitation. In conjunction with these early activities, some ditching and diking were done to regulate tidal flooding. However, these impacts were minor compared to those that followed the Industrial Revolution (1850s) when, with increased mechanization, marshes were dredged for marinas, filled for development, ditched for mosquito control, filled with dredge spills and tidal-gated in order to prevent upland flooding. The subsequent invasion of *Ph. australis* in disturbed coastal wetlands resulted in outcompeting of the native plant species and desiccation as a result of strong transpiration out of its habitat (Bertness 1999). Wetland protection laws since the 1970s, and no-net-loss policies have led to restoration efforts. Moreover, with the past decade Open Marsh Water Management has been widely practised using biological control which favours small fish to control mosquitos and simultaneously promotes restoration (Niering 1997b).

13.4.2 Erosion, reclamation and embankment

Salt marshes have been eroding rapidly in south-west England during the past 150 years, and particularly in the past few decades. The mechanisms of erosion include landward recession of the marsh edge, wave erosion of the marsh surface, internal dissection due to enlargement and coalescence of tidal creeks and mud basins, and direct removal due to human activities. Increased wind and wave energy is supposed to contribute most strongly to erosion. Increased mean sea level and tidal range are underlying factors leading to coastal 'squeezing' of salt marshes between the sea and seawalls (Pye 2000, Cooper *et al.* 2001). Establishing seawalls on the intertidal flats makes them more vulnerable to erosion. In artificial salt marshes the abandonment of accretion works results in retreating of the marsh edge (Esselink 2000).

Salt-marsh erosion is predicted as an effect of increased sea-level rise. An experiment was carried out at the Wadden Sea island of Ameland, the Netherlands. As a result of gas extraction, soil subsidence of 10 cm had taken place within 15 years, affecting both the low and high salt marsh. The net elevation of low-marsh plots did not change, indicating that subsidence kept pace with sedimentation. In contrast, the net elevation of the high-marsh plots decreased by 10 cm, indicating that no extra sedimentation took place. The vegetation changed in neither the lower nor the upper plots of the grazed salt marsh (Dijkema 1997, Eysink et al. 2000).

Most losses in salt-marsh area are caused by embankments. The area of mainland salt marshes has greatly declined during recent centuries because the accretion of new marshes has not kept pace with the rate of embankment in the northern Netherlands (Dijkema 1987).

13.4.3 Reduction of tidal amplitude and salinity

Another important cause of losses of salt marshes is coastal protection by shortening the coastline of estuarine coasts. This often coincides with the creation of storage basins. Discharge of runoff water from the hinterland can be hindered by prolonged high water during storm periods.

Desalinization causes the transformation of salt-marsh communities into communities adapted to freshwater conditions. The changes are faster in former salt marshes than in former intertidal flats (van Rooij & Groen 1996). Continued grazing by livestock retards the losses of halophytic plant species (Westhoff & Sykora 1979, van Rooij & Drost 1996). Moreover, grazing benefits short turfs that are favoured by winterstaging geese, as in the former salt marshes. Undisturbed succession results in scrub and forest with characteristic bird species (van Rooij & Drost 1996, van Wieren 1998). Erosion of the coast by wave action may result from fixed water levels.

Not all estuaries are dammed. In 1986 a sluice-gate barrier was completed in the mouth of the Ooster-

schelde estuary, the Netherlands. It can be closed during storm surges. Although the barrier allows tidal exchange, the tidal flow has been restricted. This caused a 26-cm decrease of the mean high tide and hence a decreased inundation frequency of the marsh. The effect on the vegetation was recorded in permanent plots in the period 1982–90. The initial high rate of vegetation change slowed down in 1989 and 1990, suggesting that a new equilibrium had established. Most plant species had moved down along the elevational gradient (de Leeuw et al. 1994).

More small-scale processes took place in New England, USA. During the past century, about 2000 ha (30%) of Connecticut's tidal marshes have been degraded or lost through coastal development. Tidal flow to many marshes was restricted by the construction of impoundments, producing microtidal environments in which Phragmites australis or less frequently Typha angustifolia became established at the expense of typical tidal marsh communities. In addition to these human influences, P. australis has also invaded brackish tidal marshes in the lower Connecticut River system where salinity levels are often reduced by freshwater inputs (Fell et al. 2000).

13.4.4 Cessation of livestock grazing

As a result of decreased agricultural exploitation and a non-interference policy in several national parks, large areas of salt marshes are no longer grazed by livestock. Long-term (> 25 years) exclosures in back-barrier marshes in the Wadden Sea revealed that the variation in plant communities along the elevational gradient decreased, and the community of Atriplex portulacoides at the lower marsh and the communities of Artemisia maritima and E. athericus at the higher marsh took over. Especially at the mid- and higher marsh, the plant species diversity declined (Bos et al. 2002). Similar changes were recorded in long-term ungrazed artificial marshes. Exclosures at further distances from the intertidal flats experienced lower rates of sedimentation and less spreading of E. athericus (Bakker et al. 2002b). At sites with fast colonization of E. athericus in an artificial salt marsh, the typical zonation of entomofauna communities along an elevational gradient disappeared (Fig. 13.4) and characteristic halobiontic species were replaced by common

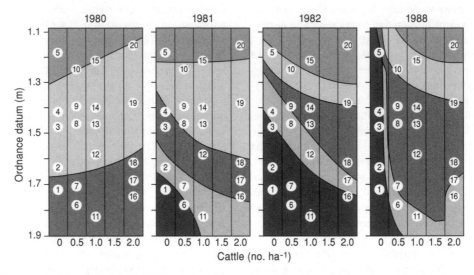

Fig. 13.4 Differentiation of the 20 pitfalls for invertebrate fauna along an elevational gradient (Ordance datum) by cluster analysis for the different stocking rates installed in 1980 and continued during subsequent years on the salt marsh of Leybucht, Germany. After Andresen *et al.* (1990).

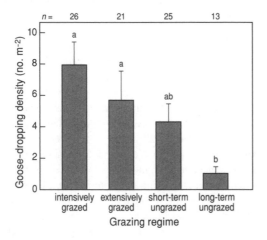

Fig. 13.5 Average goose grazing pressure at transects in the entire Wadden Sea from the seawall to the intertidal flats in relation to livestock grazing regime for all transects that were paired within the same site. Different letters indicate significant differences ($P < 0.05$). After Bos *et al.* (2005a).

inland species of tall forb communities (Andresen *et al.* 1990). The grazing intensity of winterstaging geese was less in long-term ungrazed than in grazed salt marshes in the Wadden Sea (Fig. 13.5; Bos *et al.*

2005a). Although geese numbers declined, especially in autumn, in the 10-year-ungrazed part of the Hamburger Hallig, Germany, the numbers of some breeding birds increased (Stock & Hofeditz 2000, 2002).

13.5 Restoration of intertidal flat communities

To restore damaged intertidal flat communities to a more natural and healthy state, the reduction of pollution and eutrophication are relatively straightforward, although not necessarily easy, issues to deal with. Once their influx into intertidal flat ecosystems has stopped, we predict a gradual return to the original state, all else being equal. But other things do not usually remain equal. For example, many intertidal-flat ecosystems in the world are now invaded by exotic species, of which some, such as the Pacific oyster, have the tendency to dominate the communities in terms of space and resources (Carlton 1999). Pacific oysters are rapidly becoming a pest in Dutch intertidal areas where they continue to increase their coverage of considerable expanses of intertidal flat. It has been suggested that in order to restore the original situation and to free up space and resources for native blue mussels (*Mytilus edulis*), such oysters should be removed by mechanical dredging. However, considering the many negative side effects of mechanical dredging, the manual harvest of these oysters would be more advisable. Perhaps, during particular times and tides, the general public should be encouraged to take these oysters by hand and use them for personal consumption. The concern that this would rapidly lead to development of businesses, generating a process towards commercial dependence on the harvest of Pacific oysters, might be a case for the appointment of professional oyster removers.

As we have seen, the direct and indirect effects of bottom trawling and dredging on intertidal and subtidal biota is a major concern (see section 13.2.1 and review in Dieter *et al.* 2003). The probability that damaged intertidal flat communities will recover has not been determined, but is likely to be even lower than the recovery of fish stocks after overfishing (Hutchings 2000). There is little doubt that the time course of any recovery will be a function of the spatial scale on which the disturbance took place (Lenihan & Micheli 2001; Fig. 13.6). The time constant of general processes in the coastal zone are quite tightly correlated with their specific spatial scale (Fig. 13.6a). Given the extent to which intertidal flats are dredged for cockles in the Dutch Wadden Sea, the time to any recovery would be rather longer than a year (see also Fig. 13.2; Piersma *et al.* 2001). Using data from the literature for minimum recovery times after disturbance for intertidal and subtidal soft-sediment habitats, a preliminary quantitative assessment of time to the beginning of a recovery can be made

Fig. 13.6 (a) Timescales of natural and human-induced processes with different spatial scales based on processes in the Wadden Sea. (b) Relationship between the minimal time to recovery (or some aspect of it) and the spatial scale of the disturbance studied. This graph includes observations on natural and man-made disturbances, some of which took place in the Wadden Sea. The vertical shaded bar indicates the average extent of the intertidal area affected by cockle-dredging in 1998–2001 in the Dutch Wadden Sea, the horizontal line indicates the calculated time to recovery. The **return time** is the time for mechanical cockle-dredgers in the Dutch Wadden Sea to revisit the same intertidal location, given random site selection. It is given as a function of the spatial extent of the fisheries and is calculated relative to the total intertidal area available. As the point where the return time line crosses the vertical shaded bar is much lower than the minimum recovery time, the present cockle-dredging practice is highly unsustainable. After Versteegh *et al.* (2004).

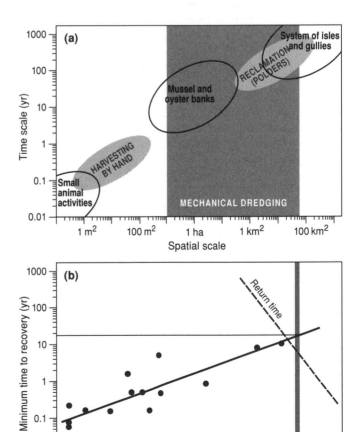

(Fig. 13.6b); this is semi-quantitative, as only data from studies where recovery was positive could naturally be included (Versteegh *et al.* 2004). Affecting up to 100 km² of intertidal flats, the mechanical dredging for cockles alone is predicted to require recovery times of 20–30 years. We regard this as the most optimistic assessment, as recovery times may be much longer if mechanical dredging has moved the intertidal ecosystem to a different, and much less biodiverse and productive, stable state (Scheffer *et al.* 2001).

One way to push a damaged intertidal flat system, devoid of eelgrass meadows, oyster beds or mussel banks due to dredging and trawling, towards the original, more biodiverse state, is to plant seagrass, build oyster reefs or create mussel beds. Subtidal oyster reefs (*C. virginica*) in Pamlico Sound, North Carolina, USA, do not function well when they are dredged and lose height. With increasing depth of the overlying water there is greater incidence of anoxia and there are lower water-flow speeds (Lenihan & Peterson 1998, Lenihan 1999). By increasing the height of such damaged oyster reefs, the performance of the oysters in terms of growth and survival increased considerably (Lenihan 1999).

In the Dutch Wadden Sea, the creation of artificial mussel beds by dumping dredged mussel on barren intertidal flats has met with some success and failures. A mussel bed of c.0.5 ha, created in June 1987 by dumping 20,000 kg (fresh mass) of blue mussels (*M. edulis*) on intertidal flats (Ens & Alting 1996) has survived several years and attracted many mussel predators. Eventually, this mussel bed was removed again by commercial dredging for seed mussels (J.B. Hulscher, personal communication). Other attempts to create mussel beds by dumping mussels on intertidal flats have met with mixed success. The creation of five experimental mussel beds in the Dutch Wadden Sea in October–November 2001 was a failure in that only two of the five sites still had mussels left after the winter (A.C. Smaal, personal communication). However, the amount of mussels deposited may have been too small during a time of food scarcity for shellfish-eating shorebirds in the Wadden Sea. We think that the artificial establishment of intertidal mussel beds on barren flats by carefully planned and supervised dumps of sufficient magnitude could help to reverse the process of ecological erosion caused by intense and large-scale dredging (see Box 13.2).

> ### Box 13.2 A vicious circle
>
> Removal of dominant stocks of filter-feeding and faecal-pellet-producing bivalves in combination with the reworking of originally rather stable sediments and an increase in tidal prism can lead to a cascade of effects, to a negative biodepository spiral, that can turn silty intertidal mudflats into much more sandy habitats (Fig. 13.7). According to this hypothesis, the mechanical removal of the large filter-feeders kicks off sedimentary changes that then automatically lead to the disappearance of other filter-feeders that also produce biodepositorily important (pseudo-)faecal pellets, *Macoma balthica* being just one example (Risk & Moffat 1977). In this scheme of things, winter storms are not the causal agent but are simply the executors of processes that have their beginnings in human perturbations of the apparently fragile balance of the intertidal sedimentary systems.

As the restoration of such eelgrass meadows happens to be one of the targets of Dutch environmental policy (e.g. van Katwijk *et al.* 2000), quite some experience has now accumulated with the transplantation of seagrass (*Z. marina*) meadows on the intertidal flats of the Dutch Wadden Sea (van Katwijk *et al.* 1998, van Katwijk & Hermus 2000). To reduce the effects of water turbidity and sediment mobility, exclosures were successfully applied at sites where light availability (a function of depth and turbidity) were adequate. To increase transplantation success sheltered locations were recommended by van Katwijk and Hermus (2000). At a local scale, a stable mussel bed could provide such a shelter, but the provision of temporal, biodegradable, dam-like structures would also provide refugia from where a successful transplantation of *Z. marina* could expand.

Thus, it is clear that restoration of biodiverse intertidal flats that were rich in biogenic structures such as mussel beds and eelgrass meadows, but were then turned into barren flats inhabited only by a few small polychaete and crustacean species as a result of large-scale mechanical dredging, probably requires more than many years without disturbances (Fig. 13.6). Using an integrated approach, the artificial re-establishment of ecosystem-engineering species like

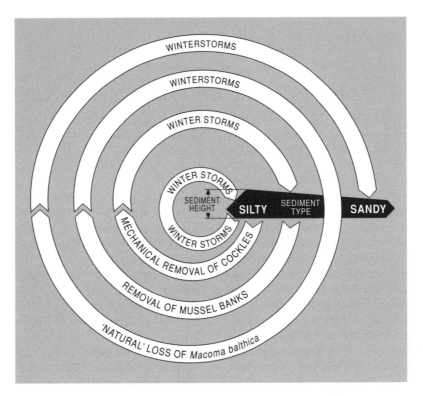

Fig. 13.7 Graphical model of the negative biodepository spiral by which mollusc-rich and reasonably silty sediments in the lee of a mussel bank, through the successive losses of cockles due to mechanical harvesting and the mussel bank itself, transform into permanently sandy, highly dynamic and lower-lying intertidal flats where even the Baltic tellin (*Macoma balthica*) is unable to maintain its population (processes as they appear to have taken place in the western Dutch Wadden Sea in 1988–98; see Piersma *et al.* 2001). Note that in the absence of human activities, the winter storms do not affect the balance of the sediments: the changes have to be started by the (mechanical) removal of the stocks of large filter-feeding and faecal-pellet-producing bivalves (blue mussels and/or common cockles) but can then turn into a self-perpetuating process with the additional loss of small filter-feeders. From Piersma and Koolhaas (1997).

mussels and eelgrass could comprise a key ingredient for restoring such intertidal flats to their original, biodiverse state.

13.6 Restoration of salt marshes

13.6.1 Setting targets

The ecological frame of reference for natural coastal systems in western Europe might be the Pechora river delta in the European north-east of Russia (van Eerden 2000). This river delta resembles the western European coast before coastal engineering activities. Present western European salt marshes seem natural ecosystems being governed by tides, but they are more or less affected by human activities as shown above. The cessation of accretion works along the mainland coasts will result in erosion of many artificial marshes in front of the present seawalls. When the area of the marshes is maintained, the geometric drainage patterns may remain for centuries. Only removal of the marsh

Table 13.4 Outline of the different landscape types from natural to artificial landscape features and examples for land use.

Drainage type	Erosion protection	Type of landscape	Land use
Creeks	No groynes	Natural	No
Creeks	No groynes	Semi-natural	Grazing/cutting/no
Ditches	Groynes	Semi-natural	Grazing/cutting/no

up to several metres deep seems to result in a natural drainage pattern during the fast sedimentation process. Salt marshes have been grazed since their emergence. Livestock grazing is beneficial for winter-staging birds, but not for many breeding birds. Authorities in charge of restoration should be aware that striving for natural salt marshes is often a contradiction. Moreover, targets can ask for contradictory measurements. Different approaches are currently practised and/or proposed to counteract the aforementioned changes in salt marshes. We discuss approaches including the interaction of sedimentation and vegetation. Hence, mitigation projects such as on dredged spoil are not taken into account (Zedler & Callaway 1999, Zedler & Lindig-Cisneros 2000).

When we adopt the aim to develop a diversity of salt-marsh vegetation, reflecting the geomorphological conditions of the habitat, the following definitions may be useful. **Natural landscapes** feature geomorphological conditions that are not affected by humans. They show a natural drainage system with meandering creeks and levees with higher elevation than the adjacent depressions. Natural landscapes occur in sandy back-barrier conditions or in parts of former Wadden Sea bays along the mainland coast. In contrast, **semi-natural landscapes** have a wide-stretched creek system but are affected in their geomorphological conditions by artificial drainage and/or by measures to enhance livestock grazing or cutting. Semi-natural landscapes are found on islands, in foreland clay marshes and in marshes with sedimentation fields and an artificial drainage system, i.e. ditches. They also include de-embanked summer-polders having an artificial drainage system without groynes. Clay pits or the establishment and mainten-

ance of sedimentation fields without an artificial drainage can result in a natural drainage system with meandering creeks and a naturally developing elevation structure.

Characteristic salt-marsh plant species can be present in all landscape types. However, their abundance in typical salt-marsh vegetation types and their spatial arrangement in the vegetation structure can be affected by land use. Geomorphological conditions may change in the long term – decades – whereas the effects of changes in land use can occur within a few years. This hierarchical arrangement implies that changes in land use cannot result in, for example, a transition from an artificial marsh into a natural or semi-natural marsh (Table 13.4).

13.6.2 De-embankments

Embankments have not only caused an interruption in salinity but also in sedimentation. The unembanked marsh in front of the newly created polder featured a continuous rise in net surface elevation. Differences in soil level in front of and behind the seawall or summer-dike will be greater when the polder is intensively drained for agricultural purposes. This will also hold for coastal systems with accumulation of peat where great shrinkage can take place (Roman et al. 1995). For the sake of coastal protection and the costs of seawall maintenance it is assumed that a well-inundated tidal marsh with a good rate of sedimentation in front of the seawall or summer-dike is better than a low-lying polder without sedimentation. Polders established at the edge of the salt marsh or even at the intertidal flats suffer from erosion of the seawall, hence the latter will be removed for reasons of coastal protection. Coastal defence and nature conservation might be combined by de-embankment of polders and subsequent restoration of these former tidal marshes, described as 'managed retreat' or 'managed realignment' (Boorman 1999). For this and other forms of restoration of halophytic communities two prerequisites need to be fulfilled: adequate abiotic conditions and the establishment of target species.

After de-embankment of a summer-polder, the renewed contact with the sea results in a quite fast re-establishment of the former abiotic conditions (Erchinger et al. 1994). A restoration is also expected to be quickly successful, for birds, as they have few dispersal problems. However, there might be constraints for plants. Are tidal plants still available in the community pool as persistent seeds in the soil seed bank as a historic record of the former marsh vegetation? A recent study in natural salt marshes indicated that most salt-marsh species have a transient or short-term persistent seed bank (Wolters & Bakker 2002). This suggests that restoration cannot rely on a persistent seed bank of salt-marsh species. Apparently dispersal of diaspores to de-embanked sites may not pose a problem. Percentages of target species, as related to the regional species pool, established in 70 de-embanked sites in north-west Europe, may amount to 70%, but most sites show lower figures (Fig. 13.8; Bakker et al. 2002b, Wolters et al. 2005).

In organogenic coastal areas such as the Baltic with a negligible tidal range, removal of the summer-dike might result in permanent stagnant pools containing slightly brackish water without any vegetation, after die-off of the freshwater community. These events were indeed found at Karrendorfer Wiesen on the German Baltic Coast (Müller-Motzfeld 1997). De-embankment of an estuarine summerpolder revealed that Ph. australis took over in the absence of livestock grazing, whereas P. maritima became dominant in a grazed site (Bakker et al. 2002b). These differences emphasize the effect of management regime on the outcome of de-embankment. This is a general phenomenon. The sites that were revealed to be most successful with respect to restoration of halophytic plant species after de-embankment were all grazed (Wolters et al. 2005).

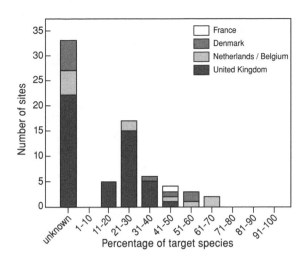

Fig. 13.8 Frequency distribution of scores of percentage of target species related to the regional species pool of over 70 de-embanked sites in northwest Europe. Note the absence of monitoring in about 50% of the sites. After Wolters et al. (2005).

13.6.3 Increase of tidal amplitude

In Connecticut, USA, tidal flows were restored by placement of a 1.5-m-diameter culvert in an impoundment. In 1988, 10 years after the start of the restoration, T. angustifolia had declined from 75 to 15%, whereas S. alterniflora had increased from < 1 to 45%. In addition, high-marsh species had re-established and covered 20%, but Ph. australis had also spread. After 1988 Phragmites declined and salt-marsh vegetation had increased to cover 85%. Although the restoration exhibits a striking result, the restored marsh only moderately resembles the pre-impoundment marsh (Fell et al. 2000).

The polder Beltringharder Koog, Germany, was embanked in 1987, including 845 ha of intensively grazed salt marsh and 2450 ha of intertidal flats. The tidal range was reduced from 3.4 m to zero immediately after embankment. Former salt marshes were covered by Cirsium arvense and Epilobium hirsutum stands, and former intertidal flats by Suaeda maritima, Spergularia spp. and Salicornia spp. In order to compensate for the losses, a lagoon system of 845 ha was established after separating it from the new polder by dams and allowing reduced tidal influence through two sluices. From 1990 onwards a 0.3-m tidal amplitude was established with stormflood simulations of 0.8-m tidal amplitude twice a month in the non-breeding period of birds. After the introduction of reduced tidal influence, the invasion of glycophytes stopped. A typical salt-marsh zonation established, including a *Salicornia* spp. zone, a *P. maritima* zone with some *Ph. australis* and an *E. athericus* zone, whereas the highest parts did not change except for the establishment of shrub species, in the absence of livestock grazing (Wolfram *et al.* 2000).

13.6.4 Changes in grazing regime

The resumption of livestock grazing in abandoned back-barrier salt marshes revealed that the changes in the vegetation after abandonment are reversible within 10 years at a relatively high stocking rate. The number of plant communities along the elevational gradient increased again. At different elevational levels the number of plant species per unit area increased, mainly because of the removal of the tall grass *E. athericus* (Bakker 1989, J.P. Bakker *et al.* 2003). However, at the low marsh the number of species decreased. The soft bottom is trampled by cattle and only annual pioneer species can be maintained (Bakker 1989). It is seen as a problem to restore artificial brackish marshes by resumed grazing, once *Ph. australis* has established (Esselink *et al.* 2002). It is estimated that the numbers of brent geese in the entire Wadden Sea in May can be a factor of 4–8 higher when all salt marshes are grazed by livestock, than in the absence of grazing (Bos *et al.* 2005b).

The overgrazed sub-arctic marshes might be restored by excluding them from grubbing by geese. However, exclosure experiments showed that once the sediment is hypersaline, it is still devoid of vegetation in the absence of geese even after a period of 20 years (F.H. Abraham, R.L. Jefferies and R.T. Alisauskas, unpublished observations). The success of restoration depends on the way species may survive the hypersaline conditions for a longer period. *Puccinellia phryganodes* and *Carex subspathacea* spread vegetatively. In contrast, dicotyledenous species associated with these graminoids depend on seed

banks for regeneration. Loss of vegetation and hypersaline soil conditions also had a deleterious effect on the soil seed bank (Chang *et al.* 2001). Even with the application of mulch and fertilizer, attempts at revegetation using soil plugs and intact plants are difficult under these conditions (Handa & Jefferies 2000). To prevent expansion of the overgrazed area in the breeding sites, the governments of the USA and Canada in 1999 introduced a spring hunt to reduce numbers of snow geese in the mid-continent population. The number of geese actually harvested since the introduction of the expanded seasons has increased from 0.6 million in 1998 to 1.4 million in 2001 (Kruse & Sharp 2002).

13.7 Concluding remarks

Understanding the historical context of the development of intertidal flats and tidal salt marshes during a period of several centuries is a prerequisite for any fruitful discussion about the perspectives of nature conservation and restoration in these systems. The intertidal flats, unless subject to long-term impact by human exploitation, can still be characterized as near-natural systems, because they were capable of recovering as long as humans use them as natural resources. As soon as these ecosystems were considered as areas for economic exploitation, mainly during the past few decades, the degree of disturbance included the risk of irreversible change. Restoration will then take a period of decades, if it is possible at all. Tidal salt marshes, on the other hand, have experienced centuries-long land use, and the majority nowadays are artificial ecosystems. This implies that these systems require continued management, livestock grazing being an inherent part of the scene. It has become clear that de-embankments, which aim at restoring salt marshes from polder areas, can quite easily result in re-establishment of the former abiotic conditions, but that the plant communities can become different from the reference communities due to species-specific dispersal of propagules, which may have consequences for small herbivores as well. This phenomenon will be further discussed in Chapter 16 in this volume, in terms of newly emerging ecosystems.

14

Restoration of Mediterranean woodlands

Ramón Vallejo, James Aronson , Juli G. Pausas and Jordi Cortina

14.1 Introduction

The Mediterranean Basin contains about 2.3 million km² of land, including part or all of 20 mountain ranges within 150 km of the sea, 40,000 km of coastline, 5000 islands and islets, a dozen or so peninsulas, and a highly complex geology and human land-use history (Suc 1984). The so-called Sea among the Lands was the principal matrix of exchange for successive civilizations over most of the last 10 millennia of history (Fig. 14.1).

Human density is very high in the Mediterranean region, particularly in the summertime. Known for its rich history, its dazzling panoply of archaeological and historical monuments and emblematic landscapes of western civilization (Grove & Rackham 2000, Allen 2001), the region has also long enjoyed the dubious distinction of being the premier tourist destination in the world. But two sharp geographical gradients occur. First, water and land are running short in coastal areas where tourists, working families and retired people seeking the sun all tend to congregate.

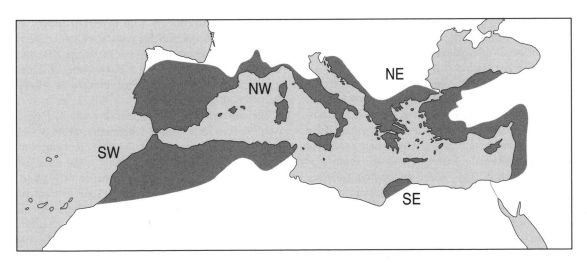

Fig. 14.1 Distribution of the Mediterranean vegetation (in grey) in the Mediterranean Basin. SW, south-west; SE, south-east; NW, north-west; NE, north-east. Modified from Blondel and Aronson (1999).

Secondly, the contrast between the northern and southern banks of the Mare Nostrum is striking and stark: ever-growing agricultural abandonment and rural exodus in southern Europe contrast like night and day with north Africa and the eastern shores of the Mediterranean, where land-use intensification and an unabated population increase in coastal and inland areas alike create a very different scene (Etienne *et al.* 1998, Blondel & Aronson 1999).

14.1.1 Mediterranean particularities

The Mediterranean is one of the world's 18 biological 'hot spots' (Myers *et al.* 2000), where exceptional concentrations of biodiversity occur, and where much of that biodiversity is in danger of depletion or extinction. The region is home to over 25,000 species of vascular plants, whereas only 6000 plant species occur in Europe outside the Mediterranean Basin despite its being an area three to four times greater in size. Approximately 247 tree-like woody species (capable of reaching at least 2 m in height) occur in the canopies of the Mediterranean forests and woodlands, whereas only 135 tree species occur in all of non-Mediterranean Europe (Quézel *et al.* 1999). Compared to northern and central Europe, unusually high species richness is also found among Mediterranean insects, mammals, birds and other groups of animals, fully matching the botanical richness referred to above. There is also an extraordinary richness of wild relatives, ancient varieties and landraces of a huge variety of domesticated plants and small livestock (Zohary & Hopf 1993).

The climate, geology and biogeography of the region have contributed to the unusually high biodiversity of the Mediterranean; both it and the adjoining Near East have long been a perhaps unparalleled nexus of exchange and interaction among contrasted biotas and cultures – among European, south-west Asian and African flora, fauna and human societies. But there is an important historical element as well. The Mediterranean Basin's very high spatial heterogeneity is still today amply mirrored, or reflected, by linguistic, legislative, cultural and agricultural diversity, especially in the various Mediterranean mountain regions (McNeil 1992).

Like the other four Mediterranean-climate regions (MCRs) in the world, the Mediterranean Basin's climate combines cool or cold and wet winters, and long, hot and dry summers. Summer drought is of variable duration, but frequent periods of drought can occur at any time of the year (Vallejo *et al.* 1999). Water is the key limiting factor (Noy-Meir 1973) for plant and animal growth, and for human societies. There is also a strong overall gradient of aridity in the Mediterranean, from the north-west to the south-east. As mean annual precipitation declines, the coefficient of variation of annual rainfall increases (Le Houérou 1984).

Furthermore, the basin is unusual among MCRs: it is the only one of the five belonging to the Old World, where human beings have for 10 millennia been living, consuming resources and transforming natural landscapes and ecosystems to their own ends. Thus, here alone among the MCRs, many plants and animals have had ample time to adapt to the human presence; those that did not adapt have, for the most part, disappeared. Thus it is comprehensible, here perhaps more than anywhere else, to speak of humans co-evolving with landscapes (Naveh 1990) rather than appearing as mere 'parasites' of the biosphere (Odum 1996).

As elsewhere in the Old World, the great and continuous density of historical layers in the Mediterranean region renders difficult the selection and use of a precise historical reference system, such as is frequently sought in New World settings (Egan & Howell 2001). This feature, which is shared with much of Europe, contrasts sharply with other continents, and has major consequences for restorationists, conservationists and ecosystem and land managers. The longstanding alteration by humans (see section 14.1.2) has lead to transformation and, often, irreversible degradation of natural ecosystems.

The combination of the great diversity of physical conditions – for example topography, geology, soils and the large number of possible pathways and stages of succession (Fig. 14.2) occurring after various disturbances as abandonment, fire or logging – imparts to Mediterranean landscapes a particular kaleidoscopic or patchwork pattern. The alternative steady states and successional pathways found in Mediterranean landscapes require that a choice be made by restorationists and land managers depending on their objectives.

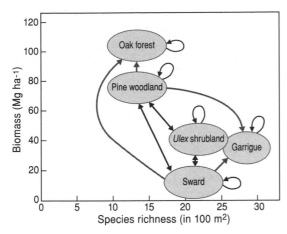

Fig. 14.2 Some of the possible successional pathways (including post-fire autosuccessional trends) in relation to above-ground biomass of mature communities (Mg ha^{-1}) and species richness (in 100 m^2) in eastern Spain (very simplified). The various stages (alternative steady states) are dominated by *Quercus ilex* ssp. *ballota* (Oak forest), *Pinus halepensis* (Pine woodland), *Ulex parviflorus* (*Ulex* shrubland), *Brachypodium retusum* (Sward) and *Quercus coccifera* (Garrigue). Variability in biomass and richness within each successional stage is not indicated, but can be large. Grey lines represent slow or rare pathways.

14.1.2 A combined approach: restoration, conservation and sustainability

Our general approach to restoring Mediterranean ecosystems is based on the need to repair, rehabilitate and restore ecosystem health and ecosystem services, while also assuring greater attention to the interrelated environmental and economic issues of biodiversity on the one hand and sustainability science on the other. This approach can be defended on purely economic grounds (Balmford *et al.* 2002), but also finds ample justification on social, cultural and ecological grounds as well. Given the ancient history of human use and transformation, the restoration of Mediterranean ecosystems necessarily follows a functional and **landscape-oriented** approach (Hobbs 2002). Furthermore, given that we are concerned with an array of cultural, semi-natural and natural landscapes that

are all more or less 'perturbation-dependent' (see Vogl 1980), and for which special treatment is required (Naveh 1991, Farina 2000), we are concerned with conservation, management for sustainable use and – where necessary – restoration of damaged, degraded or destroyed ecosystems.

Furthermore, we focus on restoration of vegetation, and especially of large woody species, and leave the restoration and re-introduction of animals aside in our discussion. Large, woody plants and trees are the main structural component in the Mediterranean Basin, and in other MCRs, and they play an important, indeed vital, role in the functioning of these ecosystems. The annual occurrence of a long, dry season at the hottest time of the year imposes a severe constraint on all plant life (Joffre & Rambal 1993), and the presence of a tree or trees greatly modifies microclimate and biological conditions both above and below the ground. Further, in all areas long modified by humans, trees and large shrubs can play an important role – as bio-indicators – in forest and vegetation history, and should also be used for determining the best options, goals and procedures for ecological conservation and restoration (Aronson *et al.* 2002). Thus, we proceed from the notion that large, woody plants provide a fundamental framework for natural and for historical cultural or semi-cultural Mediterranean ecosystems, and that therefore they must not be underestimated in the present context.

In line with the most recent definition of ecological restoration (SER 2002; www.ser.org), the basic objectives of restoration actions in the Mediterranean Basin are, or should be as follows.

- To stop degradation, especially desertification processes affecting the most sensitive Mediterranean ecosystems.
- To promote improved ecosystem and landscape function and structure, taking into consideration that both groups of attributes do not relate in simple or unique ways (e.g. Fig. 14.2).
- To assist secondary succession through stimulating natural regeneration, by
  1 making use of recognized succession trajectories (Fig. 14.2), that offer a referential multi-attribute system of potential restoration trajectories in terms of improving structure and function;

2 using known reference systems, such as known or attributed original vegetation and semi-natural, humanized landscapes;

3 fully exploiting the potential of native species, ecotypes and provenances.

• To increase ecosystem resilience, especially in relation to the most threatening disturbances, such as wildfires, extreme drought events and gradual drift to hotter and drier climates (Pausas & Vallejo 1999, Pausas 2004).

• To promote self-regenerating systems that will be as independent as possible from further subsidies, and to ensure ecosystem sustainability and health for both natural and semi-cultural systems (such as the dehesas, described in Box 14.1).

14.2 Disturbance and land-use changes

14.2.1 Ancient history of human impacts

There are natural drivers for degradation in the Mediterranean area (Grove & Rackham 2000); however, human impacts are long-standing features of the Mediterranean Basin. Environmental degradation in the Mediterranean is ancient (Thirgood 1981, Wainwright 1994). Palaeolithic people used fire deliberately to facilitate hunting and food gathering (Stewart 1956, Trabaud 1998) and, since then, millennia of severe pressure resulting in burning, cutting and grazing non-arable lands, and clearing, terracing, cultivating and later abandoning arable portions, have created a vast array of strongly human-modified landscapes. Still today we see a kaleidoscopic tapestry of managed cultural and semi-cultural landscapes and agroecosystems which reflects the historical, cultural and legislative diversity of the region.

As compared to the Persian, Greek, Egyptian and Carpathian civilizations, the Roman empire was one vast city-building enterprise requiring huge amounts of wood (Hughes 1982). In Spain, the human population was already close to the theoretical carrying capacity of traditional Mediterranean agroecosystems during the second century AD (Butzer 1990). In the Middle Ages, the Arab empire also made large inroads into the forests of the north African littoral and mountains, and those of Iberia. The recognition of land degradation and the declaration of the intention to pro-

Box 14.1 Dehesas

Over five centuries or more, people in southern Spain, Portugal, Italy and parts of north Africa have fashioned and maintained a two- or multi-tiered, semi-natural ecosystem from within the matrix provided by natural woodlands (Joffre et al. 1988, 1999, González-Bernáldez 1992). These silvopastoral or agrosilvopastoral systems make multiple use of trees (shade, fruits, bark, firewood, etc.) and the herbaceous stratum. Grazing by domestic livestock of annual and especially perennial herbs and grasses is invariably an important element in dehesa (Spanish) and montado (Portuguese) management; concurrently shrubs and small trees are regularly removed. However, in Sardinia, where the system is called pascolo arbolato, and in north Africa the shrubs are usually present. Of particular interest are the 2.2 million ha of cork oak (Quercus suber)-dominated open woodlands in north-western Africa and south-western Europe where periodic harvests of natural cork provide an important cash supplement to the annual revenues derived from animal husbandry and, in southern Iberia, from sown crops such as cereals.

Dehesas are thus artificially opened and managed woodlands that are simplified compared to natural woodlands. They have the virtue of mimicking natural Mediterranean ecosystems and are thus highly attractive alternatives, from both ecological and economic perspectives, as compared to other dry-farming or irrigated systems where all or most trees are eliminated. Unfortunately, these formations are widely threatened with extinction (e.g. Mellado 1989), either through intensification, or extensification of their usage, both of which tend to be deleterious for the resilience and productivity of the systems. They require new inputs and new ideas to promote natural regeneration and to re-introduce lost species while maintaining economic viability in an era of profound socio-economic change. Many of them require active restoration following a careful diagnosis and cooperative dialogue with all stakeholders.

mote reforestation by the various ruling regimes also came relatively early, and developed especially since the Modern age (Marsh 1871). In short, we can see the Mediterranean Basin as a moving mosaic where various land uses have moved in space and in time.

This long history of human pressure in the landscape had several important consequences.

- Original vegetation (e.g. virgin and old-growth vegetation) is mostly absent in the Mediterranean Basin, and semi-natural forest is confined to remote and inaccessible, uncultivable zones. No forests appear in flat and fertile soils, and most mid- and low-altitude vegetation communities were transformed to agriculture or degraded long ago to low and sparse vegetation due to a long, indeed millennial, period of overuse (grazing, fire, cutting). The composition and structure of semi-natural forests have been greatly modified by long-term uses (terracing, fuelwood gathering, charcoal manufacture, animal husbandry and grazing, etc.). However, due to the complex geology and diverse topography of the Mediterranean Basin, there is still a significant portion of semi-natural vegetation, which is higher compared to central Europe, that houses an important faunal diversity (including bear, lynx, wolf and large raptors).
- Abuse and overuse have also affected the soils. One general outstanding feature of dry Mediterranean soils is their low organic-matter content. As a consequence, low levels of microbial activity and low aggregate stability are common. This produces a high risk of soil compaction, surface sealing and crust formation in silty soils, which greatly increases runoff and soil erosion when plant cover is scarce (Vallejo *et al.* 1999). In addition, soils developed from calcareous substrates, very common all over the Mediterranean, tend to have low P availability.
- Where water is an important limiting factor, as land degradation increases, soil structure and loss of soil through surface erosion lead to decreased efficiency of rainfall capture, and thus to regression of vegetation cover (Thornes 1987, Thornes & Brandt 1994, Whisenant 1999). This hydro-pedological negative-feedback loop is common in the Mediterranean Basin, especially in the drier regions.
- People have practised an 'artificial negative selection' on wild plants (Burkart 1976). This term describes the short-sighted practice whereby people selectively remove the most useful phenotypes and genotypes of woody plants (and other organisms), both within and among species and genera, in a progressive fashion generated by some kind of positive-feedback cycle. Only inferior genotypes, phenotypes and, ultimately, species are left to reproduce and contribute to the seed bank in the areas subject to this short-sighted mining and mismanagement. An exception to this rule is the dehesa system described in Box 14.1.
- Wars, imperial and colonial appropriation, plagues and other catastrophes leading to social disintegration may have contrasting effects on land use. They favour destruction and intensification of uses, but they may also have the opposite effect by precluding other uses, affecting human population density and distribution.

In summary, as a consequence of human activities over millennia, carried out in a seasonally dry, unpredictable climate, many Mediterranean ecosystems were, and still are, affected by more or less irreversible desertification, especially in the long-inhabited and cultivated transition zones in the hotter, drier parts of these MCRs.

14.2.2 Recent changes and conflicts

During the past century, with the advent of industrial and tourist development, European Mediterranean countries have experienced an important change related to the abandonment of rural livelihoods and the sprawl of cities in coastal areas. These recent changes have had strong ecological implications.

- Abandonment of large areas of former agricultural and pastoral land and reduction of grazing pressure, fuelwood gathering, fibre cropping, etc. Abandoned fields and ungrazed land have been recolonized by early secondary successional species or, frequently, planted with pines. This results in homogenization of large areas with flammable even-aged stands of trees or shrubland and accumulation of litter creating a continuous fuel bed and increasing fire hazard (Pausas & Vallejo 1999, Pausas 2004). Under dry to semi-arid conditions, colonization by late-successional species seems to be rather slow (Francis 1990, Albaladejo *et al.* 1998). Furthermore, the collapse of unmaintained terraces results quickly in increased soil losses.
- A shift from exploitative to recreational use of wildland, including urbanization of rural areas by

people that do not live from the land and the increasing urban–wildland interface. These factors have increased the potential fire-ignition points and the risk of damage to structures and humans.

- Urbanization along the coast favoured by rural exodus and tourist pressure, and a heavy reduction and fragmentation of coastal and littoral ecosystems. Flatlands are particularly vulnerable, while rocky hills and outcrops are more resistant to this trend.
- Especially important losses of concern are located in coastal wetlands, which (i) are an important nexus of biodiversity, (ii) have an important influence on water fluxes and flows (Millán 1998) and (iii) play a key role in landscape-scale nutrient dynamics and pollution control.

Furthermore, European Mediterranean countries are suffering from ongoing intensification of agricultural production in the most productive areas. The Common Agricultural Policy of the European Union is subsidizing certain agricultural products (cereals, sunflower, rape, olives) and livestock husbandry, especially in less-favoured regions of southern Europe. The same policy is subsidizing set-aside of marginal agricultural lands to reduce stocks, promoting the conversion of these lands into forests.

During the past century, different socio-economic, demographic and political trends have occurred between the north (European) and south (African) rim of the Mediterranean Basin. The processes outlined in the above list for European Mediterranean systems can, to some extent, be detected (locally) in the southern rim. However, southern Mediterranean landscapes still suffer from over-exploitation, especially overgrazing (after clearing) and the associated degradation problems outlined in section 14.2.1.

In summary, the main land-degradation problems in the Mediterranean Basin include increased soil erosion, fire hazard and (locally) overgrazing, water scarcity, soil salinization, urbanization and reduced ecosystem stability and diversity. These degradation processes have a strong geographical component driven by land use and its changes that were very dynamic in the past decades. In southern Europe, recent land-use changes have produced two opposite trends, i.e. intensification in coastal areas and land abandonment in inland marginal lands. Among these dynamic changes, traditional land uses still remain

in the less-developed regions, as in north African countries. In some cases, ecosystem degradation in the Mediterranean semi-arid region is relictual (Puigdefábregas & Mendizabal 1998); that is, degradation forces are not acting any more. The questions for these cases in the semi-arid region, where vegetation is interspersed with patches devoid of vascular plants, are to what extent this heterogeneity is functional, and whether open areas are necessary for the survival of vegetation patches, assuming that the resulting heterogeneity of soil surface conditions yields optimum equilibrium with the prevailing climate and environment. Alternatively, degraded patches may produce losses of resources, which are especially related to extreme events that might be repaired by improving soil surface properties and introducing woody species. Degradation processes, either recent or relict, may not be reversed spontaneously, once one or more ecological thresholds have been crossed, and this degradation can only be reversed by human intervention in the form of restoration actions. In the next section we review some of the restoration approaches currently used and discuss how we can go further towards new, efficient restoration techniques.

14.3 Restoration approaches in Mediterranean conditions

14.3.1 Restoration priorities

Within our general framework and objectives outlined before (section 14.1.2), we here focus on ecological restoration of large-scale ecosystems, degraded most often by long-term over-exploitation and wildfires. The specific objectives of restoration differ widely among the different Mediterranean ecosystems, although the general and priority objective is soil and water-cycle conservation (Cortina & Vallejo 1999). Specific objectives vary depending on the degree of degradation, and climatic, biotic and socio-economic constraints (Table 14.1). In the case of woodland restoration in Mediterranean ecosystems, which is a common objective in the Mediterranean, we consider the following priorities.

1 Soil and water conservation as the main priority, for reducing or preventing soil losses and for regulating water and nutrient fluxes.

Table 14.1 Framework for the restoration of Mediterranean ecosystems. Drivers for restoration are identified, as well as actions that can be undertaken to attenuate them, and available techniques to implement these actions. Each driver must be offset to ensure successful restoration.

Driver	Action	Technique
Persistent stress (disturbances, unwanted species)	Release stress	Limited access to people, herbivores, etc. Fire prevention, windbreaks Species control (fire, herbicides, clearing)
Low propagule availability	Artificial introduction	Seeding, planting
	Promote dispersion	Bird-mediated restoration, frugivory-mediated restoration (artificial perches, catches, habitat amelioration)
Adverse environmental conditions	Reduce soil losses	Emergency seeding, mulching, sediment traps
	Ameliorate soil properties	Amendments, nutrient immobilization, mulching, drainage, soil preparation
	Improve microclimate	Shelters, mulching, microsite selection

2 Improving the resistance, and especially the resilience, of ecosystems with respect to human- and non-human disturbances: to ensure sustainability of the restored lands we aim to promote plant, animal and microbial communities resilient to current and future disturbance regimes.

3 Increasing mature woody formations, both forests and shrublands, depending on the environmental conditions of the site, in order to improve ecosystem and landscape quality and to increase carbon storage under scenarios of global warming and CO_2 build-up.

4 Promoting biodiversity and fostering the re-introduction of key species that have disappeared because of past land uses.

14.3.2 Passive restoration

Relatively inexpensive, passive restoration techniques are preferable where ecosystem structural/functional damage is relatively limited and resilience is high. This is the case in some overgrazed ecosystems, where the exclusion, or severe restriction of livestock grazing for some years is sometimes sufficient to promote self-recovery (e.g. Floret *et al.* 1981, Wesstrom & Steen 1993). This may also apply to some post-fire conditions, where innate resilience combined with grazing

exclusion may be sufficient to ensure satisfactory post-fire regeneration (Ne'eman 1997). In some cases, rodent control may also be required to reduce seed predation and promote establishment of new plantings. However, in most cases, simple grazing exclusion or even tighter control of grazing will evidently not be sufficient to achieve restoration of ecosystems in mature stages of development. In such situations, active restoration will be required, with relatively large inputs and typically a number of synergistic techniques – both biotic and abiotic – being applied concurrently. We will deal with these issues below.

14.3.3 First-aid restoration

In some cases, a quick restoration action is urgently needed before the degradation process reaches or exceeds a certain threshold beyond which restoration becomes almost non-viable or prohibitively expensive. This may be the case in some sensitive systems after wildfire, where fire eliminates most vegetation, leaving an unprotected soil (Robichaud *et al.* 2000). In conditions of steep slopes, erodible soils and poor regeneration capacity retained in the vegetation, post-fire rainfall can generate acute erosion processes (De Luís *et al.* 2001). In eastern Spain, for example, many plant communities dominated by obligate

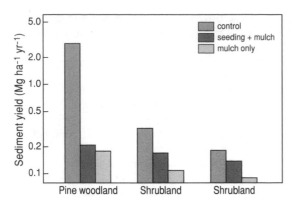

Fig. 14.3 Total sediment yield (Mg ha^{-1} yr^{-1}; log scale) in three different burned slopes (one pine woodland and two shrublands) in Benidorm, south-eastern Spain, subject to three different treatments: control (untreated), mulch only (200 g m^{-2} of straw) and mulch plus seeding (grasses and legumes). The area was burned in August 1992 and the sediment yield was recorded from May 1993 to November 1994. Elaborated from Bautista *et al.* (1996).

seeders (species unable to resprout after fire), growing preferentially on marl substrates and, especially, on south-facing slopes, commonly show low early post-fire recovery (Pausas *et al.* 1999) and thus higher erosion and runoff risk.

To provide rapid soil protection, two main (non-exclusive) techniques can be applied:

(i) **seeding** with herbaceous species and (ii) **mulching**; that is, protection of soil surface with various kinds of materials. Both seeding and mulching reduce soil losses (Fig. 14.3), surface crusting and water evaporation, and enhance water infiltration. For instance, in a post-fire pine woodland in eastern Spain, control plots had 7.2-times-higher soil losses than plots with a straw mulching treatment (200 g m^{-2} of straw; Bautista *et al.* 1996; Fig. 14.3). In an experiment with both seeding (legumes and grasses) and mulching (straw), the reduction in soil depletion was, on average, 12 Mg ha^{-1} yr^{-1} (Vallejo & Alloza 1998), which is well above the level of tolerable soil loss for shallow and erodible soils with low rates of soil formation (2–5 Mg ha^{-1} yr^{-1}; Smith & Stamey 1965, Arnoldus 1977). There are many different types of mulch that can be used (e.g. straw, wood and bark chip, shredded wood, rock fragments, paper sheets or other organic materials). For post-fire conditions, on-site slash may be a good and cheap alternative. In some cases, especially when erosion risk is high and seedlings are not likely to stop it, so-called first-aid or jump-starting restoration must include site stabilization by above-ground obstruction structures (e.g., rocks, logs, branches, brush piles, etc). These obstruction structures also retain organic matter, nutrients and propagules, and thus promote plant establishment and regeneration (Ludwig & Tongway 1996, Tongway & Ludwig 1996). For instance, site stabilization is sometimes needed in steep mountain slopes when old terraces collapse, often occurring as a result of the loss of vegetation cover after a wildfire.

Seeding species mixtures usually include commercially available seeds of native or naturalized herbaceous species, combining perennials with annuals, and grasses with legumes. Annuals show rapid germination, whereas perennials allow longer persistence. Differences in life-history traits, such as rooting depth and the potential for N fixation in legumes justify the use of mixtures. In the experiment mentioned above (Vallejo & Alloza 1998), seeded plots showed plant recovery within 2 months of application. However, field observations and measurements taken 6 and 18 months after seeding showed a short-lived, transient increase in plant cover, especially significant under semi-arid conditions, with almost all introduced species having disappeared after 18 months. Therefore, no inhibition of native-species regrowth was observed in those plots. Hence, the technique proved to be efficient in protecting vulnerable ecosystems after fire.

Seeding and mulching can be used on a large scale or in remote areas by aerial applications, although their efficiency and cost-effectiveness in these situations can be questioned (Robichaud *et al.* 2000).

14.3.4 Woodland restoration

To ensure long-term restoration, in most cases it is necessary to introduce woody species. We now briefly review some of the most prominent and promising of physical and biological tools being tested in

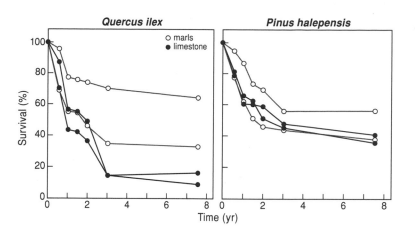

Fig. 14.4 Survival of *Quercus ilex* ssp. *ballota* and *Pinus halepensis* seedlings planted in the same year (1992) and on different bedrocks: marl colluvium (○) and limestones (●); different sets of the same symbol type refer to different plots. Data from CEAM (Fundación Centro de Estudios Ambientales del Mediterráneo; Mediterranean Centre for Environmental Studies) experimental plots in eastern Spain.

Mediterranean ecosystems. The mixing, adaptation and application of these techniques will of course vary from one context to another. Restoration projects are expensive and usually cannot be justified by direct production benefits. Economic justification relies on indirect benefits, mostly services to the society (i.e. external factors). Restoration investments must be kept as reduced as possible, but this is highly dependent on the specific socio-economic and cultural context. For example, irrigation is seldom used in restoration projects of southern European countries because the scarce water available is prioritized for other uses (mostly agriculture and tourism). However, irrigation is not unusual under semi-arid and arid conditions elsewhere (Allen 1995, Lovich & Bainbridge 1999, Bainbridge 2002), and it is commonly applied for seedling establishment in northern Africa.

To ensure sustainability, selection of woody species for restoration should be based, as much as possible, on the natural late-successional vegetation of the area, and on the environmental characteristics of the site. Traditionally pines were planted in many areas in the Mediterranean Basin for catchment protection and sand-dune fixation. Pines have high survival and growth rates, allowing a relatively quick re-vegetation success. However, extensive pine plantations provide an excellent fuel bed for large, devastating fires. Furthermore, pine woodlands commonly show low resilience to recurrent fires because Mediterranean pines do not resprout after fire. Early

attempts to introduce broad-leaved resprouting species (e.g. *Quercus* species; Fig. 14.4) met with high mortality (Mesón & Montoya 1993) and, until recently, techniques for introducing broad-leaved or evergreen, late-successional resprouting species in Mediterranean conditions were poorly developed. A relevant research question concerns whether it is possible to artificially skip stages in natural succession in this fashion, as suggested in Fig. 14.2.

As mentioned already, water-use efficiency is a key factor affecting plant survival and growth in Mediterranean conditions (Fig. 14.5). Water availability and plant water-use efficiency may be manipulated by different restoration techniques, such as seedling preconditioning, soil preparation, fertilization, protection by tree shelters and use of nurse plants. In short, Mediterranean restoration efforts generally aim at maximizing water-use efficiency and water availability (Table 14.2). Furthermore, other techniques based on the recognition and use of the ecological site diversity may also be applied, such as microsite selection and bird-mediated restoration. Each species ensemble and technique should be applied to the different landscape parts as appropriate. Restoration strategies should optimize the available resources and processes of the degraded site in its current state and context.

As many woody late-successional species do not form a permanent seed bank, seeding woody species is an attractive technique to re-introduce target species

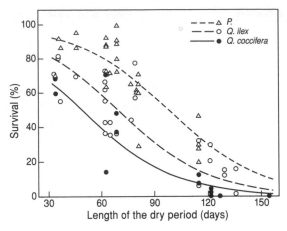

Fig. 14.5 Proportion of plants surviving in relation to the length of the dry period during the first post-plantation year, for three species (*Pinus halepensis*, *Quercus ilex* ssp. *ballota* and *Quercus coccifera*) planted in eastern Spain. Lines are significant logistic fits (*P. hakpensis* < 0.0001). Dry period refers to the maximum number of consecutive days with precipitation less than 5 mm. Elaborated from Alloza and Vallejo (1999).

Table 14.2 Mediterranean restoration techniques concerned with water.

Objective	Technique
Increase water-use	Selection of drought-tolerant species and ecotypes
efficiency	Seedling preconditioning
	Improve below-ground performance
	Improve nutritional status
Increase water supply	Soil preparation and amendment
	Irrigation
	Microsite selection
Reduce water losses	Tree shelters
	Mulching
	Microsite selection
	Control of competing species

owing to its low cost, the low impact of field operations and improved possibilities for treating remote areas through aerial seeding. However, high predation risk represents a serious managerial constraint for the direct use of seeding (Hadri & Tschinkel 1975, Bergstern 1985). Uncertain seed germination and seedling establishment under dry conditions are also important considerations.

Seedling quality

The best procedures for characterizing seedling quality are still under discussion (Mattsson 1997). Following the concept of quality for purpose, in the context of the restoration of the Mediterranean ecosystem, seedlings should be able to withstand unfavourable growing conditions (transplant shock, summer stress, cycles of drought), and still take advantage of favourable climatic periods to achieve sustained growth. Prior to the advent and general dissemination of modern nursery practices, woody seedlings were commonly overstressed and frequently showed abnormal growth (e.g. root spiralling) that strongly compromised their survival and field performance. This was a consequence of technical limitations and of the prevalent idea that seedlings hardened in this way were acclimatized to adverse field conditions. Results did not contradict these ideas in the short term, probably because of the resistance and narrow range of species used, as well as careful tending. However, with the use of a wider array of plant species and the occurrence of extremely dry years it has become increasingly evident that seedling quality must be ameliorated to increase restoration success.

Performance of high-quality seedlings, in terms of size, nutritional status, morphology, etc., is now commonly much improved (Seva *et al.* 2000, Villar *et al.* 2000). Most of these studies are based on nursery manipulations simultaneously affecting several plant morpho-functional traits. This is of practical use, but hampers a thorough understanding of the traits that are responsible for plant performance in the field. Seedling size together with other, mostly visually assessed characteristics, are used to define acceptable stocks. However, above-ground size may not be a good indicator of seedling quality under semi-arid conditions (Seva *et al.* 2000). Root-system structure is equally important, with beard-like root systems produced by aerial root pruning being an important technique to apply for some species such as pines.

Restoration techniques mostly aim at minimizing water stress of introduced plants. In many cases, a key factor in plantation success is the transplant shock; that is, the initial short-term stress experienced by seedlings as they are transferred from favourable nursery conditions to the adverse field environment. Nursery techniques to avoid transplant shock include manipulations of the watering regime and radiation environment for improving seedling quality; that is, for preconditioning the seedling to unfavourable field conditions. Preconditioning has four main objectives (Landis *et al.* 1989):

1 to manipulate seedling morphology and to induce dormancy;
2 to acclimate seedlings to the natural environment;
3 to develop stress resistance;
4 to improve seedling survival and growth after outplanting.

Drought-preconditioning has been tested for various Mediterranean species (e.g. Nunes *et al.* 1989, Ksontini *et al.* 1998, Vilagrosa *et al.* 2003), and although it is an attractive technique for Mediterranean conditions, it shows poor to moderate results. Species response may be related to their drought-avoiding strategy, and it may be necessary to design species-specific, drought-preconditioning techniques in accordance with these traits (Vilagrosa *et al.* 2003).

Many methods are currently available to manipulate seedling traits, including type of container, growing medium, fertilization and irrigation regime, irradiance, atmosphere manipulation, hormones and hormone-like compounds, mycorrhizae, above-ground and root pruning, etc. The range being so wide, experiments focusing on particular plant traits and meta-analysis of product tests could be of much use to elucidate the interest of present and future eco-technological tools. Studies on the long-term performance of introduced seedlings are scarce, and thus there is very little information on their capability to respond to a favourable climatic period after a long drought: very often seedlings can survive long droughts but show very poor growth and vitality. This

lack of information is quite surprising, as climatic variability is the norm in dry and arid ecosystems.

Microsite selection and nurse plants

Another way to increase the success of seedling establishment is by recognizing the spatial variability of water-availability conditions and soil properties in the site: the existence of 'safe sites' for germination and survival (Harper 1977). An easy way to locate favourable microsites for the introduction of target species is to identify facilitative interactions among plants. In other words, selecting a suitable microsite, close to a nurse plant, can facilitate the survival and growth of introduced seedlings (Callaway 1995). Facilitative interactions are especially important in semi-arid conditions, where isolated vegetation patches act as 'resource islands', mainly in terms of shade and soil fertility (Breshears *et al.* 1998, Schlesinger & Pilmanis 1998). It is generally assumed that the balance between competition and facilitation shifts towards facilitation in stressful environments, such as in arid areas (Bertness & Callaway 1994). The patterned landscape of semi-arid tussock steppes (*Stipa tenacissima*) is a clear example of this, as tussock microsites have higher organic matter, higher water availability, lower temperatures and lower penetration resistance than inter-tussock patches (Maestre *et al.* 2002). These environmental modifications facilitate the development of bryo-lichenic communities (Maestre 2003), and of introduced woody plants. The survival and physiological status of planted seedlings of several shrubs (*Medicago arborea*, *Quercus coccifera* and *Pistacia lentiscus*) is improved in the tussock microsites as compared to the inter-tussock microsites (Maestre *et al.* 2001). In other systems, the nurse plant can be a spiny shrub, as it protects planted seedlings from grazing (Gómez *et al.* 2001). For example, in the Sierra Nevada, southern Spain, 4 years after planting, seedlings of *Quercus*, *Pinus* and *Acer* spp. introduced under the non-spiny *Salvia lavandulifolia* and under spiny shrubs such as *Berberis vulgaris*, *Prunus spinosa* and *Rosa* spp., had, on average, rates of survival three times higher than those planted in open microsites (Castro *et al.* 2002). In contrast, under semi-arid conditions *Pinus halepensis* may not be capable of facilitating the

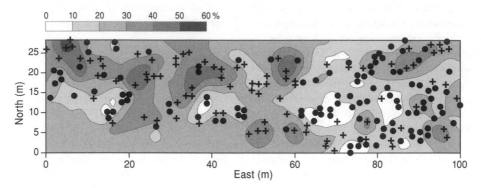

Fig. 14.6 Small-scale spatial distribution of the amount of bare soil covering planting holes (as shown by the grey scale, in %; darker areas had more bare soil), and of seedling survival 1 year after planting. Crosses (+) and circles (•) are dead and alive seedlings, respectively. There is a significant negative relationship between the amount of bare soil and the survival (logistic regression; $P < 0.001$). Elaborated from Maestre *et al.* (2003b).

establishment of woody shrubs (Maestre *et al.* 2003a). Therefore, the balance between competition and facilitation is complex and drawing generalizations about it is unadvised at this point. Biotic crusts, such as lichens, cyanobacteria, algae and mosses thriving in surface soil, may also improve nutritional status and growth of vascular plants and facilitate their establishment (Belnap *et al.* 2001).

There are other factors affecting heterogeneity in biotic and abiotic conditions in dry areas that may be relevant to restoration success. Large-scale changes in exposition, bedrock, slope, etc. can substantially affect the outcome of restoration. But subtle small-scale changes in soil properties and microtopography can also be important. Patches of plant survival in apparently homogeneous areas are frequently associated with small differences in soil depth, stoniness, texture or nutrient availability (Maestre *et al.* 2003b; Fig. 14.6). As plant responses to these factors may not be linear it is very important to identify thresholds that may explain such contrasts in plant performance. Nevertheless, small-scale heterogeneity may often not be discernible by direct visual observation. Thus, other indicators are clearly needed to bring patchiness into management and restoration plans and programmes.

As our knowledge on the biotic and abiotic drivers of seedling establishment progress, we will be able to incorporate them in routine restoration practices. However, degraded ecosystems may not be structured

enough to provide favourable sites for establishment. Or we may want to further promote factors favouring seedling establishment. In these cases we can use a wide array of ecotechnological tools and techniques (see below), which in many cases mimic biotic and biotic interactions.

Soil preparation

Soil-preparation techniques have been developed to improve water supply to planted seedlings and ameliorate soil physico-chemical properties. Runoff harvesting aims to intercept runoff and redirect water to the planted seedling, and can be performed by subsoiling (deep regolith drilling) or by the creation of small runoff collection areas up-slope to direct water to the plantation hole (microcatchments). Successful results have been obtained in arid areas of several countries (e.g. see Whisenant 1999). In arid areas low-infiltration surfaces allow runoff concentration in vegetated patches and increase the productivity of the whole system (Hillel 1992). Compacted soils show poor aeration, low water permeability and high resistance to root penetration, thus reducing the effective soil depth for roots. Thus subsoiling/ploughing has also been used for reducing soil compaction.

Terracing has traditionally been used in Mediterranean countries for agricultural purposes and later for reforestation (ICONA 1989). The effectiveness of this technique varies greatly with available soil depth, climate and accuracy of the works. However, visual impact and modification of the soil profile are strong reasons for limiting the use of this technique nowadays, despite the emotional attachment – or nostalgia – that many people still have for Mediterranean terraces.

Soil fertility and amendments

Forest soils of the Mediterranean Basin are frequently poor in soil organic matter and low in phosphorus availability (Vallejo *et al.* 1998). Furthermore, land uses and disturbances such as wildfire or erosion commonly result in decreased soil fertility (Bottner *et al.* 1995, Vallejo *et al.* 1998). To what extent soil nutrient impoverishment is hampering restoration is a matter of discussion. Soil organic matter intervenes in many soil processes affecting plant growth, but especially in soil structure (i.e. stability versus soil crusting and erosion, and water-holding capacity) and soil fertility. A soil organic matter concentration of 1.7% has been used to identify soils in a predesertified stage (Montanarella 2002). Seedling establishment does not seem to be related to soil organic matter content at levels between 2 and 4% (Maestre *et al.* 2003b). Further research on potential threshold values in organic-matter content and other soil properties could be of great help in optimizing restoration practices.

Response to resource additions is considered an indicator of limitation. Introduced seedlings commonly respond to inorganic and organic fertilizers (Roldán *et al.* 1996, Valdecantos *et al.* 1996, 2002). Negative and null responses are being associated with metal toxicity, salinity and increased above- and below-ground competition (Valdecantos 2001, Valdecantos *et al.* 2002). There is a wide range of organic residues available for improving soil fertility. They have traditionally been used for the restoration of old quarries (Bradshaw & Chadwick 1980, Sort & Alcañiz 1996), but they can be applied in other contexts (Roldán *et al.* 1996, Valdecantos *et al.* 1996, Navas *et al.* 1999).

Tree shelters

Tree shelters or protective tubes are used to modify the physical environment of the planted seedling (acting as mini-greenhouses). If designed conveniently, they can help to reduce evaporative demand and

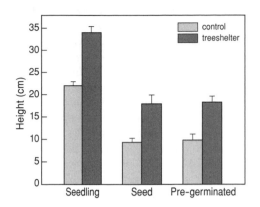

Fig. 14.7 Effect of tree shelters on growth of *Quercus ilex* ssp. *ballota* 4.5 years after planting in eastern Spain using three different techniques: standard 1-year-old seedlings (Seedling), directly sown acorns (Seed) and acorns pre-germinated 1 week prior to planting (Pre-germinated). Source: J.P. Seva, unpublished results.

improve overall seedling performance (Bergez & Dupraz 1997, Bellot *et al.* 2002). The use of tree shelters has gradually increased, being readily adopted by practitioners. Tree shelters provide protection against herbivory and extreme climatic conditions. Ventilated tree shelters help to avoid excessive warming. They can improve seedling survival and growth (Carreras *et al.* 1997, Bellot *et al.* 2002; Fig. 14.7), but the use of this technique alone does not guarantee plantation success (Oliet & Artero 1993, Navarro & Martínez 1997, Peñuelas *et al.* 1997, Grantz *et al.* 1998). The root-to-shoot ratio of protected seedlings is often lower than in unprotected seedlings. Thanks to the slow growth rates, Mediterranean seedlings growing inside tree shelters may gradually acclimatize to adverse climatic conditions outside the shelter and improve shoot growth. When this does not happen, unbalanced growth in stem height may occur. Still, a reduced root-to-shoot ratio could negatively affect seedling potential to withstand drought.

Bird-mediated restoration

The role of plant–animal mutualisms has been suggested for cost-efficient restoration plans (Handel 1997). Bird-mediated restoration is based on the observation of succession of abandoned woody crops, as olive trees, carob trees, almond trees and vineyards. In these conditions, the natural succession is significantly faster than in non-woody crops thanks to the role of the trees as perch sites for frugivorous birds. These birds defecate seeds of late-successional, bird-dispersed species that germinate around the perching tree forming a nucleus of advanced succession (Verdú & García-Fayos 1996). Furthermore, the perch tree may also create a favourable microsite for germination and survival. Dead trees may also be used by birds as caches (e.g. for acorns; Mosandl & Kleinert 1998). These bird-mediated facilitation processes inspired a restoration technique based on providing bird perches (e.g. dead trees, artificial woody structures) in old-field sites to accelerate colonization rates and ecosystem restoration. Although being an attractive and inexpensive technique to help succession, it has seldom been applied to Mediterranean ecosystems and most examples come from elsewhere, mainly from tropical ecosystems (e.g. McClanahan & Wolfe 1993). Because in many areas of extensive old-fields there is a very low seed availability of late-successional species, this technique could be appropriate. A potential limitation for the use of these techniques may be the lack of suitable dispersers (Alcántara et al. 1997) and of a close source of target species seeds; however, its application clearly needs to be explored and tested under Mediterranean conditions.

14.4 Concluding remarks

Several critical issues in ecological restoration for Mediterranean ecosystems deserve further development in the near future.

14.4.1 A landscape approach

In many Mediterranean systems, and due to the large and long-standing human impacts, degradation processes are not local and large heterogeneous areas need to be restored. Restoration needs to be viewed and approached at the landscape (and/or regional) scale. Different combinations of the above-mentioned restoration techniques may be required for different purposes, but also for different parts of the landscape.

Thus, landscape-restoration programmes should be diverse, adaptive, self-organizing and able to face the ecological realities of change (Whisenant 1999, 2002). The selection of the species (often a combination of species) for each part of the landscape, and the arrangement of the restoration patches at the different scales may determine the sustainability of the restoration and the self-sustaining recovery process.

14.4.2 Evaluation and monitoring

In order to improve our understanding of the success or failure of restoration actions, there is a need for long-term monitoring and evaluation of restoration actions in Mediterranean landscapes. Evaluating ecological restoration success on the ecosystem and landscape scales can be performed using indicators (e.g. Tongway & Hindley 1995, Aronson & Le Floc'h 1996a, Whisenant 1999), although widely accepted standard protocols are not yet available.

14.4.3 Extreme and unpredictable dry conditions

Restoration techniques for Mediterranean conditions have greatly improved in recent decades, thanks to the inputs from disciplines such as community ecology, ecophysiology and soil science. However, we still lack well-tested and reliable techniques for restoring degraded ecosystems in arid or semi-arid regions, and thus further research is needed in this context. Moreover, nursery production needs to be diversified in order to provide high-quality seedlings for different purposes and conditions.

14.4.4 Economics

The benefits of restoration projects are indirect and long-term, and thus they do not have a market value under prevailing economic systems. For this reason, restoration actions in the Mediterranean are strongly dependent on subsidies – mostly from the European Union – and changes in the existing subsidy policies could have strong impacts on our landscapes.

Community involvement and sustained commitment are also essential to foster and nurture restoration.

Acknowledgements

We thank J.A. Alloza, F. Maestre, S. Bautista and J.P. Seva for providing helpful information for the elaboration of this chapter and J.P. Bakker and J. van Andel for many helpful comments on the manuscript. Elaboration of this review has been financed partially by the Generalitat Valenciana, BANCAIXA, the European project REDMED (ENV4-CT97-0682) and the Spanish project FANCB (REN2001-0424-C02-02/GLO, CICYT). J.A. also thanks the Wallace Research Foundation (USA).

15

Restoration of alpine ecosystems

Bernhard Krautzer and Helmut Wittmann

15.1 Introduction

15.1.1 Historical overview, state of the art

Mass movement and erosion in a young high mountain range, such as the Alps, are utterly natural processes. A decisive regulating mechanism to counteract this natural instability is intact vegetation (Tappeiner 1996). Humans have interceded in this sequence of events for around 7000 years. Overgrazing, deforestation and technical interception over hundreds of years have repeatedly provoked erosion, resulting in slight to catastrophic effects (Stone 1992). Nevertheless, many alpine regions would have been beyond settlement had people not undertaken protective measures at a very early date against the results of erosion. Thus, a more or less stable balance was achieved in which agrarian use and mining were in the foreground of human interest.

With the conclusion of the Alpine Convention in 1991, the region of the Alps was defined for the first time and borders were established at a community level. According to this definition the Alps cover an area of 191,287 km^2 and the number of inhabitants is 13 million (CIPRA 1998). It was the aim of this convention on the protection of the Alps to pursue a comprehensive policy for the preservation and protection of the Alps by applying the principles of prevention, payment by the polluter and cooperation between member states, concerned regions and the European Community, concisely to balance ecology and economy (EC 1991).

Permanent changes have taken place in the entire region of the Alps during the course of the last 50 years. Wide areas used for agrarian purposes have been reduced or abandoned. On the other hand, there has been widespread opening of power stations and intensive road building, torrent and avalanche barriers, as well as extensive infrastructural measures especially for winter tourism. Some 40,000 ski runs, amounting to 120,000 km in length, have been built in recent decades in the Alps and are used annually by 20 million tourists (Veit 2002).

All of the measures described lead to intensive building each year, which then requires the restoration of the areas burdened by the intrusion. But at increasing altitudes restoration becomes increasingly more difficult due to the rapidly worsening climatic conditions. Due to cost, restoration continues to be relinquished in some areas of the Alps, but a combination of usually cheap restoration procedures and cheap and alien seed mixtures are relied upon. The resulting ecological and often economic damage is comprehensive: soil erosion, increased surface drainage, inadequate vegetation cover, the high costs of ecologically dubious fertilization measures and management, and flora falsification are some of the effects that follow.

For 15 years, intensive research has been carried out by various institutes to break this negative circle of events. In various research projects (e.g. Urbanska 1986, 1997, Wild & Florineth 1999, Wittmann & Rücker 1999, Florineth 2000, Krautzer *et al.* 2003) it has been proved that a combination of high-quality application techniques and site-specific vegetation or seed has led to stable, sustainable and ecologically adapted populations of high value for the protection of nature. Fertilization and management measures can be clearly reduced, which makes these methods useful in the medium term, as well as being economical.

The following depictions should offer a brief overview of the restoration problems in alpine environments as well as the possibilities with and necessity for site-specific restoration measures, and also the limits of what is possible.

15.1.2 Concepts and terms

Appropriate to the climatic changes at specific altitudes, vegetation in the Alps and many other landscape elements and processes are divided into altitudinal zones (Veit 2002). The change of these factors, according to altitude, leads to a vertical sequence in various climatic areas (Ozenda 1988). The high zones are separated by borders that are fairly easily recognizable: the montane zone is separated from the subalpine zone by the forest line, the subalpine zone from the alpine zone by the tree line, the alpine zone from the subnival zone by the grassland border.

The following depictions of the restoration of alpine ecosystems relate to the subalpine and alpine zones and are thus limited to the zones between altitudes of 1300 and 2400 m (Ellenberg 1996). In lower zones, overcoming the power of erosion is easier by degrees. At extreme altitudes, over 2400 m, satisfactory restoration is no longer possible according to the current state of technical awareness.

In the *Richtlinie für standortgerechte Begrünungen* (which translates to *Guidelines for Site-Specific Restoration*; ÖAG 2000) the important terms with respect to restoration measures are defined exactly: vegetation is site-specific when after generally extensive agricultural use or non-use it is enduringly self-stabilizing, and when the manufacturing of agricultural products is not a prime target for this plant society. This site-specific vegetation, with the exception of finishing and development management, or possible intensive agricultural use, requires no further management measures.

Vegetation created by humans is then site-specific when the following three criteria are fulfilled.

1 **Site-adapted:** the ecological amplitudes (the demands) of the applied plant species should be in accord with the characteristics of the site.

2 **Indigenous:** the plant varieties used are to be seen as indigenous when they are found in the geographical region (e.g. Val d'Aosta, Hohe Tauern), at least in the same region in which restoration takes place, and are evident, or have been evident, at appropriate natural sites.

3 **Regional:** the seed or plant material used should originate from the immediate surroundings of the project area and from the habitats which with respect to essential site factors are appropriate to the type of vegetation to be produced. Due to a lack of availability of regional seed, the regional criterion should be aimed at, but is not obligatory.

15.2 Consequences of a change in land use

15.2.1 Change of agricultural use

Comprehensive deforestation measures undertaken in the high zones of the Alps – especially after the beginning of the Middle Ages – to create grazing areas on the one hand and to supply the enormous timber needs for salt mines and mining operations on the other, have repeatedly caused ecological crises in the Alps, which were counteracted partly by restrictive forestry laws (e.g. in Austria, Bavaria and Switzerland). In the last 150 years, agrarian usage in high zones and other less-productive areas has clearly receded, which has led to a corresponding expansion of forested areas (Cernusca et al. 1996). As a result, the exploitation of alpine meadows intensified or was, on occasion, completely abandoned. In the subalpine zone, during the transition process from the lavish and intensively cultivated alpine meadow areas to the original forest vegetation, a creeping destabilization of the ecological systems of high alpine-meadow regions can arise. Summer precipitation flows increasingly on the surface and erosion that cannot be restored immediately – as in alpine-meadow cultivation – leads to extensive landslides and the formation of shell-shaped erosion scarp (Stahr 1996, Tasser et al. 2003). Up to the time at which the original vegetation – forests – is re-established there is a clear increase of natural erosion processes, which requires repeatedly more restoration and preventative measures (Gray & Sotir 1996).

15.2.2 Opening for tourism

Contrary to agricultural use, summer and winter tourism in the last few decades has led to extensive opening of high locations. There are already more than 13,000 lifts, cable cars and other transport facilities throughout the Alps, which are used mostly for tourism. Taking the calculations made in the middle of the 1990s as a starting point, the actual area of ski runs and lift facilities is more than 110,000 ha, of which 10,000 ha are already covered by artificial snow (CIPRA 1998). Even if an exact estimation is not possible, at least half of these areas are in high zones. As before, thousands of hectares are levelled annually as part of the opening for tourism and infrastructural improvements, and these areas now require restoration. Necessary measures for the protection of the facilities (above all, torrent and avalanche barriers) also require large areas each year.

15.3 Specific alpine characteristics

15.3.1 Alpine climate

Plants at high altitudes are often subject to frequent and often rough change of climatic factors. The transition of the seasons takes place very quickly. With increasing altitude, the vegetation period is around 1 week less per 100 m of altitude (Reisigl & Keller 1987). The differentiation of the macroclimate from the microclimate, dependent on altitude and broad location is important. The most important difference from sites in valley locations can be briefly characterized by the following factors (see Fig. 15.1).

- Temperature decreases in the air and in the deeper levels of the Earth by an average of 0.6°C per 100 m of altitude. Frost is a possibility at all times of the year in high zones; at the beginning and end of the vegetation period an interchanging frost climate generally predominates (Arenson 2002). The climatic vegetation period with average daily temperatures of over 10°C is around 67 days at an altitude of 2000 m, which is one-third of the vegetation period in valley areas (Krautzer *et al.* 2003).
- The deep-ground temperatures in the mountains strongly reduce the activity of micro-organisms. Reduction of dead organic mass and thus the provision of basic mineral nutrition is inhibited. The subterranean habitat is thus limited – contrary to the grasses in warmer, lower zones – to the most strongly warmed, humus-rich, generally acidic and intensively rooted upper layers of the ground.
- Precipitation increases with altitude; in addition, on the fringes of the Alps, where clouds from adverse weather fronts – coming mainly from the Atlantic but

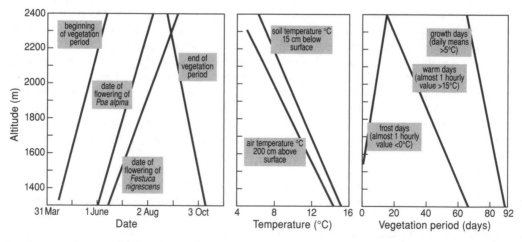

Fig. 15.1 Average changes of important phenological and climatic parameters according to altitude. After Krautzer *et al.* (2003).

partly also from the Mediterranean Sea – accumulate, precipitation is 800–1000 mm more than in the heart of the Alps. Evaporation rises. Critical situations in the water balance of plants, however, are rare, other than in special locations (those of strong radiation, high temperature and high wind).

- Wind increases in frequency and strength with altitude. This strongly influences the distribution of snow in winter and thus the length of the snow-free period and the water balance, which makes a strong erosion effect possible in exposed locations. Various inclinations of the sun's radiation create differing degrees of heat.

- The difference between a north and south aspect is increasingly greater with altitude. The micro-climate, however, can surpass the influence of the macroclimate and altitude.

15.3.2 Loss of the mother soil, loss of the established species, and erosion

Removal of the topsoil, as is common in technical inter-ception, means the destruction of the mother soil specific to a site. Without this, ecologically adapted grassland can no longer thrive. Only pioneer com-munities can develop on stony ground lacking fine soil, which is usually created within the sphere of re-cultivation.

The species in the high zones of naturally present grassland communities are adapted to an optimum degree to the soil and weather conditions of their habitat (short vegetation period, low provision of nutrition). Insofar as they were previously available, these species as well as the natural seed bank and the vegetative organs of renewal contained in the mother soil are generally lost during building activities.

The mastering of erosion, with all of its resulting effects, such as increased surface drainage and loss of topsoil, and up to the formation of karst, is one of the main problems in re-cultivation work in alpine environments. Average slope gradients of 30–45% in the area of ski runs, and far above in the area of natural erosion zones and avalanche barriers, make adequate erosion protection a prerequisite for successful restoration. Only sufficient vegetation cover stabilizes the topsoil and reduces soil erosion to an acceptable degree (Markart *et al.* 1997). Results of several

assessments indicate that at altitudes of between 1800 and 2400 m a minimum vegetation cover of 70–80% is required to avoid erosion (Stocking & Elwell 1976, Mosimann 1984, Tasser *et al.* 1999, Peratoner 2003). Therefore, a sufficient combination of application technique and adapted seed mixture, or plant mater-ial, reaching the minimum requirement of sustainable vegetation with 70–80% cover within the first two vegetation periods, must be the main target for restoration at high altitudes. Assessments of the veg-etation cover of the sites showed that, under average conditions at high altitudes, the necessary minimum demand of cover can be achieved in the second vegetation period at the earliest. This requires applica-tion techniques with sufficient protection for topsoil for the first two vegetation periods.

15.3.3 Ski runs and agricultural utilization

No resulting burden normally occurs in the recul-tivated areas in the course of building protective measures (torrent and avalanche barriers). But areas used as ski runs in winter and/or alpine pasture in summer are subject to very special site factors, such as extended snow cover that is generally longer than usual at the given altitude. The generally dense snow cover on ski runs, often with deposits of ice layers, hinders the exchange of oxygen between the plant cover, the ground and the atmosphere in winter. Mechanical disturbance factors, such as the effect of the steel edges and chains on ski-run preparation machinery, can have a destructive influence on vegetation on rounded hills and steep slopes.

Low herbal vegetation growing on ski slopes also decisively increases the influence of surface runoff following large volumes of precipitation in high zones. In contrast to forests and areas stocked with dwarf shrubs, only technical measures (diagonally running and open drainage channels) can effect a safe disposal of the surface water.

Large areas used for agriculture are mostly burdened by grazing in summer. This burdening can lead to a multiplicity of damage (trampling and pressure dam-age by animals, humans and machines, soil density, etc.), which cause erosion and then require expensive restoration measures. Nevertheless, the task of the work-intensive management of alpine meadows in high zones, with regular measures taken, such as the removal of stones and bushes and the rapid restoration of points where erosion has begun in the course of a possibly long process of reforesting, can also lead to the massive formation of shell-shaped erosion scarp. Mostly affected are pastures below the potential forest line with deeply crevassed soil and a slope gradient of over 30°. Those areas showing shell-shaped erosion scarp are mainly those that have not been utilized (i.e. mowed or grazed) for many years (Bernhaupt 1980). Plant communities, in which the formation of shell-shaped erosion scarp appears, are marked by a high share of biomass-rich grasses and high-growing herbs. The root horizon is relatively low and uniform. The formation of a sliding horizon is fostered in this way. Stable plant communities, in comparison, comprise somewhat loose swards and low growth. They form relatively small amounts of biomass above the surface, but relatively high amounts of root biomass.

15.3.4 Plant growth and vegetation

In most cases, above all in the alpine zone, natural development back to the original vegetation units within human planning periods is not to be expected. With respect to sufficient protection against erosion, the first target function of restoration in high zones is usually the achievement of stable grassland. Previous damage mostly leads to the destruction of the vegetated earth or affects the eroded areas with a lack of humus groundcover. The results are mostly more or less a lack of fine soil, stony restoration bases with slight water-storage capacity and a high tendency to erode (Chambers 1997). To achieve sufficient protection against erosion restoration measures must take place as quickly as possible.

15.4 Approaches for sustainable restoration in alpine environments

15.4.1 General

Restoration at high altitudes is subject to limits. With increasing altitude, restoration following building measures always becomes more difficult. Irrespective of whether terrain corrections have been made in the course of constructing ski runs, or forest and alpine-meadow trails, measures for the improvement of tourism infrastructure or torrent and avalanche barriers, only a combination of high-quality plant material or seed adapted to the site with the optimum restoration technique will bring permanent success. The use of seed mixtures is only sensible in high zones within which the species can reproduce. This border is generally achieved at altitudes of 2300–2400 m. The planning of restoration at extremely high altitudes (over 2000 m) is to be carried out by appropriately trained experts with sufficient experience. The methods used are suitable for the climatic and geobotanical conditions and are principally for the restoration of alpine ecological systems. For the production of vegetation cover that is similar or identical to nature in areas with no primary agricultural use, the methods given here can be recommended for all altitudes in the subalpine–alpine zones.

Natural re-vegetation

Currently to be seen as site-specific and restorable according to the latest advances in the field are areas subject to human influence, somewhat nutrition-rich plant communities as the basis for various grazing lands, former cattle resting places and fields of high perennials and shrub communities. The recultivation of areas not subject to human influence, exposed alpine grasslands (curved sedge, cushion sedge, high-altitude formations of horst sedge grasslands, etc.) is currently hardly possible, as is that of windy ridge communities (the community of the three-leaved rush, etc.). There is no commercial seed available of the characteristic species of these types of vegetation (some are also beyond production). Moreover, the majority of these natural grasslands cannot be planted; many of them die shortly after transplantation.

In principle, it is possible in high zones to permit natural successive procedures. Due to specialized considerations for nature protection, this strategy is sometimes (with appropriately favourable conditions and awareness of the danger of erosion) favoured over other restoration methods, or at least combined with them. The substratum and site conditions will define the development of the vegetation. The spreading of mulch materials or geological textiles, facilitating so-called safe sites (Urbanska 1997), can foster the habitat and the germination of wind-blown seed.

Above all, on lime-rich restoration sites with sufficient fine soil and diaspora material, vegetation development is relatively rapid and richly blossomed pioneer stages can occur. On silicate sites in high zones the redevelopment of the vegetation takes place extremely slowly, for which reason this method is not to be used at such sites.

In some areas in the subalpine–alpine zones it is utterly legitimate to re-cultivate with rubble, meaning rock material of various granular sizes containing no vegetation. Such habitats are a part of the natural surroundings and it is possible to gain similarity to the areas to be re-naturalized. Such procedures have proved especially useful where, through the combination of restored areas and open-rubble areas in the course of site restoration, the mosaic of restored sections typical of the high zone and vegetation-free sites can be obtained.

The opportunities for natural re-vegetation are very limited in alpine environments, which necessitates the application of other methods.

Demands on the application technique

The available topsoil should be carefully collected and left at the site at the beginning of building activities. The diaspora material it contains and the remaining parts of vegetation make resettlement possible with vegetation from the original site. This is important for an enrichment of indigenous plants deriving from restoration, because their seed cannot be obtained on the market or is very expensive. Even after a partial mixing of the top soil with mineral layers, which are devoid of a seed bank, seed densities can be considerable (Diemer & Prock 1993). Therefore, conservation and redistribution of the topsoil should be considered a very important task when planning restoration. Discarding the topsoil represents a waste of valuable autochthonous plant material, which is available, on-site, for site-specific, low-impact restoration (Peratoner 2003).

On inclined areas, a sufficient cover of the topsoil with mulch or geo-textiles is a prerequisite for minimizing surface drainage and soil erosion. With the exception of intentionally initiated successive areas, recultivation techniques are to be used exclusively,

which guarantee sufficient protection for the topsoil. This includes seeding processes combined with a cover of the topsoil with a layer of mulch, netting or matting, as well as hay-mulch seeding. With the use of hay-flower seeding, the necessity for additional cover is to be decided on by an expert.

Ski runs

Ski runs are emphasized as an independent focus because above all the re-cultivation of bare terrain following the erection of skiing facilities has been, and is often, insufficiently carried out. Owing to this situation, the important economic and tourism policies have acquired a somewhat negative image. Numerous terrain corrections at high altitudes are recognizable as extensive vegetation-free areas with a high erosion potential after decades of operation and, despite some restoration attempts, are seen as wounds in the landscape. Above all, in relation to the re-cultivation of such skiing areas the latest advances in the field are not being utilized, and the lack of contractual and realizable legal nature-protection criteria and guidelines is especially flagrant. In numerous cases, permanent site-specific restoration was agreed upon in decisions supporting legal nature protection as well as in tenders, but was never realized. In this respect, it is to be maintained that where sites are beyond restoration, according to the latest advances in the field, terrain-changing measures are to be used.

Fertilization

Restoration in the area of ski runs is generally only successful with the use of seeds or plants interacting with proper fertilization. A poor volume of minerals available to plants is mostly found in areas following levelling. Rapid development of the seeding to a full grass cover is also necessary in site-specific restoration for rapid erosion protection at such sites. A single fertilization of such areas with a suitable fertilizer is generally sufficient to achieve establishment (Krautzer et al. 2003). If there is insufficient cover in the second year of vegetation, further fertilization measures to achieve a sufficient grass density is necessary (Holaus & Partl 1996). These measures can also be combined with seeding-over with a site-specific seed mixture. With the achievement of a relatively dense

Table 15.1 Main and secondary components for site-specific seed mixtures for restoration of alpine ecosystems (ÖAG 2000, Krautzer et al. 2003).

Species	Distribution	Vegetation belt			Parent rock		Moisture		Against fertilization	Tolerance	
		Montane	Subalpine	Alpine	Silicious	Calcareous	Dry	Wet		Against cutting	Against trampling
Main components (≥ 60% of weight)											
Grasses											
Avenella flexuosa	Worldwide	+	+	+	+	–	+	(–)	(–)	–	(–)
Bellardiochloa variegata	Middle/south Europe	–	+	+	+	(–)	(+)	(+)	(+)	(+)	(+)
Deschampsia cespitosa	Worldwide	+	+	+	+	+	(–)	+	+	(+)	+
Festuca nigrescens	Europe	+	+	+	+	+	+	+	+	+	+
Festuca picturata	Middle Europe	–	+	+	+	(–)	+	(+)	(+)	+	(+)
Festuca pseudodura	Middle Europe	–	(+)	+	+	(–)	+	(–)	(+)	–	(+)
Festuca supina	North/middle Europe	–	+	+	+	(–)	+	(–)	(–)	(–)	–
Festuca varia s. str.	Europe	(–)	+	+	+	+	+	–	+	–	+
Phleum hirsutum	Middle/south Europe	(+)	+	+	(–)	(+)	+	(–)	+	+	+
Phleum rhaeticum	Middle/south Europe	(+)	+	+	+	+	(+)	+	+	+	+
Poa alpina	Europe/Siberia/N.Am.	(+)	+	+	(+)	+	+	(+)	+	+	+
Sesleria albicans	Europe	+	+	+	+	+	+	–	(–)	(–)	(–)
Leguminosae											
Anthyllis vulneraria ssp. alpestris	Europe	+	+	+	(–)	+	+	–	(+)	(–)	(+)
Trifolium alpinum	Middle/south/west Europe	–	(+)	+	+	–	(+)	(+)	+	+	+
Trifolium badium	Europe/Siberia	(+)	+	+	+	(+)	+	+	(+)	+	+
Trifolium pratense ssp. nivale	Middle/south Europe	–	+	+	+	(+)	(+)	+	(+)	+	+
Secondary components (≤ 40% of weight)											
Grasses											
Agrostis capillaris	Euroasia	+	+	(–)	+	(+)	(+)	+	+	+	+
Agrostis stolonifera	Euroasia/North America	+	(+)	–	+	+	(+)	+	+	+	+
Briza media	Europe	+	(+)	–	+	+	+	(+)	(+)	+	(+)
Cynosurus cristatus	Euroasia	+	+	–	(+)	+	(+)	(+)	+	+	+
Festuca rubra ssp. rubra	Euroasia/North America	+	+	(+)	+	+	+	+	+	+	+
Koeleria pyramidata	Europe	+	+	–	(–)	+	+	–	+	+	(+)
Phleum pratense	Worldwide	+	(+)	(–)	+	+	(+)	+	(+)	+	+
Poa pratensis	Euroasia/North America	+	(+)	–	+	+	+	+	+	+	+
Leguminosae											
Anthyllis vulneraria	Middle/south Europe	+	(+)	–	(–)	+	+	–	(+)	(–)	(+)
Lotus corniculatus	Worldwide	+	(+)	–	(–)	+	+	–	+	(–)	+
Trifolium hybridum	Europe/Siberia/N.Am.	+	(+)	–	(+)	+	(+)	+	+	+	+
Trifolium repens	Europe/Siberia	+	+	(+)	+	+	(+)	+	+	+	+
Herbs											
Achillea millefolium s.l.	Europe/Siberia	+	+	(+)	(+)	+	(+)	(+)	+	+	+
Leontodon hispidus s.l.	Europe	+	+	+	(+)	(+)	(+)	(+)	(+)	(+)	+

+, Very good; (+), good; (–), poor; –, very poor; N.Am., North America.

grass cover, the measures can be limited to unsatis-factory patches within the area.

Fertilizer of a slow and permanent effect should be used, which promotes the build-up of humus and has good plant tolerance. Attention should be given to achieving a balanced nutritional relationship (Heer & Körner 2002). The use of roughage-promoting or unhygienic fertilizer should be avoided. Where possible, organic fertilizer such as well-rotted farmyard manure, composted fertilizer or certified biological compost (according to the existing legal regulations) should be used. The use of fluid and semi-solid sewage should be avoided. The use of organic mineral fertilizers with the appropriate characteristics (slow, permanent release of nutrition) is possible. Their use should be limited to the necessary degree in relation to the positive additional effects of the organic fertilizer (multiple effects, deposit effect, herb tolerance, build-up of humus).

15.4.2 Restoration with seeds

Seed mixtures

The conventional high-zone mixtures available on the market mainly comprise high-growing, non-site-specific plants originally bred for grassland economy in valley locations or as grasses for sporting events. These species are adapted to lower, warmer locations and are generally not suitable for restoration in high zones (Florineth 1992). The high nutritional needs of these species require long-term, expensive fertilization measures to achieve the necessary grass density. These species also show relatively high biomass production, which again requires regular cutting, grazing or removal of the materials arising. This is because in the short vegetation period that occurs at high altitudes the additional biomass that grows does not decompose sufficiently, with the result that the vegetation's stigmas would be choked. In many cases, further use or management of the restored areas is not wished for or possible.

Site-specific subalpine and alpine plants are adapted to an optimum degree to the high-zone climate. They produce little biomass, but with an appropriate choice of species they do produce high-quality feed. Seeding with site-specific seeds generally requires only small amounts of nutrition, and short-term management measures lead quickly to natural, generally extensive self-maintaining grass, which has high resistance against subsequent uses for tourism and agriculture. With the use of site-specific seed mixtures, the required sowing volumes commonly used in practice can be reduced from 200–500 to 80–160 kg ha^{-1}. Grasses and legumes were selected within the sphere of several international research projects, which are suitable for seed production in valley locations and can be used in various site-specific alpine seed mixtures (Krautzer et al. 2003, Peratoner 2003). In the meantime, the ecological species suitable for high-zone restoration will multiply over a broad area and can then be graded according to altitude, original rock and usage, packaged as high-quality restoration mixtures and then marketed. The use of such site-specific seed mixtures should be obligatory when sowing in high zones.

Standards for seed mixtures

Minimum standards for site-specific seed mixtures for high-altitude restoration include the following. First, when faced with a lack of indigenous plant material, seed mixtures need to be used for restoration at high altitudes. To avoid errors when using such mixtures, alternative recipes are used that fulfil the following criteria. Due to their natural area of distribution, site-specific, high-altitude mixtures are divided into main components and secondary components (see Table 15.1). The main components encompass the species currently commercially available. A further series of site-specific species (grasses, legumes and special herbs) are more or less offered commercially and regularly in small amounts, which according to definition may also be used as high-altitude seed. High-altitude seed must comprise at least 60% main-component seed, by weight. The remaining 40% of the weight can be made up with secondary components (Table 15.1). Secondly, mixtures must comprise at least five species. The share of a single species must not exceed 40% of the weight. Leguminosae must comprise at least 10% of the weight of high-altitude mixtures. Thirdly, for high altitudes (> 2000 m) site-specific, ready-made mixtures are generally usable to only a limited extent. A special combination of site-specific mixtures made up by experts is necessary.

Finally, according to the degree of availability, and the choice of restoration method, improvements with the inclusion of further species, and the reduction of cultivation varieties or their use through local ecological types, are possible and desired.

15.4.3 Plant clippings and nets

Mulch seeding

In the mulch-seeding process, soil and seeds are covered and protected with various organic materials. For optimum growth the depth of the layer of mulch should not be more than 3–4 cm and should be pervious to light. The most common mulch materials are hay and straw. To avoid the inclusion of undesired seed, in principle only hay of the second or third cut should be used.

With the simple hay or straw seeding methods, a 3–4 cm straw or hay cover is applied over the seeding. The prerequisite for this restoration method is sites that are protected against the wind and not too steep. The material expenditure is 300–600 g m^{-2} in a dry state.

At steep points, especially above the tree line, the black-green seeding method is suitable. Seeds and fertilizer are applied into the 3–4 cm straw layer and an unstable bitumen emulsion sprayed over it (not to be used in drinking-water-protected areas). Hay is not as suitable for spraying with bitumen because it is compressed; due to thinner stalks and better cohesion, hay-cover seeding alone is more stable than straw. Hay and straw can also acquire sufficient cohesion through light, organic gluten.

Hay-mulch seeding

With the availability of appropriate areas, the seeds can also be won by special mowing in suitable donor areas. The areas to be mown should generally bear site-specific vegetation that is appropriate to the aims of the areas to be restored. Mowing is undertaken at staggered intervals (with two or three mowing dates) to include the broadest possible spectrum of species in a mature state. These mowing dates should be determined by an expert. The plants to be harvested should not be in an overly ripe state because a slight loss of seed can take place. With the intermediate storage of the hay, which often requires the selection of several mowing dates, sufficient drying is necessary to hinder the attack of mould. The ratio of winning to restoration areas is generally 1 : 1 or 1 : 2. The hay won in this way, and the seeds it contains, is to be applied to the restoration area in a uniform layer to a maximum depth of 2 cm. Over-intensive application is to be avoided to prevent anaerobic decomposition processes in the distributed seed.

Geological textiles

A number of geological textiles are available commercially. This netting of jute, coconut fibre, synthetic fibres or wire can be used for all restoration processes described above. When possible, the use of synthetic fibres and wire netting as a planting aid in site-specific restoration should be avoided (galvanized-iron netting and synthetic netting have lifespans of around 30 years and are not biologically degraded). Geological textiles are used predominantly where there is a clear danger of erosion or extreme site conditions (e.g. on very steep, ridged banks). They offer the possibility for stronger surface protection and, according to the materials used, are more or less stable in the face of natural forces such as falling rock, snowdrifts and precipitation. According to the material, site conditions and altitude, the netting rots within 1–4 years, leaving no residue.

Hay-flower seeding

Required for this method is the availability of the seed-rich remains from threshing floors in hay barns, which above all at high altitudes is still mostly of sufficient quality. This material should come from hay that is not older than 1 year or maximally 2 years. A further prerequisite is that the hay must be cut sufficiently late, which promotes the forming of mature seeds in many field grasses and herbs. Sieving is often recommended to acquire an appropriately high seed concentration. The hay flowers (0.5–2.0 kg m^{-2}) are sown with their stalks to a maximum depth of 2 cm. An additional layer of mulch is only necessary when sieved material has been used. To hinder loss through scattering by the wind, seeding should only take place on wet soil, or if the hay flowers are watered after seeding. At high altitudes, seeding weighted with steel building grids, wire netting or coconut netting, which can be removed after a few weeks, has proven successful. A certain degree of protection of the soil against mechanical interference is achieved with the mulch layer, and microclimatic conditions are improved. The additional use of a cover crop has proven useful. If the germination capacity of the hay flowers is insufficient, important seed components can be additionally purchased and sown.

15.4.4 Restoration with plant material

Sod clippings

Shoots or rosettes (mostly mechanically separated vegetation turfs) are loosely distributed. Distribution can take place mechanically in areas that can be driven on. In this way, a much larger area can be restored with well-established vegetation than with grass swards. Restoration, however, is significantly more sparse, and the danger of erosion higher.

Grass turfs

Available and natural vegetation is above all the best substance in the alpine zone for enduring restoration identical to nature. Therefore, extreme care should be taken when using such vegetation because destruction or a lack of re-use must be strictly avoided.

Grass turfs (also known as grass swards) or larger pieces of vegetation won during levelling or path construction are grouped together following completion of the work. They are very suitable for the rapid and site-specific restoration of damaged areas. On steeper banks, the grass turfs must be fixed with wooden nails. Wherever possible the planting of grass turfs should take place before shooting or after the start of the autumn vegetation pause, just after the melting of snow or immediately before the coming of snow in winter. At these times the success of planting, even in the extreme high zones, is very good.

Before levelling begins, the available grass or pieces of vegetation are lifted together with the rooted soil and laid again after levelling. Depending on whether the turfs are cut manually, or lifted mechanically, their size is 0.15–0.5 m². If required they can be stored in pits or stacked on pallets (maximum of 1 m wide × 0.6 m high) to hinder drying out, stifling and rotting.

The storage period should not exceed a maximum of 2–3 weeks in summer. Following the end of levelling the grass turfs or pieces of vegetation are again laid out and pressed in lightly.

With appropriate planning of the building process, the direct use of vegetation turfs is possible without intermediate storage.

Potted plants

The plants and seeds are pre-cultivated in nurseries and planted with a well-developed root stock at the restoration site. Site-specific species with a good vegetative growth are used for this (Grabherr & Hohengartner 1989). One can also turn to mother plants or seeds taken directly at the site by experts. With the appropriate choice of species, excellent results can be achieved at extreme sites in this way. The supporting use of this method as a post-improvement measure against sparseness in the restoration area is favourable.

Ready-made sward (sod rolls)

Sod rolls with site-specific vegetation are already available in small amounts for differing starting substrates. Sod rolls are produced at specialized firms over a period of around 12 months until the sufficient development of site-specific altitude species is ensured. According to need and restoration aims, certain grass mixtures can be produced beforehand. The grass is then harvested to order and transported to the restoration area. Thus a complete cover of restoration areas is possible in the shortest possible time. This method is especially interesting in restorations following small-area interception and in extreme locations.

15.4.5 Combined techniques

Vegetation transplantation: combined seed-sward process

In this special restoration technique, the covering with grass swards, or other pieces of vegetation, is combined with dry or wet seed. The grass swards used must be appropriate to the desired site-specific type of vegetation and are generally acquired from the project area or the immediate vicinity at the beginning of building work. There can therefore be cases of an interception in the vegetation sphere beyond the immediate project area to achieve optimum success through the division of available vegetation. The area to be restored is therefore often larger than the original project sphere.

The grass swards (0.2–0.5 m²) are placed in groups in dry locations, to prevent them from drying out, or in a grid-like pattern in areas subject to high precipitation. Site-specific seed is applied to sparse patches between the swards. This seed has a stabilizing effect on the vegetation-bearing layer. Due to the short distances between the covered grass swards, it is possible for well-established vegetation to move into the intermediate spaces (Fig. 15.2). In this way, these patches will also be restored and inhabited in a natural way by species that are not available as seeds.

This method has been tested at altitudes of at least 2400 m and according to the latest advances in the field. Especially suitable are moderately nutrition-rich plant communities subject to little human influence, such as those found on grazing land (of the most differing types), high-growing perennials and green alder bushes.

The conception of this restoration technique, and above all the selection of grass-donor areas, is only to be undertaken by appropriate experts. In steeper areas (with a gradient of over 30%), and in terrain endangered by erosion, the use of geological textile matting or similar is planned for securing the covered vegetation or for the protection of the topsoil against erosion.

15.4.6 Management

Constant cultivation is not obligatory or necessary following the use of site-specific seed mixtures or plant material. With the appropriate composition of the seed mixtures or the use of appropriate plant materials, a restoration area can be left to itself, which is greatly desired for the restoration of areas prone to erosion, and those containing torrent and avalanche barriers.

Management of ski-run restoration is in most cases also necessary in areas not used predominantly for

Fig. 15.2 Seed production of site-specific species (harvest of *Poa alpina*), a precondition for ecological restoration of alpine environments.

farming. Management takes place in the form of extensive grazing or annual mowing (Persson 1995), with or without the removal of organic material (only small amounts of biomass).

Above all, in the first years of seeding with fertilization, ski areas must be managed. Until the achievement of sufficient grass density, at least over the first two vegetation periods, no grazing or trampling is to take place (Klug *et al.* 2002). Annual mowing is necessary following the appearance of appropriately lush growth. This mowing removes biomass and thus hinders the stifling of the growth in winter. Tillering of the plants is also stimulated and promotes grass density. If necessary, grazing should be hindered by fencing on steep and footfall-sensitive areas, in favour of mowing.

When cover is less than 50% in the year following restoration, further necessary measures are to be taken, such as reseeding or replanting with a site-specific seed mixture (30–50 kg ha^{-1}) or plant material. When necessary, appropriate improvement work must be undertaken in small areas.

15.5 Concluding remarks

A state of acceptance is normally not given before the site being restored shows signs of development that ensures the achievement of the restoration aims or is appropriate to the same. Confirmation of the work carried out and the achievement of an acceptable state of development is in certain cases to be executed through proper and successful care until completion. Therefore, an exact evaluation of success and failures is an important foundation.

15.5.1 Evaluation of success

Evaluation of the success of restoration requires special criteria in high-altitude sites. The primary aim of every restoration is sufficient protection against erosion following restoration until the vegetation has developed to be able to fulfil the task adequately. This immediate protection against erosion is enabled by the use of recommended application techniques with the covering of the topsoil. As an essential limit for sufficient erosion protection created by the developing vegetation, ground cover of 70–80% is considered appropriate by experts (Stocking & Elwell 1976, Mosimann 1984). With a site-specific choice of species, the vegetation can be considered stable from this point in time. Restoration created by seeding should form vegetation that is as uniform as possible, which when left uncut, unless otherwise agreed, must show at least 70% of the projected ground cover. In justified cases, a divergent ground cover can be agreed upon. Vegetation-free areas of over 20×20 cm^2 are not permitted in cases where potted plants are used. Up to 60% of the projected cover should be comprised of the species specified in the seed mixture, or as deter-

mined by the restoration aims in terms of vegetation type. The species-specific annual condition of the plants is to be taken into account when defining the degree of cover. Nursery vegetation and alien vegetation should not be a part of the required degree of cover. Divergent degrees of cover or states of decrease, above all in the restoration of difficult sites, are to be agreed upon and taken into account during evaluation.

The sown or planted vegetation in high locations must have survived two rest periods and two frost phases before evaluation can be conclusive. In special cases (e.g. re-introduction projects) individual evaluation criteria are necessary.

15.5.2 Failures

Restoration can easily fail in extreme locations or at extreme altitudes. The most common causes for such a lack of success are listed below.

False restoration methods

The more extreme the conditions, the more specific must be the planning of the restoration or rehabilitation measures. The securing of valuable pieces of vegetation, the gathering, restoration, intermediate storing and expert reapplication of the topsoil, the subsequent prevention against erosion, the use of special restoration methods and the choice of donor areas for the combined seed/sward technique or for hay-mulch seeding, require planning by appropriately experienced experts. Successful high-location restoration at over 2000 m has always been planned and maintained by trained experts.

False seed

A common mistake, even in less than extreme conditions, is the choice of unsuitable seed. Not only the use of lowland seed in the subalpine and alpine zones, but also the lack of attention given to decisive criteria, such as the degree of acidity of the soil, or the availability of nutrition, are causes for insufficient restoration success. Also valid here is the maxim that the more extreme the conditions, the more necessary is the use of trained experts.

False fertilization

As already mentioned, fertilization at the restoration site and the restoration method are to be mutually adapted. Too little as well as too much can hinder success. In this way, with the combined seed/sward technique, heavy fertilization can destroy vegetation of the replaced swards and the natural seed slumbering in the soil. The slightest failure in this respect can be caused by small doses of slowly working and long-term fertilizer.

Inexpert work

Grass swards as well as seeds are living materials and therefore careful handling and expert attention is indispensable. Badly stored grass swards, inexpert fixing of the sward in the soil, a lack of adequate bedding-in and the connected drying-out phenomenon can even destroy restoration undertaken with high expenditure. Above all, under difficult conditions one must call in a competent restoration expert.

Lack of subsequent management

In many cases, a certain degree of subsequent management is required for the success of restoration: when mowing is to be undertaken an exactly dosed post-fertilization, additional seeding or necessary fencing against grazing animals is required for the achievement of the projected level of restoration. All of these measures are essential elements of restoration that must not be forgotten if one wishes to achieve appropriate success.

15.5.3 Prospects for the future

Above all, restoration at extreme altitudes has made great progress in recent years. Whereas the restoration of areas above 2000 m back to a natural state was considered impossible 20 years ago, there are now fine examples of restoration at altitudes up 2400 m. Even when the techniques used are comparably extensive, they nevertheless create at these high altitudes management-free units of vegetation that are identical or almost identical to nature.

The current awareness of technology for the restoration of alpine ecosystems in various neighbouring alpine states is defined very differently and the knowledge of special restoration methods is insufficient. The legal sphere dedicated to extensive restoration methods also lacks uniformity. What is common in some countries is strictly forbidden in others. Above all, due to the manifested prohibitions, mostly given in nature-protection laws, the use of vegetation alien to such sites is in practice often ignored due to a lack of the knowledge about alternatives. Although permission for building projects at high altitudes is obligatory in almost all of the affected states, the protection laws are less than strictly controlled. There is also a lack of information among the authorities concerning what is technically possible. The drawing up of binding guidelines for site-specific restoration at high altitudes within the region of the Alps, which reflect the latest awareness of technological advances, is needed urgently. Specialized experts from within the region of the Alps should participate in the drawing up of such guidelines.

PART 4

Challenges for the future

16

Challenges for ecological theory

James Aronson and Jelte van Andel

16.1 Introduction and overview

Restoration ecology is in full bloom. The number and percentage of scholarly papers reporting on ecological restoration research published in the *Journal of Applied Ecology* has grown steadily over the past 40 years (Ormerod 2003). In the last 10 years especially there has been an expansion – not to say an explosion – of scientific and popular articles, newsletters, websites and books on the subject, all around the world, and in all parts of Europe. This intellectual and academic flowering of the science of restoration ecology is the reflection of a parallel evolution in the *practice* of ecological restoration in nations around the world. Public policy and new legislation have led to budgets being created, and full-time planning and administrative departments becoming established, to work on ecological restoration. Germany, the UK, the Netherlands, Spain, Canada and USA are among the countries taking the lead in the northern hemisphere, while South Africa, Costa Rica, Australia, New Zealand and India, among others, take the lead in the geopolitical south. The **Society for Ecological Restoration International** (SER International), which brings together scientists and non-scientists of all sorts, now has members from over 35 countries. SER International's Science and Policy Working Group (SER 2002; www.ser.org) has produced the first Primer of ecological restoration, providing criteria for distinguishing this discipline from other endeavours of environmental improvement, and standards for the evaluation of ecological restoration and the attributes of so-called restored ecosystems. Yet the scientific field of restoration ecology still has a long way to go in clarifying the basic concepts, models, hypotheses and

workaday definitions of ecological restoration, and how it fits into a broader picture of environmental management and problem-solving. This chapter aims to address these questions.

Whereas we now have some information on how different kinds of terrestrial and aquatic ecosystems actually develop and interact over time, we still have many difficulties in predicting how they, in specific cases, can be expected to respond to intentional interventions, different levels of exploitation, and various drivers of local change such as global climate change, biodiversity loss and biological invasions. Thus, as Bradshaw (1987) and Harper (1987) have argued, restoration can provide a crucial test for ecological models and theories. Additionally, restoration ecology is just one part of something bigger and indeed unprecedented: that is, an international, transdisciplinary effort addressing sustainability in a scientific fashion (Kates *et al.* 2001). The 'science of ecosystem services' and, more generally, a 'pragmatic ecological science' (Palmer *et al.* 2004) are also both beginning to emerge and take shape, in response to the gigantic and unprecedented ecological and environmental problems facing us today as a global society. Central and vital to all these efforts is the integration of ecology and economics, taking full cognizance of varying social and cultural values and constraints occurring in each different country. Trans-disciplinary and holistic evaluation also are needed to facilitate the planning, practice and coordination of ecological restoration and related activities at national, regional and international levels or scales. The need to understand complexity, and to embrace uncertainty and unpredictability in the development of complex adaptive systems, is becoming more widely recognized, and at least two theoretical approaches have been proposed (Roe 1998; see also Holling 2000, 2001, Gunderson & Holling 2002, van Eeten & Roe 2002).

For ecosystem and restoration ecologists dealing directly with specific systems in specific places, the planning and indeed all conceptual and practical efforts in restoration need to be cognizant of two new facts: (i) the quickly growing importance of so-called emerging ecosystems, arising as a result of biotic migrations, introductions and invasions that occurred in the last century and a half, and as a result of rapidly changing land-use patterns related to urbanization, industrialization and globalization of commerce (Milton 2003) and (ii) the realization that a great number of existing ecological systems around the world are better conceived as socio-ecological systems than as natural systems. This is because the deeply transforming impact of sustained human activities has been in motion and having impact – often irreversible – for many millennia. In most parts of the world it is only in the last 100–150 years that human impact has been deep and broad enough to be noticeable – from outer space, for example – as an indelible ecological footprint and, consequently, for our present era to be labelled, somewhat ominously, as the 'Anthropocene Era' (Crutzen 2002). During this same period, the phenomenon of emerging ecosystems has grown gradually in scale and abundance until now it may be considered of dramatic importance. Emerging ecosystems are of course not new, but are now arising in large numbers and on broader scales. While emerging ecosystems may not sound too menacing, it is clear that increases in global commerce and migrations have led to a vastly increased spread of pests and diseases that cause great harm when they are transported long distances and thus divorced from their natural predators and pathogens. The notion of socio-ecological systems is a particularly powerful tool in the dual effort to develop stronger links between social and ecological systems in the 'real' world (Berkes & Folke 1998), and to forge stronger linkages among researchers and educators in the biological sciences and the social sciences, as well as the broader non-scientific, non-academic portion of our local and global societies. As a new paradigm they also help us to take biocomplexity into account and to strive to restore ecosystems that will be adaptive and resilient to local surprises and to global change.

Before further exploring the way forward in restoration ecology, a crucial preliminary question to ask is, why restore ecosystems? Why invest the important amounts of time, energy and other resources required to actively restore, or assist in the self-restoration of, ecosystems that have been degraded, damaged or destroyed? Who will benefit? And, who will be willing to pay? What are the alternatives – so-called designer ecosystems, unmanaged emerging ecosystems, abandoned ecosystems, etc. – and are they socially, legally, politically and financially acceptable as substitutes?

Simply put, three fundamental reasons to restore ecosystems can be cited:

1 preservation of native biodiversity;
2 maintenance or improvement of sustainable economic productivity; and
3 protection – or augmentation – of our stock of **natural capital** providing a flow of ecosystem goods and services which benefit us, and all other species, in more ways than we can count or evaluate in simple monetary terms.

Actually, the third reason is a combination of the first two.

The first of these reasons is justifiable first and foremost by the inherent, innate values related to biodiversity. Biodiversity is a synonym for life, and thus commands our respect, stewardship and protection *per se*. In addition, biodiversity plays many roles in the adaptive evolution and consequently in the functioning of all life-supporting ecosystems.

The second reason – sustainable economic productivity – can be readily justified as well, especially in developing countries of the south, to use the term in its current geo-economic sense – where most people still live directly on the land, and on natural capital, as opposed to human-manufactured economy. Here, the primary issues are human welfare and economic livelihoods, as opposed to biodiversity *per se*. As in the industrialized north, socio-ecological systems are common, but it is in the south that the largest expanses of more or less natural ecosystems, and the bulk of biodiversity, also persist despite human use and exploitation. In the geo-economic north, including western Europe, this second justification of restoration, which is pragmatic and economic, has rarely been evoked as a reason to restore, at least until recently. Indeed, restorative activities undertaken in this pragmatic spirit are most often referred to as rehabilitation, in the sense given to this term in Chapter 1 and in the SER Primer (SER 2002). But this should assuredly not be taken as a hierarchical level of lesser importance or difficulty as has often been done in the past. Indeed, from a global-society point of view the contrary is probably true.

The third reason to restore encompasses the first two reasons and seeks to reconcile them. Ecosystem goods and services, the life-sustaining products of our stock of natural capital (Jurdant *et al.* 1977, Costanza & Daly 1992, Daily 1997, Clewell 2000a, Daly & Farley 2004) are best assured by ecosystems that are functioning in a healthy fashion, in the broad sense that ecosystem health was defined in Chapter 1 (Cairns 1993, Daily & Ellison 2002).

16.2 Eight hot topics

What are the major challenges to ecological modelling and theory from the twin points of view of the science and the practice of ecological restoration? In the broad context, at least eight hot topics in theoretical ecology should be mentioned, with special emphasis on the ecosystem and landscape levels, where disproportionately little work has been done to date in ecology.

1 Biodiversity and its role in ecosystem functioning.
2 Assembly rules and the structuring of ecosystems.
3 Ecotones, ecoclines and landscape boundaries.
4 Ecosystem resilience.
5 Ecosystem health and integrity.
6 Emerging ecosystems.
7 Socio-ecological systems and their relevance to the setting of restoration objectives and references.
8 Linkages, re-integration and the science of sustainability and ecosystem services.

16.2.1 Biodiversity and its role in ecosystem functioning

It has frequently been suggested that an increase of biodiversity has not only an intrinsic value (see reason 1, above), but also has positive functional or dynamic effects at the level of ecosystems. For example, an increase in species richness can contribute to an increase in ecosystem productivity (the rivet hypothesis; see Fig. 4.3 in this volume). But it is of course important to distinguish between native and introduced species, and also to consider the timeframe of most relevance when considering productivity. In restoration ecology, species richness cannot without risk be considered in neutral terms in the sense of Hubbell's (2001) neutral theory of biodiversity. Indeed, a first pertinent question of a general nature is: in a given situation, how do biodiversity changes – **losses**, at genetic, specific or ecological levels, **gains**, of invasive or introduced species, or intentionally reintroduced natives – affect ecosystem functioning and ecosystem integration in landscapes and bioregions? This is of course a relatively long-standing issue in ecology (Schulze & Mooney 1993, Ernest & Brown 2001, Kinzig *et al.* 2001, Giller & O'Donavan 2002, Loreau *et al.* 2002, inter alia), but one which can clearly be much further elucidated in the context of ecological restoration projects than anywhere else. Similarly, it is important to enquire in each specific situation where restoration is being considered, do species numbers really matter? Are some species redundant? Do **driver**, **passenger** or **keystone species** really exist, and if so how do they interact with each other? Does redundancy really exist, as some suggest, in natural or socio-ecological systems? This broad issue was touched upon in Chapter 2 in this volume. Here we shall refer to a state-of-the-art analysis presented by Lawler *et al.* (2001), who identified four unresolved issues.

1 The relationship between ecosystem functioning and biodiversity is not well-enough established to use as an argument for preserving biodiversity.
2 Ecosystem functioning forms an asymptote with species richness at a relatively low number of species.
3 Because randomly assembled communities show that biodiversity influences ecosystem functioning, species identity may be unimportant in some cases.
4 Dominant species may be solely responsible for correlations between biodiversity and ecosystem functioning.

This paper raised a number of important research topics, and pushes us to go beyond simply demonstrating that biodiversity can affect ecosystem functioning. Specifically, it suggests that – in some cases – biodiversity may be insufficient as a reason to justify the expense of restoration, at least from a purely engineering or socio-economic point of view. But, after having recognized this point of view, we need to comment on it. Firstly, productivity, frequently used as an estimate of ecosystem functioning, may be considered as an agricultural or silvicultural rather than an ecological parameter. Secondly, as we emphasized in the previous section, there are other very important, compelling reasons to conserve and, where possible, to restore biodiversity. Indeed, current biodiversity represents the evolutionary potential of the future, and in this way it at least affects all ecosystems' long-term functioning, except perhaps those that are completely designed and engineered. We therefore suggest focusing on the question of how restoration of ecosystem functioning can contribute to the restoration of biodiversity rather than putting it the other way around (see Fig. 5.1 in this volume; van Andel 1998a). Scientific challenges in this area abound and much remains to be elucidated through careful research.

16.2.2 Assembly rules and the structuring of ecosystems

Restoration of ecosystems is, as was indicated in Chapter 2 (see Fig. 2.2) a pathway towards some desired ecosystem, rather than the immediate re-establishment of such a system. Rehabilitation has much more the character of a discovery, a guided designer process that can work out even without a target set. The notion of assembly rules (see Chapter 5) suggests that the structuring of developing ecosystems may be subjected to certain general laws of species interactions and succession. This notion is obviously relevant to the **design** of ecosystems as much as to their restoration or rehabilitation. The important point, for ecologists, is to determine to what extent pre-existing ecosystems are – or can be – taken as models or references for the reparative work or redesigning work at hand. More discussion on this will be provided below.

Contrary to the Clementsian model of succession leading to some sort of 'climax', ecologists now generally embrace a flux-of-nature paradigm and accept the reality of many possible steady states, basins of attractions and above all multiple trajectories that ecosystems can experience or exhibit in the course of their development (see Suding et al. 2004). In that context, the model of adaptive cycles (Holling 1973) is a useful alternative metaphor for how ecosystems develop (see below). Nevertheless, this model or metaphor of ecosystem development should by no means be taken by the reader as being definitive or all-inclusive. In the field it is often hard to see its application and faced with the myriad forms of living systems that exist we need an abundance of modelling approaches (e.g. Ludwig & Tongway 1995, Tongway & Ludwig 2002). These models of course represent just one conceptual approach, but they have considerable power for generating new hypotheses and applications to conservation, management and restoration. Despite loss of biodiversity at genetic and species levels, and despite rampant homogenization of cultures, and the erasure of former landscape boundaries and transition zones, the diversity of ecological systems may actually be increasing at present, rather than decreasing. But the result may not be entirely to our satisfaction, or to that of future human generations.

The topic of assembly rules calls for renewed population and, especially, community ecology research to deepen understanding of the genetic, population and community dynamics within different types of ecosystem: plants, animals and all the rest. A recent book (Temperton et al. 2004) and a conceptual paper (Belyea & Lancaster 1999) should help the reader gain an overview of the field. The issue to be considered here is, to what extent assembly rules that may have shaped communities in the past can still be usefully applied to predict the development of future, newly emerging communities. One key issue to mention here is that of so-called functional groups of organisms and the extent to which they can be used in predictive or retroactive analysis and piloting (or reconstruction) of communities and ecosystems regarded as damaged, disturbed or destroyed by human activities (McIntyre & Lavorel 2001, Gondard et al. 2003, Pausas et al. 2003). How does succession take place, after all, and what are the adaptive cycles, if any, and the feedback systems, assembly rules and other inherent functional, evolutionary or simply dynamic mechanisms that make ecosystems develop and interact in one way or another? If we can sort these questions out – biome by biome – then we will unquestionably be better placed to predict how much time, energy and capital of all sorts will be required, or should be allocated, to ecological restoration and rehabilitation, for reasons 1, 2 and/or 3 given in section 16.1. Functional groups have also shown a clear usefulness in the sticky problem of scaling from species to vegetation, and higher levels of complexity (Körner 1993, Lavorel & Garnier 2001, Pywell *et al.* 2003).

16.2.3 Ecotones, ecoclines and landscape boundaries

From this point, we shall leave the subject of successional pathways aside and focus instead on the final or targeted ecosystems intended to be restored, rehabilitated or designed, and the landscapes in which they occur. Ecosystems are embedded in landscapes and they depend largely on processes at work at the regional scale. In much of Europe, and elsewhere, ecological transition zones have been deeply modified by human actions (Correll 1991). Human impact on the functioning of landscape boundaries has been profound and widespread, and yet at present those boundaries appear to be changing quickly as a result of rural exodus, agricultural abandonment and the rapidly growing cities and their complex urban/rural interfaces, among other things.

Ecotones and landscape boundaries can be diffuse or sharp (Wiens *et al.* 1985, Holland *et al.* 1991). Van der Maarel (1990) distinguished between ecotones (sharp boundaries with strong environmental fluctuations) and ecoclines (gradual differences in at least one major environmental factor, fluctuating relatively little), and stated that ecoclines are the most interesting in terms of species diversity and species rarity. Can notions of system resilience be applied here, as some researchers in the new field of biocomplexity maintain? How do the fractal properties of such zones, and landscapes in general, affect species and community dynamics (Palmer 1992, Ritchie & Olff 1999b), and how should they influence restoration efforts? In the case of ecosystems occurring as patches in a cultivated landscape, where there are boundaries rather than transition zones, the establishment of corridors between the patches may be an important restoration option or programme element, but this approach may not respond or compensate entirely for a species characteristic of intermediate or frontier zones along environmental gradients. Moreover, the usefulness for species migration among patches largely depends on the species of plant or animal (see Chapters 3 and 6).

16.2.4 Ecosystem resilience

Resilience has been defined as the time taken to return to the steady state of a stable ecosystem after a temporarily disturbing event (Holling 1973; see also Fig. 2.3). This is clearly a crucial and very complex issue as the innate complexity and adaptive nature of ecosystems must nowadays cope with increasingly complex multiscalar disturbances all around the globe. What happens to an ecosystem's developmental trajectory when species disappear locally, and/or new ones appear and persist (Pahl-Wostl & Ulanowicz 1993, Pahl-Wostl 1995)? Faced with these or other environmental challenges, does the ecological system show sufficient adaptive capacity or resilience (Gunderson 2000, Carpenter *et al.* 2001, Walker *et al.* 2002) for it to remain recognizable as a system or, rather, does it somehow mutate, so to speak, or shift to a new domain of attraction? Can we recognize, design or choose among alternative ecosystem trajectories and feedback systems (Odum *et al.* 1979, Suding *et al.* 2004)? What are the risks of ecological threshold crossings (Knoop & Walker 1985, Hobbs & Norton 1996)? What are the opportunities that adaptive ecosystems can exploit? How many alternative states can a given ecosystem move among in a given configuration, and what is its resistance to the crossing of an apparently irreversible threshold which pushes it into a new configuration of alternative states, in a new basin of attraction? If an ecological system exhibits resilience, as defined by Holling (1973), what are its limits and response times to various influences such as those described in the three previous sections? How do these features vary from one system to another? To what extent does the position of a system on a transformation (or degradation) gradient affect its resilience and resistance? To what extent can we identify threshold crossings in the past, or predict them in the future (Friedel 1991, Agnew 1997)? Do 'early warning' or 'immune systems' exist in ecosystems (Levin 2001)? Can a 'healthy' system invent and experiment (Holling 2001) in an ongoing context of self-organization within a 'panarchy'; that is, a nested hierarchical set of adaptive cycles (Gunderson & Holling 2002)? The reader interested in resilience, self-organization and related topics should be aware of the highly influential model of adaptive cycles introduced by C.S. Holling (1973) and developed in numerous arenas by the internet-based Resilience Alliance which publishes the online journal *Ecology and Society* (www.ecologyandsociety.org; formerly *Conservation Ecology*).

These questions are all clearly fundamental to the conceptual advancement of the science and the practice of restoration, especially if the motivation is the broad reason 3, above; all the natural goods and services flowing from natural capital. At the heart of these issues is the notion of resilience (see also Chapter 2), which of course is a term originating in physics, where it is used to describe the degree of mechanical resistance of a material subjected to an impact. A notable cross-disciplinary interest in this notion exists today, which underlines the fact that resilience is an excellent vehicle or focal point for the study of systems at many nested levels of complexity, and at various spatial scales. For example, the term resilience is much discussed in psychology and psychiatry today, referring to the capacity of a child or adult to return to a healthy or normal life despite experiencing a serious wound or stress, or the capacity to continue learning and growing even in environments or situations that should be debilitating (Cyrulnik 2001). Notably, Walker *et al.* (2002), of the Resilience Alliance, have recently redefined resilience as 'the degree to which an ecosystem expresses capacity for learning and adaptation'. This definition is highly reminiscent of the terms used by the psychiatrist Cyrulnik mentioned just above. Although the rapprochement should be made with caution, it does seem increasingly clear that the sciences of human health on the one hand, and of ecosystem health on the other, have much to give one another. To become truly scientific, however, the study of ecosystem health, like that of ecosystem resilience, must move steadily beyond the realm of metaphor to that of measurement (Carpenter *et al.* 2001).

16.2.5 Ecosystem health and integrity

Health does not equal integrity, but the two are interrelated. The SER Primer (SER 2002) distinguishes between these terms by defining **integrity** in terms of biodiversity – particularly species composition and community structure – whereas **health** is defined as an ecosystem's overall dynamic state at a given time based on ecosystem functioning. Indeed, Ulanowicz (1997) and Mageau *et al.* (1995) argue that ecosystem health is something to be measured at a given point in time, while integrity can only be evaluated over a longer period. Kay and Regier (2000) argue that ecological integrity is 'about maintaining the integrity of the process of [and the capacity for] self-organisation', over time and under a variety of environmental conditions or contexts. A major research goal is to refine these concepts and methods of analysis and measurement (Costanza *et al.* 1992, Wu & Loucks 1995, Rapport *et al.* 1998, Ernest & Brown 2001, Müller 2003; see also Chapter 2). A major challenge today is to experimentally and practically apply both concepts – health and integrity – to natural and socio-ecological systems at both regional and global scales (Crabbé *et al.* 2000).

A relevant point can be made here about health, and it is one that may help in communication between scientists and non-scientists, and among biologists, social scientists, physicians and psychologists. Health for humans can be defined as the absence of known diseases, parasites, etc., and a general sense of well-being. Of course, ecologists cannot readily use the notion of 'a *sense* of well-being' when considering the health of ecosystems (but see Kay & Regier 2000, pp. 133–4). Yet, as we have shown already, an essential element of both human and ecosystem health is resilience or adaptive capacity. From this it follows that reason 3 of why to restore, above, is by far the most important, since by focusing on protecting, managing and, where necessary, restoring natural capital, we have the best chance of maintaining long-term, intergenerational resilience and overall health of life-supporting ecosystems. In this light, the notion of ecosystem or natural capital should be understood as referring not only to economic goods and services (reason 2), and not only to biodiversity (reason 1), but also to health, both of life-supporting ecosystems and of all living creatures, including human individuals and societies. Popular wisdom around the world, and the common sense that comes with age and maturity, both tell us that a person's greatest and most abiding wealth is in fact his or her **health**. From the preceding discussion however, it should be clear that restoration ecology must be concerned with both the integrity and the health of ecosystems.

16.2.6 Emerging ecosystems

The concept of emerging ecosystems – as explained in Milton (2003) and Lugo and Helmer (2004) – deals primarily with land units partially or totally transformed by people for agricultural, commercial or industrial use that have an uncertain, and above all unprecedented, ecological and socio-economic trajectory. Indeed, 'ecosystems that develop after changing social, economic and cultural conditions, so change the environment that new biotic assemblages colonize and persist for decades with positive or negative social, economic and biodiversity consequences'. For information on emerging ecosystems from the United Nations Educational, Scientific, and Cultural Organization (UNESCO), see www.unesco.org/mab.

It is this concept of emerging ecosystems that led us earlier to suggest that ecosystem diversity may actually be increasing worldwide, and it will be central to our discussion in the remaining sections. However, even if emerging ecosystems increase habitat diversity and regional species diversity let us not forget that they probably reduce species diversity at the global level. As Sax and Gaines (2003) point out, we have not created any new species, we have caused the extinction of many, and those that remain have been generously spread around the planet. The outcome of this mixing and matching is still obscure and a major issue for restoration science and practice.

We propose to consider ecosystems receiving restoration efforts as belonging to the large and growing number of 'emerging ecosystems'. Designer ecosystems belong there too, even if they are usually not designed to be autogenic or self-organizing. Socio-ecological systems (see below) can be of many sorts, depending on many factors, but in today's changing world they should be seen as emerging too. In all emerging ecosystems, we argue, a critical issue

is that of ecological models or references, and from where they are taken or constructed. For complex adaptive ecosystems there is no fundamentally right state or community. But for both scientific and socio-cultural-economic reasons, some states or better still trajectories are certainly preferable to others, especially with regard to socio-ecological systems.

16.2.7 Socio-ecological systems and their relevance to the setting of objectives and references

We all know it is literally impossible to fully restore 'original' ecosystems of the past. If you play the tape of life back, so to speak, it will never come out the same. However, if in a given situation the goal is restoration – as opposed to large-scale gardening, landscaping or fully engineered, designer ecosystems – for reasons 1–3 listed above, then one useful approach lies in carefully selecting an ecosystem or landscape of reference to serve as a standard or yardstick for the evaluation of restoration endeavours (Aronson *et al.* 1993a, 1995, Aronson & Le Floc'h 1996a, 1996b). But how can one select an **ecosystem of reference**? This topic, which was discussed in Chapters 1 and 2 of this volume appears in a new light when we introduce the notions of emerging ecosystems and socio-ecological systems. If an historical orientation to references is adapted – as is most often the case – reconnecting some sort of ecological and cultural continuity, or reconsolidation with some stage or period of the local past, ranks high on the list of desiderata. But how do we know to what extent the period we are choosing, say the 16th century for the Americas, or the 18th century for Australia, was itself the most natural, or desirable, in terms of biodiversity, economic productivity or natural-capital value, the three reasons or motivations for ecological restoration given above? Quite simply, we do not and, probably in most cases, cannot. In Chapter 1 this was called the moving-target syndrome.

The Canadian anthropologist and past chairman of SER International, Eric Higgs has argued (Higgs 2003) that an historical 'range of variability' approach should be adopted in situations where reference conditions are difficult to discern or historical information is mostly absent. The goal in this case would be to bring an ecosystem back within an historical range, 'or closer to a long-term average' than is currently the case, as a result of ill-advised over-exploitation or ill-management by previous generations. The alternative approach is that of considering a restored ecosystem as a special category of 'emerging ecosystems' and therefore to relax our concern with historical authenticity and strict references to focus instead on pragmatic issues and what Clewell (2000b) – also a founder and past chairman of SER International – has called 'natural authenticity'. Many authors seem to be converging on this approach, while others remain adamantly in favour of historical references (see Egan & Howell 2001). This issue will be discussed further in Chapter 17, but for now we simply re-iterate that answering the question, why restore?, first of all is the first step in deciding on the best way to proceed.

An under-studied axis of comparison: European versus neo-European perspectives

Before continuing, and completing, our list of eight hot topics, let us pause here to consider if there is anything special or different about restoration ecology in Europe? After all, nearly all European ecosystems and landscapes are emphatically not wilderness or pristine ecosystems of any sort such as North American and Australian conservationists, for example, are largely concerned with, even though they seldom represent more than 5–10% of their national or regional territories. Indeed, ecosystems and especially landscapes in Europe are almost all strongly cultural, or semi-cultural, and should probably best be described as long-standing, but still evolving, socio-ecological systems, given the predominant influence of human activities on vegetation, wildlife and nutrient, energy and water flows over many millennia (see Chapter 1). This contrasts sharply with the situation in the USA, Canada and other 'neo-Europes' (Crosby 1986); that is, Australia, New Zealand and parts of southern Africa and South America, those temperate (or extra-tropical) regions of the planet where, since the 16–19th centuries, there has been imposed the form of overwhelming human domination that Alfred Crosby called Europeanization of landscapes, resources, biota and of course ecosystems. In evolutionary and ecological terms, this contrast is an important one that

merits our attention here, and indeed merits much more comparative research (Hobbs & Hopkins 1990, Hobbs & Saunders 1992, Hall 1998, 2000, Aronson *et al.* 2002). Nevertheless, in Africa, south-east Asia and Oceania, human influence has had a major impact not just for millennia but for 100,000 (Australia) or millions of years. Europeans should keep this in mind when evoking a possible specialness of the European perspective. Still, there is a major difference of intensity, as indeed in the contrast between the entire geopolitical north, and the geopolitical south, a difference called the 'digital divide' in a pivotal paper by Kates *et al.* (2001) on sustainable development and sustainability science (see section 16.2.8 and Chapter 17).

Returning to the question of reference areas and ecosystems, we note that it may be easier in Europe, most of Asia and tropical Africa, than in any of the neo-Europes – or the few remaining remote parts of the world where the 'human footprint' (Janzen 1998) is still relatively shallow – to accept the idea that a full return or restoration to the past is simply not possible. This idea is expressed in Fig. 16.1, which revisits Fig. 1.6, in light of global ecological flux (global and demographic changes in the so-called Anthropocene Era), and the growing preponderance of emerging ecosystems. The notion of restoration to the future implies a paradox, but this paradox is an inherent part of the problem of ecological restoration, and of restoration ecology as well. Note that degraded systems in Fig. 16.1 differ from emerging ecosystems in that their autogenic succession is generally more or less blocked. The feasibility of a return to the past is questioned, as in Chapter 1, because of the moving-target syndrome. Restoration (or rehabilitation), or designing ecosystems for the future, will require jump-starting investments and sustained investments, and also dedication, on the part of local populations and institutions. But is the representation in this model really useful in our current state of planetary flux? At least it keeps us aware of the need to extrapolate our limited current knowledge of biocomplexity towards future developments, rather than to seek to restore the past in a strict sense. Figure 16.2 recasts the model in a different light. The notion of sustainability includes both ecological and socio-economic aspects and challenges restoration ecology to become a multidisciplinary science, or a key

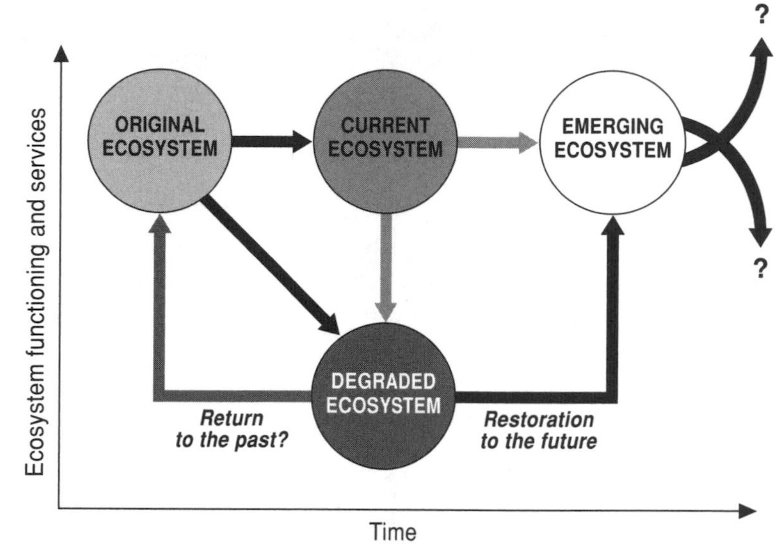

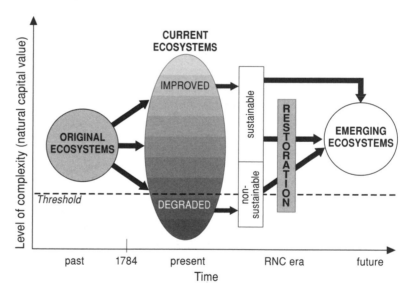

Fig. 16.1 A new restoration ecology dynamics model in a world of emerging ecosystems, ever-deeper human footprint and global ecological flux. Various possible trajectories exist which are not linear, nor even entirely predictable.

Fig. 16.2 A new approach to ecosystem development in an unprecedented era when humans dominate all planetary ecosystems. The date 1784 was suggested by Paul Crutzen (2002) as the year marking the beginning of the Anthropocene Era, as in that year James Watt completed the design of the double-action steam engine that made possible the so-called Industrial Revolution and also greatly accelerated the global process of urbanization.

component of what is frequently called sustainability science. Human transformation of ecosystems can have various effects on natural capital value, and restoration of some ecosystems may not be deemed necessary. Yet, in the current era of massive change and transformation, virtually all human-dominated or managed systems should be seen as emerging and as having many possible future trajectories.

16.2.8 Linkages, integration and sustainability science

The last hot topic we wish to discuss is the most important of all, for society, and the most challenging for researchers as it implies, and requires, transcending traditional and academic boundaries. As noted already, natural capital value (Fig. 16.2) is a notion

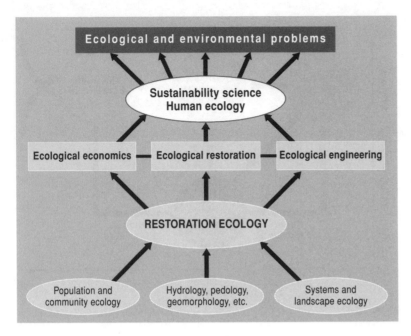

Fig. 16.3 Numerous scientific disciplines feed restoration ecology which in turn aids in the orientation of applied endeavours and enterprises confronting specific ecological and environmental problems at all levels of complexity.

corresponding to the sum of the values of all the goods and services that ecosystems provide to individuals and to society. For those goods and services that are traded in explicit markets, such as minerals, timber, fuelwood, oils, resins, wild fruits and berries, medicinal plants and mushrooms, it is possible to obtain their market prices. Indeed, to calculate their respective monetary values at a given time is often a straightforward exercise, since market prices provide a great deal of the necessary information. However, there are many other ecosystem goods and services, for example clean air, erosion protection, carbon sequestration, aesthetic amenities and recreational facilities, for which well-defined property rights do not exist, which are not traded in explicit markets and, therefore, do not have explicit market prices. This is a crucial challenge for the emerging integrative or sustainability science, mentioned above. Restoration ecology and ecological restoration are vital components of this new approach or paradigm, but they must be integrated with conservation and long-term regional and 'planetary ecological management' for well-being and sustainability (Prescott-Allen *et al.* 1991, Gadgil 1995, van Eeten & Roe 2002, Carey 2003, 2004).

In Fig. 16.3 we summarize these ideas schematically. Here it becomes clear which role restoration ecology has to play. Ecological theories, as treated in Chapters 2–6, are now confronted with problems they did not have to confront when they were first developed. Large datasets are available, and significant progress has been made in understanding alternative stable states, species interactions, cross-scale effects and metapopulation dynamics, to mention only a few. The challenge now is to make such knowledge applicable to the task of predicting and piloting the development of future, mostly non-equilibrium, emerging ecological systems, not only in terms of statistical probabilities, but in specific cases and in a context of pragmatic problem-solving and sustainability science. This approach indicates the way forward.

16.3 The way forward

Restoration ecology is a branch of science that is not only in the forefront of theoretical and applied ecology today, but also at the cusp of integrative sustainability science, ecosystem science and human ecology (Hardin 1985, 1991, Kates *et al.* 2001, Müller 2003). Its practitioners, researchers and students get involved, in the field, in both private and especially public domains of action, problem-solving and intervention. They try to communicate, at national, continental and global levels with decision-makers, legislators and industries to help them take ecological and other quantitative data provided by scientists into account when taking decisions on major ecological and environmental issues (Hollick 1981, Palmer *et al.* 2004). In addition to the three reasons to restore given at the outset of this chapter, we must recognize that we – as a global society – are quite simply running out of fertile land, potable water, renewable energy and other vital resources. We need to find new and sustainable ways to coexist with other species within the limits of a finite, human-dominated world (Hardin 1993, Sanderson *et al.* 2002, Postel & Richter 2003, Rosenzweig 2003). We need to learn to live within limits and in recognition of the Earth's and each specific bioregion's carrying capacity (Arrow *et al.* 1995). We also need to control – or reduce – our 'ecological overshoot' (Wackernagel *et al.* 2002) and instead adopt a truly long-term stewardship approach to our planet's limited resources, paying careful attention to ecosystem resilience and landscape health.

Looking backward, restoration ecology has grown out of theoretical and academically applied ecology and conservation biology, and still relies heavily on them. Looking forward, it is part of an emerging human ecology, an eco-economy paradigm that builds on truly ecological economics, and is the vital challenge of securing sustainability for our burgeoning global society and the biosphere. All three of the reasons for ecological restoration cited at the outset – biodiversity, sustainable productivity and long-term stewardship of natural capital – as well as the over-arching problem or challenge mentioned just above, should drive societies towards much-increased investment in research, development and training in restoration ecology, in the broadest and widest context of the global pursuit for survival and sustainability. In light of the recent report from the Ecological Society of America (Palmer *et al.* 2004), we hypothesize that in many or most situations **restored ecosystems** offer better value and greater promise for the long run than newly invented designer systems. The same is probably true of the aggregates of ecosystems we recognize as landscapes. Both types are necessary, no doubt, but we would suggest placing top priority on the former.

The ecological and environmental problems and challenges we face today are planetary, local and regional all at the same time (Dasgupta *et al.* 2000). The ultimate framework in which to teach and study ecological restoration is that of our lonely planet under pressure and human ecology in the immediate locale where teaching and study are taking place. Restoration ecologists and practitioners in Europe, and elsewhere, should take up the challenge to play an active, pragmatic role in trans-disciplinary research, in actual restoration projects, and in advising the public and public policy-makers in the critical years and decades to come.

Acknowledgements

The authors warmly thank James Blignaut, Andy Clewell, Sue Milton and Edouard Le Floc'h for long-standing collaboration and especially for their constructive comments on earlier drafts of this chapter. We also thank Juli Pausas and Ramón Vallejo for helpful discussions and comments during the chapter's long period of gestation.

17

Challenges for the practice of ecological restoration

James Aronson and Ramón Vallejo

17.1 Introduction

In preceding chapters, a wide range of ecological contexts and problems have been encountered, representing most of the major ecoregions of Europe. One important context, however, has not been treated, that of spreading peri-urban areas growing like fractal fairy rings around big towns and cities all across Europe and, indeed, around the world. A compelling need clearly exists for concerted conservation, management and restoration (CMR) programmes specifically for the peri-urban areas where so many different needs and actors come into play. In this chapter we shall illustrate the context of peri-urban demands for restoration, and CMR, through the example of a study carried out in conjunction with massive urbanization and engineering actions undertaken in a river delta near to Barcelona. This case is a good example of the effects of intensive urban development creating pressure on the remaining natural or semi-natural ecosystems in close proximity to urban centres. For the development of new human ecological paradigms, such settings are instructive and useful.

Additionally, a very different setting or situation will be considered in this chapter, that of compliance with international conventions developed to protect environmental and social services derived from natural capital.

In this chapter we will also address the challenge to develop integrative techniques for diagnosis, monitoring, evaluating and adjustments or so-called fine-tuning of projects undertaken in various scales of space,

bio- and socio-complexity, and time. These issues have been discussed in previous chapters in specific contexts. Here we identify some of the key theoretical issues and methodological aspects that need special attention, giving special emphasis to the need to balance and integrate ecological and socio-economic criteria, constraints and desiderata within both the planning and evaluation techniques. Specifically, we discuss further the issues related to references, and the need to conduct ecological restoration with an integrated landscape perspective and awareness of the enormous ecological and environmental problems facing us. When we say us, once again, we refer to each country and region in the European Union (EU), and throughout the rest of our global society as well. The notions of natural capital, socio-economic ecological systems and emerging ecosystems, all of which were introduced in Chapter 16, will come into discussion here as well.

17.2 Emerging demands for restoration

In the coming years, two distinctly different contexts will emerge for most of the societal demand for ecological restoration in Europe. The first category is one of relatively small-scale, high-resolution settings, where land and other resources have huge socio-economic value and are used evermore intensively. Ecological transformation, including unequivocal degradation in some portions of these areas, has been severe, and ecological restoration and rehabilitation are and will be increasingly called for, in close co-ordination with biological conservation of specified sites, protection of natural capital and more-integrated use and management of land units allocated to supplying goods and services to the cities. As explained above, we refer to the tentacular and fractal peri-urban landscapes that are mushrooming in all countries, even as the human footprint deepens and broadens worldwide (Sanderson *et al.* 2002). In these urban–rural interfaces, people and nature meet face to face, so to speak, and conflict resolution requires unusual diplomacy, civic spirit and conceptual clarity if public as well as private interests are to be defended adequately (Riley 1998, Benfield *et al.* 2001, Postel & Richter 2003).

Concurrently, at higher scales, but lower resolution, a context now exists where international conventions on environmental issues of global concern oblige European states to address very diffuse and large-scale issues for which no clear market values have been established and no private ownership or sovereignty is overtly expressed. At these spatio-temporal scales, resources and much of the land affected have low (short-term) economic value, as defined by prevailing economic models and accepted market practices which flagrantly neglect the value of natural capital and the cost – to all of us – of its reduction or destruction. Much of the territory concerned by this set of contexts may have low or medium use and intensity of transformation/degradation and the demand for public investment and legislation is driven by such long-term, broad-scale issues as biodiversity loss, climate change and desertification. But at least some decision-makers in the European Union clearly recognize that our current situation is one of ecological 'overshoot' (Wackernagel *et al.* 2002) and multiple crises, and that it requires our serious and immediate attention and engagement at the planetary and all other scales.

In all the above-cited international issues, ecological restoration and rehabilitation are expected to play a critical role, in one way or another, but at present that role is far from being explicit in the relevant texts. Furthermore, an administrative and legalistic device known as mitigation often permits private interests, and some public administrations, to side-step the critical issues and provide window-dressing instead of meaningful restoration, conservation and management. Yet, taking the long view, European and indeed global society attitudes and policies concern-

ing nature have been changing steadily since the 1940s at least, when the first international treaties aimed at the protection of biological resources were signed; for example, the first whaling convention of 1946, and the Ramsar Convention on wetlands established in 1971 (www.ramsar.org). In 1972, the first United Nations (UN) conference on environmental matters took place in Stockholm and following the disastrous famines in Sahelien Africa in the late 1960s the very widespread menace of desertification was recognized at the Nairobi Conference in 1977. In 1994, a United Nations Convention to Combat Desertification (UNCCD) was adopted by the officially affected countries (UNCCD 1994; www.unccd.int). Global organizing for transnational protection of nature, and nature's services, took further shape with the creation of the UN's World Charter for Nature (UN 1982), the Rio Declaration on Environment and Development (UN 1992a), Agenda 21 (UN 1992b) and the worldwide adoption of the Convention on Biological Diversity in 1992–3 (CBD 2003; www.biodiv.org). In 1994, the UN Framework Convention on Climate Change was adopted (UNFCCC 1994; www.unfccc.int) and the controversial Kyoto Protocol was launched in 1997 (UNFCCC 1997). All of these conventions place constraints on the contracting parties (i.e. the signatory countries) and in each nation politicians and administrators have to make tough decisions regarding balancing the short-term needs of people and society, and the general, long-term need for sustainable management, conservation and restoration of natural capital for future generations.

Land-use change is one of the major factors affecting ecosystem degradation and it should be one of the drivers in increased investment in restoration. Although land-use changes might be affected by climate change in the long run, socio-economics is the main recognized driving force for land-use change and accepted levels of intensity and transformation. The European scenarios for land-use change are quite diverse, namely World Markets, Global Sustainability, Provincial Enterprises and Local Sustainability (Parry 2000). Those scenarios yield very diverse outputs affecting emissions, land-use changes and policies, and their interactions. People's sensitivity to issues of nature conservation and restoration will inevitably evolve as well. Assuming that current trends of land use are maintained over the next two or three decades, we can expect the continuation of extensive marginal agricultural land abandonment in formerly densely populated rural areas, with intensification of agriculture in the most productive areas. The EU's Common Agricultural Policy has already strongly influenced the management and ecological trajectory of many marginal fields, for example through EEC regulation 2080/92 (EC 1992b) promoting afforestation of set-aside fields. Consequently, set-aside fields offer numerous, propitious settings for programmes and experiments in ecological restoration, in the framework of guidelines and demands of the international treaties, and the specific contexts of local priorities, risks and sensitivities.

In relation to the Climate Change Convention (UNFCCC 1994), meanwhile, the Kyoto Protocol (UNFCCC 1997) is progressively encouraging monetary market mechanisms to participate in the regulation of greenhouse gas emissions and to finance restoration or mitigation measures to enhance carbon sequestration. Climate-change scenarios are already being considered for carbon management and they certainly merit careful reflection on appropriate strategies and techniques for ecological restoration. Given the spatial variability of expected climate change, the question is how to incorporate predicted (near-)future climate scenarios in specific ecological restoration projects. Clearly, we must anticipate ecosystem and ecoregional changes and migrations, but there are compelling cultural pressures to retain and emulate contemporary (or recent historical) references (see below) in the design of ecological restoration projects. From a certain point of view, however, such an approach would represent an attempt to resist – or deny – projected climate changes. In southern Europe, for example, the question arises, should we favour the persistence of species, or species assemblage characteristic of drier climates to anticipate the predicted intensification of drought in this ecoregion? In the framework of the UNCCD Annex IV for the northern Mediterranean Region (UNCCD 2001), the European Mediterranean countries are developing National Action Plans towards combat desertification that include afforestation and reforestation projects, and other active measures to mitigate land degradation and preventing wildfires.

The ACACIA report (Parry 2000) on the potential effects of climate change in Europe suggests that policies aiming at conserving diversity and various ecosystem services should be based on detailed regional assessment, since impacts are regionally variable with regard to both climate and land-use diversity. These various projects and reports all include ecosystem-management measures, but the economic and ecological connections among conservation, management and restoration activities are generally not spelled out. In relation to the Convention on Biological Diversity (CBD 2003), for example, habitat restoration and other practical measures for protecting species diversity and eco-management at landscape scales should be specified for the (semi-)natural agricultural and urbanized peri-urban areas.

17.3 Evaluation of whole systems

In addition to the need for greater integration of CMR, there is a clear need for great coherence and connectivity among three major components of restoration itself (Tongway & Ludwig 2002).

1 Historical, geographical and ecological assessment of ecosystem or landscape degradation (How did things get this way?).
2 Technical and institutional means responsible for responding to ecosystem damage and degradation or transformation.
3 Monitoring techniques for evaluating progress towards a desired, or a satisfactory, outcome within a certain timeframe.

We would add the following to Tongway and Ludwig's list.

4 Realistic and holistic eco-economic budgeting and accounting for damage to, and reparation of ecosystems – including socio-economic systems – which represent the stock from which societies draw natural capital.

Evaluation includes both diagnosis and monitoring, and provides the retrospective and on-going data sets that not only allow programmes to be evaluated holistically, but also allow fine-tuning by managers. We present these ideas schematically in Fig. 17.1. As van

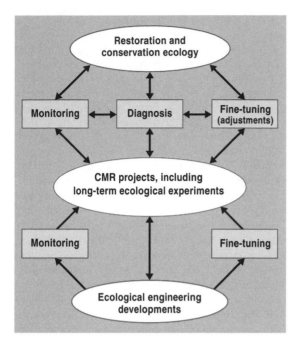

Fig. 17.1 Schematic view of the pathways and linkages among restoration and conservation ecology (top), on-going refinement of ecological engineering methods (bottom) and the central trio of conservation, management and restoration (CMR) projects and programmes. Monitoring, diagnosis (assessment) and managerial fine-tuning represent the modes of interaction between science and applied enterprises confronting specific ecological and environmental problems of all sorts (see also Fig. 16.3).

Eeten and Roe (2002) and other authors point out, ecological, engineering and resource management concerns and criteria must all be considered.

Turning now to the related issue of choice and combination of indicators, we note that there is still much to do in order to develop, test and validate practical means and criteria to evaluate success of our efforts in biological, ecological and socio-ecological, or human ecology, terms. Aronson *et al.* (1993a) and Aronson and Le Floc'h (1996a) pointed out that no single bio-indicator or 'flagship' species can provide the basis for a full, or reliable, assessment or diagnostic evaluation of the status or progress of a

restoration or rehabilitation intervention, be it conceived at the ecosystem level, the landscape level or the ecoregional level. Instead, ecological restoration projects need broad suites of relevant, reliable and complementary traits or ecosystem attributes that together describe and reflect the structural, compositional and functional dynamics of an ecosystem at any given point in time and can be monitored over time to provide a moving picture, and not just a snapshot. At the next higher hierarchical level, the same reasoning applies to the agglomeration of interactive ecosystems that ecologists label as landscapes (Aronson and Le Floc'h 1996a), except that socio-ecological or human ecological attributes must be added to the physical, pedological and biological ones. Over time, these suites of indicators should reveal ecosystem or landscape responses to new management practices, including those conceived within a CMR framework. In addition, they should help reveal when and how 'switches', 'flips' or 'threshold crossings' take place, and the dynamics of 'resistance to restoration', two important areas of active study (Hobbs & Norton 1996) where more research is urgently needed.

Obviously, we need indicators that are quantifiable, reliable and transferable, but they also should be inexpensive and cost-effective, so as not to get left out of CMR programmes for budgetary reasons. Moreover, since large numbers of attributes can rarely be monitored in a single project, it is important, early on, to identify a compact and affordable suite of the most pertinent, sensitive and reliable attributes for each experimental site, gradient or project. Ideally, they should be amenable to synchronic and diachronic comparisons, as mentioned above. Moreover, we repeat that, in the new paradigms discussed here, ecological restoration and CMR measures of diagnosis, evaluation and monitoring must address not only ecological but also social, political, juridical, economic and, in some cases, cultural aspects of ecosystem responses and resilience. They must also be simple enough to aid in clear communication transfers, and synergy for specific projects, among scientists of different backgrounds and training, and also the more numerous non-scientists, including ecosystem users, stakeholders, resource and area managers, industrialists and engineers (see Winterhalder *et al.* 2004).

17.3.1 Reference areas: pros and problems

Aronson and Le Floc'h (1996a) and Aronson *et al.* (2002) have argued for the usefulness of establishing references – areas, ecosystems, landscapes – or even a collation of reference information from various sources (see Clewell & Lea 1990, Aronson *et al.* 1993a, 1995, Egan & Howell 2001, SER 2002). This process or approach is useful, as we and others have argued, even if it is not essential to ecological restoration (see Pickett & Parker 1994, Janzen 1998, 2002). In many although not all situations there exists an array of forms, phases or avatars in which vegetation can be seen, most of which are rather removed from a pristine, undisturbed state. Accordingly, several types of **reference** and sources of reference information (White & Walker 1997) can be suitable for restoration projects, including the restoration of long-standing and still-relevant socio-economic systems, which can serve as a reference or yardstick with which to evaluate and attempt to repair damaged ecosystems today. But there are several obstacles to finding a clear and straightforward application. Firstly, in Europe, as in most parts of the world today, we often only have vestiges, and fragmented landscapes or ecological systems that could effectively serve as references for ecological restoration. Secondly, in Europe, as elsewhere, the situation is further complicated by the many layers of history and culture which render the choice of references highly arbitrary. Finally, as the reader will recall, in Chapter 16 we introduced the notion of emerging ecosystems, a notion that effectively obliges us to consider the whole issue of choice and use of reference areas in a new light. Reconciliation is possible, but the conceptual process involved is as complex as the notion of socio-economic systems. To further complicate the picture, truly cultural landscapes are prevalent, and much valued by citizens in general and many ecologists in particular (Farina 2000, Naveh 2000). Cultural landscapes can and often should provide the best reference for actual restoration or rehabilitation projects. Of course, the relative value and naturalness of a cultural landscape depends a great deal on present-day perception and may underestimate the extent to which profound and perhaps irreversible degradation processes were initiated in the past in order to produce those cultural landscapes (see Box 17.1).

Box 17.1 Food for thought

Restoration should be guided by scientific knowledge (restoration ecology principles) and accommodate social demands. As social demands are very variable in time and space, a precautionary principle would suggest that restoration actions (and especially large-scale and long-term management, conservation and land-use plans) should maximize the diversity of future management options or ecosystem uses, for example by avoiding transformations leading to pathways of no return. In other words, it is prudent to avoid irreversible processes that would require future huge investments of energy to restore the site or system if needs or priorities change at a later date. Introducing the repair cost in the business plan before initiating profound land transformations, in whatever significant units (e.g. energy, non-renewable resources, community impact indices or money), provides good perspectives; some mining companies are already progressing in this direction. The cost of the destruction or non-reversible transformation of nature should also be valued and factored in. This is the area of research and development known as ecological economics, and large numbers of books, documents and websites exist for consultation, a sample of which were cited in Chapter 16. The application of the precautionary principle is progressing nowadays in Europe; see for example the REACH (Registration, Evaluation, Authorisation and Restriction of Chemicals) proposal for a strategy on chemicals policy, from its generic recognition in the Treaty on the European Union (EC 1992c, article 130 R.2).

The first step for setting up a reference can be to create some simple models, flow charts or other graphics describing how things have become the way they are in the study area, and delineating some of the alternative states (or basins of attraction; see Anand & Desrochers 2004) that have existed, or that are possible for the systems in question (see Wyant *et al.* 1995, Yates & Hobbs 1997). In Fig. 14.2 was presented one example, showing some of the possible successional pathways (including post-fire trends) in relation to above-ground biomass of mature woody plant communities and plant species richness (in 100-m^2 plots) in eastern Spain. A much more detailed example of this kind of exercise is given in Fig. 17.2.

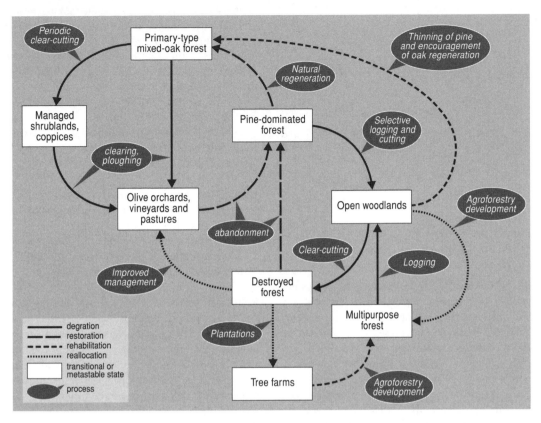

Fig. 17.2 State and transition model for a *Pinus halepensis* woodland in Mediterranean France showing degradation pathways and various possible responses. Modified from Gondard *et al.* (2003).

Note that no values are given, and no species axes or dimensions are indicated on this figure, as is typical of state and transition models. However, choices taken about which pathways to foster or favour must depend on a combination of ecological and socio-economic deliberations and negotiations, ideally with a natural-capital approach getting the attention it deserves, along with various social, ecological and cultural concerns. It is to be noted that the Society for Ecological Restoration International Primer (SER 2002; www.ser.org) states clearly that ecological restoration is intentional, but that the succession pathways labelled as restoration can either be entirely passive – that is, spontaneous regeneration – or else carefully monitored and adjusted by on-going management.

An alternative – or complement – to the idea of real, existing reference areas is provided in the SER Primer,

under the rubric 'Attributes of restored ecosystems'. When no historical reference is available or adequate, as is very often the case in Europe, this approach can help orient the ecological restoration programme. It would also apply to emerging ecosystems when present conditions do not allow the recovery of a degraded ecosystem to a known historic condition. To express this idea, and to continue the ideas given on this subject in Figs 16.1 and 16.2, Fig. 17.3 provides a schematic description of the new approach to ecological restoration we espouse, which aims to reconcile the reality of emerging ecosystems, and the growing prevalence of socio-economic systems, with the fundamental notions of ecological restoration, including the potential value of historical reference systems.

On the left-hand *y*-axis of Fig. 17.3, note that all three 'whys' for ecological restoration (see section 16.1) are combined in order to go beyond previous

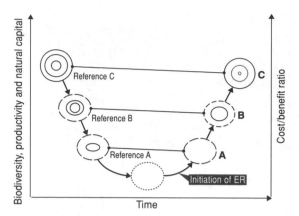

Fig. 17.3 A new approach to ecosystem restoration. See text for discussion.

schemas of this sort, wherein biodiversity, ecosystem functioning and the like were utilized separately to describe this dimension of the graph. For the sake of brevity and simplicity, the presence or absence of significant threshold-crossings (see Fig. 16.2) has been ignored here. Further, it should be noted that the ecosystem under manipulation is depicted as a simple circle, and is considered as an emerging ecosystem, moving inexorably towards the future, as mentioned above. It changes, over time, in terms of composition, rate of functioning and, above all, relative degree of ecological and socio-economic integration with its broader setting. Loss or addition of outer, concentric circles relate to the relative degree of **linkages** between the ecosystem and its surroundings. To date, the most common usage of this term, when applied to parks and other protected areas, has been geographical, physical and biological in nature. Yet the notion of linkages must be extended to a wider and more comprehensive usage, including socio-economic as well as other ecological connections.

From the point where a decision is taken to engage the process, and for each stage or phase (A, B, C, etc.) of ecological restoration, an appropriate reference can theoretically be found, or constructed from the previous or presumed time-space in which the system showed both higher degrees of linkages, and also higher levels of health and natural capital in the sense that these terms were used in Chapter 16. In this scenario, the restorationist is, like a chameleon, always look-

ing backwards, with one eye, while looking steadfastly forward with the other one! Finally, on the right-hand *y*-axis of Fig. 17.3, we present the hypothesis that the benefit/cost ratio of an ecosystem (or socio-economic system) undergoing reorientation increases with time from the initiation of restoration.

In closing this section, we suggest that despite cost and difficulty of set-up, an over-arching series of needs – for practitioners, far-sighted stakeholders and researchers alike – is to collaborate on the development and maintenance of:

- large and long-term ecological research sites;
- long-term gradient studies encompassing one or several ecotones or ecoclines; and
- demonstration and experimental sites for selection, and presentation of species selected for projects and the various field techniques available.

Ecological restoration takes much time and space, and lots of people. It is absolutely essential that local people, especially stakeholders and young people, are incorporated into the process. Similarly, appropriate spatial scales need to be embraced, with a nested hierarchical or landscape perspective to the complex problems of ecological restoration and CMR. We will address these questions further in the following section.

17.4 Missing tools

17.4.1 Specific methods and general strategies

The statement 'think globally and act locally' applies well to ecological restoration. However, ecological restoration projects are essentially local, in terms of the constraining physical and biological conditions, and especially in terms of local social and cultural linkages, or absence thereof. This fact frequently explains difficulties in communication among researchers and practitioners coming from culturally and socio-economically distinct countries. The refinement of biome-specific methods and broad strategies for fine-tuning or adjustments of a technical or engineering nature is very much required, as highlighted in almost every chapter of Part 3 of this book, and in

Fig. 17.1, above. For example, to name just a few methods for drier parts of southern Europe, we need to improve water-harvesting and -desalinization techniques, evergreen and deciduous forest tree-propagation and -establishment techniques, watershed revegetation, multi-purpose use of large and small herbivores, and various post-fire soil-stabilization measures, among others (Vallejo 1996, Vallejo & Alloza 2003; see also Whisenant 1999). Generally speaking, the issue of fine-tuning needs to be discussed and negotiated among scientists, managers and other stakeholders working together on an ecological restoration or a CMR project, since the intervention undertaken in this sense may alter the initial experimental design!

Having identified and elaborated on some of the problems that face ecological restoration as it pertains to many, but not all, socio-economic and historic situations in Europe, we shall now present a case study that illustrates the questions and problems addressed thus far in this chapter. In it, we wish especially to raise the issue of the complex urban–rural interfaces, mentioned at the outset of the chapter as being one of the key challenges for ecological restoration in the coming years. Far more than in the restoration of mires, wetlands or forests far from human population nodes, restoration actions in peri-urban and urban areas cannot avoid a plethora of socio-economic, financial and political issues – and conflicts – that must be confronted even if they are entirely reconcilable.

17.4.2 Case study: the Llobregat river delta, Barcelona, Spain

On the perimeters of growing towns and cities, the worldwide phenomenon of urban expansion is generating a kind of ecological degradation, in the urban–rural or urban–wildland interfaces where land uses often undergo rapid transformations, including abandonment of former agricultural uses and/or reallocation related to housing, commerce or transport. Concurrently, of course, human population density increases dramatically. CMR in those degraded lands can include measures ranging from simple gardening and landscaping criteria to be used more thoroughly in land-use planning and, at the other extreme, rehabilitation or full-scale restoration efforts for certain

ecosystems within the areas concerned. Indeed, these peri-urban areas constitute an important nexus for ecological restoration practice and science in the coming years, and may in fact generate a large portion of the social and political demands for ecological restoration in developed countries. Ideally, ecological restoration efforts should be planned in the context of the re-integration, reconstruction or 're-weaving' of fragmented landscapes (Saunders et al. 1993) and communities, so as to provide a positive, constructive or proactive socio-economic systems approach to the actions undertaken.

In wildland areas near large urban areas that are undergoing strong pressures for increased transport corridors and building, both conservation and restoration are increasingly engaged and performed in the context of so-called ecological compensation (van der Meulen & Salman 1996), also sometimes called mitigation, for example in North America. In theory, this process seeks to ensure the protection, improvement or restoration of high-value ecosystems and landscapes, within or in close proximity to peri-urban or protected areas that are allocated to high-intensity, urban utilization which is in most cases incompatible with ecosystem or biodiversity conservation. Therefore, where there is a strong demand on land for urban development, a trade-off is often proposed between land conservation and land occupation in nearby areas. This is done in the perhaps naïve expectation that given adequate study and planning, the two land uses will be ecologically as well as economically compatible. In many cases, these sorts of projects, if successful, represent reclamation projects or habitat enhancement rather than restoration per se. However, there are exceptions, such as that described in Clewell (1999).

The problem

The Llobregat river delta has been deeply affected by centuries-old, intensive agricultural activity. Notably, the coastal wetlands were almost entirely drained during the 19th and early 20th centuries, and industrial development has proceeded rapidly since the 1960s to the present day. However, by the early 20th century it was recognized that dune stabilization would be needed to protect nearby fields from unwanted coastal sands. Dune fixation was thus

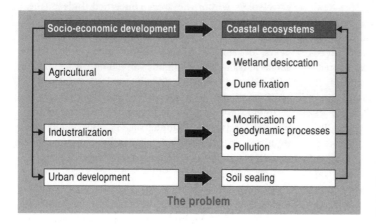

Fig. 17.4 Schematic representation
of major problems in the study case.
Original from N. Abad.

attempted through *Pinus pinea* (stone pine) plantations. Nowadays, however, those pine stands show practically no regeneration, and are declining rapidly in the first coastal line. More recently there has been an ever-growing development of tourist facilities, which contributes to pressures and conflicts over land, water and energy use (Fig. 17.4). In addition, the redevelopment of the harbour of Barcelona has led to the initiation of a large engineering project to deviate the Llobregat river in conjunction with the enlargement of the neighbouring international airport.

Since the 1960s, the enlargement of Barcelona's harbour, and dam construction along the Llobregat river watershed, have reduced the transport of sediments both from the coastal current and from the river's tidal floods. This in turn has caused regression of the coastal line. Industrial development in this period has increased air pollution and seawater pollution, especially affecting the production of aerosols rich in surfactant substances that injure coastal vegetation (Bussotti *et al.* 1995). Increased industrial and agricultural demand for fresh water has led to intensified groundwater pumping and concomitant saline intrusion. Tourist development has led to numerous secondary housing complexes in the fixed dunes areas, and the occupation of the first dune line by campsites, all of which modify the original topography and depend on the use of exogenous soils from inland areas, and various inert paving materials, which replace the original sands (Fig. 17.5). The original pre-disturbance ecosystems and communities of the coastal areas

were already absent from the site at the time that the study described in Box 17.2 (page 244) was undertaken. Instead, various cosmopolitan ruderal species were abundant, for example *Chenopodium ambrosioides*, *Portulaca oleracea* and *Conyza bonariensis*, as were exotics like *Chloris gayana*, *Carpobrotus edulis*, *Cortaderia sellowiana* and *Myoporum laetum*. Some of these are proving to be very invasive.

Restoration and rehabilitation procedure

The first step in the restoration process was to artificially reconstruct a dune topography. Unwanted soil materials were eliminated, and inter-dune depressions were created by excavation, providing sand material to build up a first dune line behind the beach. A previous study of water-table fluctuations helped establish the appropriate excavation depth corresponding to the minimum water-table depth. To reconstruct dune-formation dynamics, sand traps were installed using reed (*Arundo donax*) fences oriented perpendicularly to the dominant winds.

Recreation of coastal ecosystems along the transect was the next step. For each of the five ecosystem types recognized, species were selected according to habitat conditions, reference ecosystems and seeds/seedling availability. Most of the key species in the various habitats were re-introduced as they were no longer present in the vicinity of the site. The goal was to enhance and, if possible, accelerate secondary succession. Trying to advance inter-dune depression

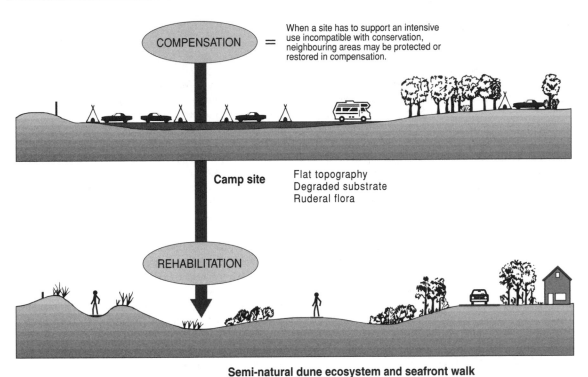

COMPENSATION = When a site has to support an intensive use incompatible with conservation, neighbouring areas may be protected or restored in compensation.

Camp site   Flat topography
Degraded substrate
Ruderal flora

REHABILITATION

Semi-natural dune ecosystem and seafront walk

Fig. 17.5 Compensation undertaken in coastal ecosystems to transform a former campsite into a restored dune ecosystem integrated in the seafront pedestrian area and urbanization programme. Original from N. Abad.

dynamics, including siltation and salinization in the medium term, two *Juncus* species and various graminoid species with different tolerance to flooding and salinization were used, including *Scirpus maritimus*, *Scirpus holoschoenus* and *Saccharum ravennae*.

Most plants were introduced in the form of transplanted seedlings cultivated in specialized nurseries, after a period of experimental trial. Some seeds were specifically collected for the project in nearby areas. Seeding was carried out for some herbaceous species to reinforce the transplantings. No post-planting care was provided, and nor was irrigation applied. Only labelling and fencing of the planted areas was undertaken.

Global evaluation: ecological and social criteria

Four years after the dune-system reconstruction, a general rehabilitation of the various habitats was considered to be successful, with some limitations. Key herbaceous species were established at high rates and most were expanding naturally (naturalizing) throughout the area. With the exception of *Pistacia lentiscus* and *Tamarix* spp., woody shrubs and pines showed poor establishment rates as a result of water stress and aerosol damage in the growing stems. The use of biodegradable tree shelters significantly improved results for these species. The construction of a rigid pathway along the top of the first dune system produced the almost immediate fixation of that dune, thereby interrupting its natural dynamics. This pathway was a requirement of the municipality for public access and use of the area. It is a good example of the kind of conflict that often occurs, and need for compromise, between ecosystem rehabilitation and social or community uses in peri-urban areas.

Trampling is one of the major disturbances for dune reconstruction in densely visited areas (Andersen 1995) that represent a limiting factor, and negotiating point, in finding the compatibility of social use and

Box 17.2 Llobregat river delta project

The project

After the abandonment of a campsite area in the 1990s, the Municipality of Gavà proposed to 'recover and rehabilitate' the seafront for public use. This is an area of high pressure along the beachfront, especially during the summer months and during weekends all year round. The mechanism adapted to promote and fund the project was 'ecological compensation'. The actors were the municipality and private companies interested in urbanizing various parts of the pine woodland area. Natural-capital enhancement was in this case supported by the municipality. A consortium of companies was authorized to urbanize and develop residential housing in a large part of the fixed dunes area and, in exchange, they took care of the cost and realization of restoration and rehabilitation in the remaining portion of the seafront area. This compensation was to consist of two components:

1 restoring sand dune ecosystems and
2 creating a pedestrian, recreational area (*Paseo maritimo*; seafront walk).

The urbanization project was designed by an architect with the support of restoration ecologists from the University of Barcelona.

Ecological objectives

First, pine woodland in the fixed dunes area was partially cleared in order to provide room for development and to promote the growth and health of the remaining trees, preserved in patches among the housing areas. Secondly, to restore the original sand-dune ecosystems by enhancing geodynamic dune-formation processes, efforts were made to recreate the original topography and to re-introduce the key plant species of the various characteristic communities. As mentioned, relevant reference communities or ecosystems were locally scarce and fragmentary. Nevertheless, from the few well-preserved dune-system remnants existing along the Spanish coast (De Bolós 1967), a beach-to-inland transect was set up to include the following ecosystems and plant communities (Fig. 17.6).

- First dune line, moving sands forming dune embryos, characterized by species supporting extremely dry and mobile substrate, like *Ammophila arenaria*.
- Fixed dunes with chamaephytic vegetation.
- Inter-dune depression with occasional flooding.
- *Tamarix africana* and *Tamarix gallica* woodlands colonizing wet soils not affected by flooding.
- Sclerophyllous shrublands, juniper woodlands and introduced pine woodlands in the fixed dunes more distant from the seaside.

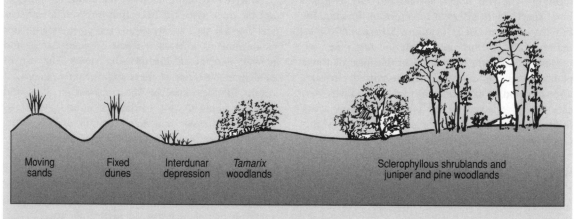

| Moving sands | Fixed dunes | Interdunar depression | *Tamarix* woodlands | Sclerophyllous shrublands and juniper and pine woodlands |

Fig. 17.6 Ideal transect of dune ecosystems taken from neighbouring reference areas. Original from N. Abad.

ecosystem restoration or rehabilitation. Although pathways were constructed to avoid trampling in the rehabilitated areas, some areas were affected by trampling, especially the areas of dune formation in the beach. This fact, together with the early fixation of the first constructed dune line, has limited the expansion of that habitat. Social use of the seafront walk has greatly increased, and the municipal authorities have a very positive evaluation of the project. Social perception of the rehabilitated ecosystem and the designed landscape is probably only partial, as dissemination of the natural values of the dune ecosystems has not been developed fully.

This project provides an example of an attempt to introduce ecological restoration in peri-urban areas. Advancing in this line would require further valorization of the ecological aspects of the project, both in the project-design phase and in the public perception through dissemination and sensitization campaigns. Ecological economics assessments should also be undertaken.

17.5 Concluding remarks

For restoration ecologists, conservationists and environmentalists in general, this is a good time for serious reflection about the world, and the role of people in the natural world. Not only are we entering a new millennium, we now see that we are more than two centuries into a new and unprecedented geological period called the Anthropocene Era (Crutzen & Stoermer 2000, Crutzen 2002). In this context, restoration ecology and ecological restoration form an essential component of what we may rhetorically call our survival strategy, and which bears the more precise term **sustainability science**, wherein nonscientists and scientists work together to imagine, develop, test and apply new methods, tools and approaches to the enormous challenges ahead. This new, integrative, broad-based science brings together the natural sciences, social sciences, economics and environmental technologies and engineering in the hope of solving problems pragmatically and promoting more healthy, durable and equitable socio-ecological systems (Palmer *et al.* 2004). (Note that we follow the definition of sustainability proposed by the US National Research Council (NRC 1999), 'meeting

human needs while conserving the Earth's life support systems and reducing hunger and poverty'.) In Chapter 16, we argued that the goal and indeed the crucible of sustainability science must be long-term, large-scale projects combining CMR. We note that the World Wildlife Fund, in its Forest Landscape Restoration programme, has adopted this same approach. The International Union for the Conservation of Nature, the International Tropical Timber Organisation and other large organizations and consortia also seem to be moving towards the same conceptual approach. It remains to be seen how and if legislative bodies at national and international levels put the CMR into practice.

For ecosystem managers and scientists, and environmental polytechnicians now in training, the challenge will be to achieve greater integration of restoration ecology models, results and principles in the actual practice and contracts carried out by the growing number of small- and large-scale operators in the industry of ecological restoration. Very often, at present, restoration (in the strict sense) is only superficially or nominally applied in projects labelled as ecological restoration or rehabilitation. This results in unspecific greening, or revegetation, where little or no reference is sought or made to pre-existing ecosystems or even plant communities, and little reflection is given to how things got the way they are and in which direction we would like to see ecosystems and landscapes moving. Often, very little systems-approach thinking or planning goes into the landscape design. Worse still, little or no rigorous evaluation, nor even any monitoring, is undertaken in many of these projects; as a result, useful feedback is few and far between. This limits the possibility of anyone having access and applying results, or lessons learned, in other projects elsewhere. Opportunities for testing theories and models are also lost to researchers and students – the decision-makers of tomorrow.

Strengthening ties and promoting feedback between the scientific community working in restoration ecology and the restoration industry is thus needed urgently, throughout Europe and globally. Great strides are being made, as reflected in the burgeoning academic and popular literature on ecological restoration, and in the largely unpublished results of innovative managers and eco-engineers. As suggested above (see Fig. 17.1), we now need to move towards the real implementation within ecological restoration projects of the already available scientific and technical knowledge. Concurrently, real-world projects need to be better designed and managed so as to provide reliable data for further research and development. Within an integrated CMR and sustainability paradigm, social scientists, ecologists, engineers, ecological economists and non-scientists of all sorts must now pull together (i) to face up to the risks and hazards of the 21st century, with its gigantic human footprint and growing ecological overshoot, (ii) to fine-tune our management of ecosystems and (iii) to repair some of the most critical ecological and environmental damage done in the past.

In Europe as elsewhere, ecological restoration has no hope of being successful unless it respects the needs of people and wildlife; that is, of the rest of nature (Prescott-Allen 1991, Rosenzweig 2003). This position – and the practical considerations it leads to – should be considered and reformulated not only in strictly biological, species-to-species terms, but also in ecological, socio-economic and cultural terms as well. In Europe, where socio-ecological systems predominate (see Chapter 16), this point should also be relatively easy to communicate, given that managed or human-dominated landscapes and ecosystems have existed there for so many centuries and human generations. We must recognize however that the idea of a dichotomy existing between people and nature is deeply embedded in most cultures and will take time to be modified and evolve into something new. Growing recognition of the fact that ours is an era of profound transformations characterized, among other things, by a growing predominance of 'emerging ecosystems' (Milton 2003) and socio-economic systems, may fasten and facilitate the kind of trans-disciplinary collaborations that will be needed to face the huge environmental and ecological challenges ahead.

When natural-capital, emerging ecosystems and socio-economic systems concepts are adopted, then, as Carpenter et al. (2001) argue, we will see our way to managing our life-supporting ecosystems 'from the inside', as stewards of evolving systems to which we belong and for which we are now fully responsible. This way of thinking should also help contribute to greater international cooperation, a *sine qua non* of effective ecological restoration, and eco-economic

conservation and management of ecosystems. Case by case, robust approaches to simultaneously addressing eco-economic and environmental concerns must be found (Costanza & Daly 1992, Prugh et al. 2003) while new environmental laws and public policies will be required. These measures and jurisprudence should be designed to ensure application of the most equitable, appropriate and cost-effective means to protect long-term, public interests, even while permitting private commercial and industrial interests to operate (Hawken et al. 1999, Heal 2000, Brown 2001, inter alia). Indeed, it appears increasingly clear that private enterprises have an important role to play in this arena (Benfield et al. 2001, Cunningham 2002, Daily & Ellison 2002), but governments will certainly need to adjust taxes, penalties and incentives so as to convince more corporations to assume their natural civic leadership role.

In Chapter 16, we noted three primary reasons to restore. While recognizing the validity of biodiversity conservation and sustainable productivity as objectives, we argued for a re-orientation of ecological restoration and, consequently, of restoration ecology towards truly integrated, transdisciplinary efforts aimed at the restoration of **natural capital** (Jurdant et al. 1977, Costanza & Daly 1992, Cairns 1993, Clewell 2000a, Milton et al. 2003). This enlarged third answer to the question of why restore encompasses the first two reasons and also incorporates the emerging notions of **ecosystem health** and **integrity**, especially in terms of resilience, self-organization and biocomplexity. Ecosystem goods and services are the dividends provided by a stock of natural capital and, theoretically, they are best assured by systems that are functioning in a healthy and sustainable fashion, in the multifunctionallity senses of health, integrity and sustainability, outlined above. Furthermore, along with Daly and Farley (2004), we note that a profound conceptual paradigm shift is required at societal levels, as related to the limits to growth, the search for sustainability and natural capital as the ultimate foundation of all economies and all societies. The science of restoration ecology and the practice of ecological restoration will undoubtedly be called upon to play major roles in the coming decades as global society comes of age and, hopefully, evolves nearer to the ideas of sustainability and social justice; this, indeed, is the new frontier.

Acknowledgements

The notes on dune restoration in a CMR context in the Llobregat delta near Barcelona were prepared in collaboration with Nuria Abad (University of Barcelona) who, together with R.V., conducted a study in this area. We also warmly thank Andre Clewell and Sue Milton for their helpful comments on the manuscript. Section 17.4.2 is based on the project reports provided by N. Abad.

References

Abbott, R.J. (1992) Plant invasions, interspecific hybridization and the evolution of new plant taxa. *Trends in Ecology and Evolution* **7**, 401–5.

Aber, J.D. and Melillo, J.M. (1991) *Terrestrial Ecosystems.* Saunders College Publishing, Philadelphia.

Abrams, P.A. (1996) Evolution and the consequences of species introductions and deletions. *Ecology* **77**, 1321–8.

Achermann, B. and Bobbink, R. (eds.) (2003) *Empirical Critical Loads for Nitrogen.* Proceedings of the expert workshop held at Bern, Switzerland, 11–13 November 2002. Report of the Swiss Agency for the Environment, Forests and Landscape SAEFL, Bern.

Adema, E.B., Grootjans, A.P., Petersen, J. and Grijpstra, J. (2001) Alternative stable states in wet calcareaous dune slacks in the Netherlands. *Journal of Vegetation Science* **13**, 107–14.

Agnew, A.D.Q. (1997) Switches, pulses and grazing in arid vegetation. *Journal of Arid Environments* **37**, 609–17.

Albaladejo, J., Martinez-Mena, M., Roldan, A. and Castillo, V. (1998) Soil degradation and desertification induced by vegetation removal in a semiarid environment. *Soil Use and Management* **14**, 1–5.

Alcántara, J.M., Rey, P.J., Valera, F. *et al.* (1997) Habitat alteration and plant intra-specific competition for seed dispersers: an example with *Olea europaea* var. *sylvestris. Oikos* **79**, 291–300.

Allen, E. (1995) Restoration ecology: limits and possibilities in arid and semiarid lands. In: *Proceedings of the Wildland Shrub and Arid Land Restoration Symposium*, pp. 7–15. USDA Forest Service INT-GTR-315, Washington, DC.

Allen, E.B. and Allen, M.F. (1990) The mediation of competition by mycorrhizae in successional and patchy environments. In: *Perspectives on Plant Competition* (eds. J.B. Grace and D. Tilman), pp. 367–89. Academic Press, London.

Allen, H.D. (2001) *Mediterranean Ecogeography.* Prentice Hall, Harlow.

Allen, J.R.L. (2000) Morphodynamics of Holocene salt marshes: a review sketch from the Atlantic and southern North Sea coasts of Europe. *Quaternary Science Review* **19**, 1155–231.

Allen, T.F.H. and Hoekstra, T.W. (1987) Problems of scaling in restoration ecology: a practical application. In: *Restoration Ecology* (eds. W.R. Jordan, III, M.E. Gilpin and J.D. Aber), pp. 289–99. Cambridge University Press, Cambridge.

Allen, T.F.H. and Hoekstra, T.W. (1992) *Toward a Unified Ecology.* Columbia University Press, New York.

Allendorf, F.W. and Leary, R.F. (1986) Heterozygosity and fitness in natural populations of animals. In: *Conservation Biology: the Science of Scarcity and Diversity* (ed. M.E. Soulé), pp. 57–76. Sinauer Associates, Sunderland, MA.

Alloza, J.A. and Vallejo, R. (1999) Relación entre las características meteorológicas del año de plantación y los resultados de las repoblaciones. *Ecología* **13**, 173–87.

Al-Mufti, M.M., Sydes, C.L., Furness, S.B., Grime, J.P. and Band, S.R. (1977) A quantitative analysis of shoot phenology and dominance in herbaceous vegetation. *Journal of Ecology* **65**, 759–91.

Alpert, P., Bone, E. and Holzapfel, C. (2000) Invasiveness, invasibility and the role of environmental stress in the spread of non-native plants. *Perspectives in Plant Ecology, Evolution and Systematics* **3**, 52–66.

Ammer, U. and Pröbstl, U. (1988) Erstaufforstungen und Landespflege. *Forstwissenschaftliches Centralblatt* **107**, 60–71.

Anand, M. and Desrochers, R.E. (2004) Quantification of restoration success using complex systems concepts and models. *Restoration Ecology* **12**, 117–23.

Andersen, U.V. (1995) Succession and soil development in man-made coastal ecosystems at the Baltic Sea. *Nordic Journal of Botany* 15, 91–104.

Anderson, E. (1948) Hybridization of the habitat. *Evolution* 2, 1–9.

Andersson, E. and Nilsson, C. (2002) Temporal variation in the drift of plant litter and propagules in a small boreal river. *Freshwater Biology* 47, 1674–84.

Andersson, G. (1988) Restoration of Lake Trummen: effects of sediment removal and fish manipulation. In: *Eutrophication and Lake Restoration: Water Quality and Biological Impacts* (ed. G. Balvay), pp. 205–14. Thonen-les-Bains.

Andrén, H. (1994) Effects of habitat fragmentation on birds and mammals in landscapes with different proportions of suitable habitat: a review. *Oikos* 71, 355–66.

Andresen, H., Bakker, J.P., Brongers, M. *et al.* (1990) Long-term changes of salt marsh communities by cattle grazing. *Vegetatio* 89, 137–48.

Angevine, M.W. and Chabot, B.F. (1979) Seed germination syndromes in higher plants. In: *Topics in Plant Population Biology* (eds. O.T. Solbrig, S. Jain, G.B. Johnson and P.H. Raven), pp. 188–206. Columbia University Press, New York.

Arenson, L.U. (2002) Unstable alpine permafrost: a potentially important natural hazard: variations of geotechnical behaviour with time and temperature. Dissertation, Swiss Federal Institute of Technology, Zürich.

Armentano, T.V. and Menges, E.S. (1986) Patterns of change in the carbon balance of organic soil-wetlands of the temperate zone. *Journal of Ecology* 74, 755–74.

Armonies, W. and Reise, K. (1999) On the population development of the introduced razor clam *Ensis americanus* near the island of Sylt (North Sea). *Helgoländer Meeresuntersuchungen* 52, 291–300.

Arnoldus, H.M.J. (1977) Predicting soil loss due to sheet and rill erosion. *FAO Conservation Guide* 1, 99–124.

Aronson, J. and Le Floc'h, E. (1996a) Vital landscape attributes: missing tools for restoration ecology. *Restoration Ecology* 4, 377–87.

Aronson, J. and Le Floc'h, E. (1996b) Hierarchies and landscape history: dialoguing with Hobbs and Norton. *Restoration Ecology* 4, 327–33.

Aronson, J., Floret, C., Le Floc'h, E. *et al.* (1993a) Restoration and rehabilitation of degraded ecosystems in arid and semi-arid regions. I. A view from the south. *Restoration Ecology* 1, 8–17.

Aronson, J., Floret, C., Le Floc'h, E. *et al.* (1993b) Restoration and rehabilitation of degraded ecosystems in arid and semi-arid regions. II. Case studies from southern Tunisia, central Chile, and northern Cameroon. *Restoration Ecology* 1, 168–87.

Aronson, J., Dhillion, S. and Le Floc'h, E. (1995) On the need to select an ecosystem of reference, however imperfect: a reply to Pickett and Parker. *Restoration Ecology* 3, 1–3.

Aronson, J., Le Floc'h, E. and Ovalle, C. (2002) Semi-arid woodlands and desert fringes. In: *Handbook of Ecological Restoration*, vol. 2. *Restoration in Practice* (eds. M.R. Perrow and A.J. Davy). pp. 466–85. Cambridge University Press, Cambridge.

Arrow, K., Bolin, B., Costanza, R. *et al.* (1995) Economic growth, carrying capacity, and the environment. *Science* 268, 520–1.

Askew, A.P., Corker, D., Hodkinson, D.J., and Thompson, K. (1997) A new apparatus to measure the rate of fall of seeds. *Functional Ecology* 11, 121–5.

Augustin, J., Merbach, W., Schmidt, W. and Reining, E. (1996) Effect of changing temperature and watertable on trace gas emission from minerotrophic mires. *Angewandte Botanik* 70, 45–51.

Austen, I., Andersen, T.J. and Edelvang, K. (1999) The influence of benthic diatoms and invertebrates on the erodibility of an intertidal mudflat, the Danish Wadden Sea. *Estuarine, Coastal and Shelf Science* 49, 99–111.

Baar, J. (1996) The ectomycorrhizal flora of primary and secondary stands of *Pinus sylvestris* in relation to soil conditions and ectomycorrhizal succession. *Journal of Vegetation Science* 7, 497–504.

Baguette, M. and Schtickzelle, N. (2003) Local population dynamics are important to the conservation of meta-populations in highly fragmented landscapes. *Journal of Applied Ecology* 40, 404–12.

Bainbridge, D.A. (2002) Alternative irrigation systems for arid land restoration. *Restoration Ecology* 20, 23–30.

Baker, H. (1937) Alluvial meadows: a comparative study of grazed and mown meadows. *Journal of Ecology* 25, 405–20.

Bakker, C., Blair, J.M. and Knapp, A.K. (2003) Does resource availability, resource heterogeneity or species turnover mediate changes in plant species richness in grazed grasslands? *Oecologia* 137, 385–91.

Bakker, E.S. and Olff, H. (2003) Impact of different-sized herbivores on recruitment opportunities for subordinate herbs in grasslands. *Journal of Vegetation Science* 14, 465–74.

Bakker, J.F. and de Jonge, V.N. (1998) Hoe veilig is de Eems-Dollard? Ontwikkelingen in enkele belangrijke verontreinigende stoffen. In: *Het Eems-Dollard estuarium:*

interacties tussen menselijke beinvloeding en natuurlijke dynamiek (eds. K. Essink and P. Esselink), pp. 47–60. Report RIKZ-98.020, Haren.

Bakker, J.P. (1989) *Nature Management by Grazing and Cutting.* Kluwer Academic Publishers, Dordrecht.

Bakker, J.P. (1998) The impact of grazing on plant communities. In: *Grazing and Conservation Management* (eds. M.F. WallisDeVries, J.P. Bakker and S.E. van Wieren), pp. 137–84. Kluwer Academic Publishers, Dordrecht.

Bakker, J.P. and Grootjans, A.P. (1991) Potential for vegetation regeneration in the middle course of the Drentsche A brook valley (The Netherlands). *Verhandlungen der Gesellschaft für Ökologie* 20, 249–63.

Bakker, J.P. and Olff, H. (1995) Nutrient dynamics during restoration of fen meadows by hay-making without fertilizer application. In: *Restoration of Temperate Wetlands* (eds. B.D. Wheeler, S.C. Shaw, W.J. Foyt and R.A. Robertson), pp. 143–66. Wiley, Chichester.

Bakker, J.P. and Londo, G. (1998) Grazing for conservation management in historical perspective. In: *Grazing and Conservation Management* (eds. M.F. WallisDeVries, J.P. Bakker and S.E. van Wieren), pp. 23–54. Kluwer Academic Publishers, Dordrecht.

Bakker, J.P. and Berendse, F. (1999) Constraints in the restoration of ecological diversity in grassland and heathland communities. *Trends in Ecology and Evolution* 14, 63–8.

Bakker, J.P. and Ter Heerdt, G.N.J. (2005) Organic grassland farming in the Netherlands: a case study on the effects on vegetation dynamics. *Basic and Applied Ecology* 6, in press.

Bakker, J.P., Bakker, E.S., Rosén, E. *et al.* (1996a) Soil seed bank composition along a gradient from dry alvar grassland to *Juniperus* scrub. *Journal of Vegetation Science* 7, 165–76.

Bakker, J.P., Poschlod, P., Strykstra, R.J. *et al.* (1996b) Seed banks and seed dispersal: important topics in restoration ecology. *Acta Botanica Neerlandica* 45, 461–90.

Bakker, J.P., Grootjans, A.P., Hermy, M. and Poschlod, P. (2000) How to define targets for ecological restoration?: introduction. *Applied Vegetation Science* 3, 1–6.

Bakker, J.P., Elzinga, J.A. and De Vries, Y. (2002a) Effects of long-term cutting in a grassland system: possibilities for restoration of plant communities on nutrient-poor soils. *Applied Vegetation Science* 5, 107–20.

Bakker, J.P., Esselink, P., Dijkema, K.S. *et al.* (2002b) Restoration of salt marshes. *Hydrobiologia* 478, 29–51.

Bakker, J.P., Bos, D. and De Vries, Y. (2003) To graze or not to graze: that is the question. In: *Proceedings of the 10th International Scientific Wadden Sea Symposium* (eds. K. Essink, M. van Leeuwe, A. Kellermann and W. Wolff), pp. 67–88. Ministry of Agriculture, Nature Management and Fisheries, Department of Marine Biology, University of Groningen.

Bakker, R. (2003) The emergence of agriculture on the Drenthe Plateau: a palaeobotanical study supported by high-resolution ^{14}C dating. *Archäologische Berichte* 16, 1–305.

Balmford, A., Bruner, A., Cooper, P. *et al.* (2002) Economic reasons for conserving wild nature. *Science* 297, 950–3.

Barendregt, A.A., Wassen, M. and Schot, P.P. (1995) Hydrological systems beyond a nature reserve, the major problem in wetland conservation of lake Naardermeer. *Biological Conservation* 72, 393–405.

Barko, J.W. and James, W.F. (1998) Effects of submerged aquatic macrophytes on nutrient dynamics, sedimentation, and resuspension. In: *The Structuring Role of Submersed Macrophytes in Lakes* (eds. E. Jeppesen, M. Søndergaard and K. Christoffersen), pp. 197–217. Springer, Heidelberg.

Bartelink, H.H. (1999) Growth and management of mixed-species stands. In: Management of Mixed-Species Forest: Silviculture and Economics (eds. A.F.M. Olsthoorn, H.H. Bartelink, J.J. Gardiner *et al.*). *IBN Scientific Contributions* 15, 186–90.

Bartholdy, J. (1997) The backbarrier sediments of the Skallingen Peninsula, Denmark. *Geografisk Tidsskrift* 97, 11–32.

Bauer, B., Fioroni, P., Ide, I. *et al.* (1995) TBT effects on the female genital system of *Littorina littorea*, a possible indicator of tributyltin pollution. *Hydrobiologia* 309, 15–28.

Bautista, S., Bellot, J. and Vallejo, V.R. (1996) Mulching treatment for postfire soil conservation in a semiarid ecosystem. *Arid Soil Research and Rehabilitation* 10, 235–42.

Becker, P.H., Thyen, S., Mickstein, S. *et al.* (1998) Monitoring pollutants in coastal bird eggs in the Wadden Sea: final report of the pilot study, 1996–1997. *Wadden Sea Ecosystem* 8, 59–101.

Begon, M., Harper, J.L. and Townsend, C.R. (1996) *Ecology*, 3rd edn. Blackwell Science, Oxford.

Behre, K.E. (1995) Die Entstehung und Entwicklung der Natur- und Kulturlandschaft der ostfriesischen Halbinsel. In: *Ostfriesland. Geschichte und Gestalt einer Kulturlandschaft* (eds. K.E. Behre and H. Van Lengen), pp. 5–37. Ostfriesische Landschaft, Aurich.

Behrendt, H. (1997) Detection of anthropogenic trends in time series of riverine load using windows of discharge and long-term means. In: Report ICES/OSPAR workshop: *Identification of Statistical Methods for Temporal Trends*, Annex 5, Copenhagen, pp. 20–9.

Bekker, R.M., Verweij, G.L., Smith, R.E.N. *et al.* (1997) Soil seed banks in European grasslands: does land use affect regeneration perspectives? *Journal of Applied Ecology* **34**, 1293–310.

Bekker, R.M., Schaminée, J.H.J., Bakker, J.P. and Thompson, K. (1998) Seed bank charactersitics of Dutch plant communities. *Acta Botanica Neerlandica* **47**, 15–26.

Bekker, R.M., Strykstra, R.J., Schaminée, J.H.J. and Hennekens, S.M. (2002) Zaadvoorraad en herintroductie: achtergronden, spectra van plantengemeenschappen en voorbeelden uit de praktijk. *Stratiotes* **24**, 27–48.

Bell, R.H.V. (1971) A grazing ecosystem in the Serengeti. *Scientific American* **224**, 86–93.

Bell, S.S., Fonseca, M.S. and Motten, L.B. (1997) Linking restoration and landscape ecology. *Restoration Ecology* **5**, 318–23.

Bellot, J., Ortíz de Urbina, J.M., Bonet, A. and Sánchez, J.R. (2002) The effects of treeshelters on the growth of *Quercus coccifera* L. seedlings in a semiarid environment. *Forestry* **75**, 89–106.

Belnap, J., Prasse, R. and Harper, K.T. (2001) Influence of biological soil crusts on soil environments and vascular plants. In: *Biological Soil Crusts: Structure, Function, and Management* (eds. J. Belknap and O.L. Lange), pp. 281–300. Springer, Berlin.

Beltman, B., van den Broek, T. and Bloemen, B. (1995) Restoration of acidified rich fen ecosystems in the Vechtplassen area; successes and failures. In: *Restoration of Temperate Wetlands* (eds. B.D. Wheeler, S.C. Shaw, W.J. Foyt and R.A. Robertson), pp. 273–86. Wiley, Chichester.

Beltman, B., van den Broek, T., van Maanen, K. and Vaneveld, K. (1996) Measures to develop a rich fen wetland landscape with a full range of successional stages. *Ecological Engineering* **7**, 299–313.

Beltman, B., van den Broek, T., Barendregt, A. *et al.* (2001) Rehabilitation of acidified and eutrophied fens in The Netherlands: effects of hydrologic manipulation and liming. *Ecological Engineering* **17**, 21–32.

Belyea, L.R. and Lancaster, J. (1999) Assembly rules within contingent ecology. *Oikos* **86**, 402–16.

Benfield, F.K., Terris, J. and Vorsanger, N. (2001) *Solving Sprawl: Models of Smart Growth in Communities across America*. Island Press, Washington, DC.

Bengtsson, J., Fagerström, T. and Rydin, H. (1994) Competition and coexistence in plant communities. *Trends in Ecology and Evolution* **9**, 246–50.

Bengtsson, L., Fleischer, S., Lindmark, G. and Ripl, W. (1975) Lake Trummen Restoration Project. I. Water and sediment chemistry. *Verhandlungen – Internationale Vereinigung für theoretische und angewandte Limnologie* **19**, 1080–7.

Benito, G., Baker, V.R. and Gregory, K.J. (1998) *Palaeohydrology and Environmental Changes*. Wiley, Chichester.

Benndorf, J. (1987) Food web manipulation without nutrient control: a useful strategy in lake restoration. *Schweizerische Zeitschrift für Hydrologie* **49**, 237–48.

Benndorf, J., Böing, W., Koop, J. and Neubauer, I. (2002) Top-down control of phytoplankton: the role of time scale, lake depth and trophic state. *Freshwater Biology* **47**, 2282–95.

Bentham, H., Harris, J.A., Birch, P. and Short, K.C. (1992) Habitat classification and soil restoration assessment using analysis of soil microbiological and physicochemical characteristics. *Journal of Applied Ecology* **29**, 711–18.

Berendse, F. and Elberse, W.Th. (1990) Competition and nutrient availability in heathland and grassland ecosystems. In: *Perspectives on Plant Competition* (eds. J.B. Grace and D. Tilman), pp. 93–116. Academic Press, London.

Berger, J.J. (1990) Evaluating ecological protection and restoration projects: a holistic approach to the assessment of complex, multi-attribute resource management problems. Doctoral dissertation, University of California, Davis.

Bergerud, A.T. and Mercer, W.E. (1989) Caribou introductions in eastern North America. *Wildlife Society Bulletin* **17**, 111–20.

Bergez, J.E. and Dupraz, C. (1997) Transpiration rate of *Prunus avium* L. seedlings inside an unventilated treeshelter. *Forest Ecology and Management* **97**, 255–64.

Bergstern, U. (1985) A study on the influence of seed predators at direct seeding of *Pinus sylvestris* L. *Sveriges Lantbruksuniversitet Rapporter* **13**, 1–15.

Berkes, F. and Folke, C. (1998) *Linking Social and Ecological Systems: Management Practices and Social Mechanisms for Building Resilience*. Cambridge University Press, Cambridge.

Bernhaupt, P. (1980) Zum Problem der Bodenerosion in Almgebieten am Beispiel der Planneralm, Wölzer Tauern, Steiermark. *Interpraevent* **1**, 291–308.

Bertness, M.D. (1999) *The Ecology of Atlantic Shorelines*. Sinauer, Sunderland, MA.

Bertness, M.D. and Callaway, R.M. (1994) Positive interactions in communities. *Trends in Ecology and Evolution* **9**, 191–3.

Bertness, M.D. and Leonard, G.H. (1997) The role of positive interactions in communities: lessons from intertidal habitats. *Ecology* 78, 1976–89.

Bertram, B.C.R. (1988) Re-introducing scimitar-horned oryx into Tunisia. In: *Conservation and Biology of Desert Antelopes* (eds. A. Dixon and D. Jones), pp. 136–45. Christopher Helm, London.

Beukema, J.J. (1995) Long-term effects of mechanical harvesting of Lugworms *Arenicola marina* on the zoobenthic community of a tidal flat in the Wadden Sea. *Netherlands Journal of Sea Research* 33, 219–27.

Beukema, J.J. and Cadée, G.C. (1986) Zoobenthos responses to eutrophication of the Dutch Wadden Sea. *Ophelia* 26, 55–64.

Beukema, J.J., Cadée, G.C. and Dekker, R. (2002) Zoobenthic biomass limited by phytoplankton abundance: evidence from parallel changes in two long-term data series in the Wadden Sea. *Journal of Sea Research* 48, 111–25.

BfN (1996) *Daten zur Natur*. Bundesamt für Naturschutz (Federal Agency of Nature Conservation), Bonn.

BfN (2000) *Map of the Natural Vegetation of Europe*, Scale 1 : 2,500,000. Bundesamt für Naturschutz (Federal Agency of Nature Conservation), Bonn-Bad Godesberg.

Bich, J.P. (1988) The feasibility of river otter reintroduction in Northern Utah. MSc thesis, Utah State University, Logan.

Bieleman, J. (1992) *Geschiedenis van de landbouw in Nederland 1500–1950: veranderingen en verscheidenheid*. Boom, Meppel.

Biere, A. and Honders, S.J. (1996) Impact of flowering phenology of *Silene alba* and *S. dioica* on susceptibility to fungal infection and seed predation. *Oikos* 77, 467–80.

Bignal, E.M. and McCracken, D.I. (1996) Low-intensity farming systems in the conservation of the countryside. *Journal of Applied Ecology* 33, 413–24.

Bijlsma, R.G. (2001) Pitrus *Juncus effusus* en Sprinkhaanrietzanger *Locustella naevia*: de discrepantie tussen theorie en praktijk. *Drentse Vogels* 14, 43–53.

Bijlsma, R., Bungaard, J. and van Putten, F. (1999) Environmental dependence of inbreeding depression and purging in *Drosophila melanogaster*. *Journal of Evolutionary Biology* 12, 1125–37.

Bijlsma, R., Bungaard, J. and Boerema, A.C. (2000) Does inbreeding affect the extinction risk of small populations?: predictions from *Drosophila*. *Journal of Evolutionary Biology* 13, 502–14.

Bill, H.C., Poschlod, P., Reich, M. and Plachter, H. (1999) Experiments and observations on seed dispersal by running water in an alpine flood plain. *Bulletin of the Geobotanical Institute ETH* 65, 13–28.

Bjork, S. (1972) Ecosystem studies in connection with the restoration of lakes. *Verhandlungen-Internationale Vereinigung für theoretische und angewandte Limnologie* 18, 379–87.

Blondel, J. and Aronson, J. (1999) *Biology and Wildlife of the Mediterranean Region*. Oxford University Press, Oxford.

Blossey, B. and Nötzold, R. (1995) Evolution of increased competitive ability in invasive nonindigenous plants: a hypothesis. *Journal of Ecology* 83, 887–9.

Bobbink, R. and Willems, J.H. (1991) Impact of different cutting regimes on the performance of *Brachypodium pinnatum* in Dutch chalk grassland. *Biological Conservation* 56, 1–22.

Bobbink, R., den Dubbelden, J. and Willems, J.H. (1989) Seasonal dynamics of phytomass and nutrients in chalk grassland. *Oikos* 55, 216–24.

Bobbink, R., Hornung, M. and Roelofs, J.G.M. (1998) The effects of air-borne nitrogen pollutants on species diversity in natural and semi-natural vegetation: a review. *Journal of Ecology* 86, 717–38.

Boers, P., van der Does, J., Quaak, M. *et al.* (1992) Fixation of phosphorus in lake-sediments using iron (III) chloride – experiences, expectations. *Hydrobiologia* 233, 211–12.

Boeye, D. and Verheyen, R. (1994) The relation between vegetation and soil chemistry gradients in a groundwater discharge fen. *Journal of Vegetation Science* 5, 553–60.

Bokdam, J. (2003) Nature conservation and grazing management: free-ranging cattle as a driving force for cyclic vegetation succession. PhD thesis, Wageningen University.

Bokdam, J. and Gleichman, J.M. (2000) Effects of grazing by free-ranging cattle on vegetation dynamics in a continental north-west European heathland. *Journal of Applied Ecology* 37, 415–31.

Bokdam, J. and WallisDeVries, J. (1992) Forage quality as a limiting factor for cattle grazing in isolated Dutch nature reserves. *Conservation Biology* 6, 399–408.

Bonn, S. and Poschlod, P. (1998) *Ausbreitungsbiologie der Pflanzen Mitteleuropas*. Quelle und Meyer Verlag, Wiesbaden.

Bonnell, M.L. and Selander, R.K. (1974) Elephant seals: genetic variation and near extinction. *Science* 184, 908–9.

Boorman, L.A. (1999) Salt marshes – present functioning and future change. *Mangroves and Salt Marshes* **3**, 227–41.

Booth, W. (1988) Reintroducing a political animal. *Science* **241**, 156–8.

Booy, G., Hendriks, R.J.J., Smulders, M.J.M. *et al.* (2000) Genetic diversity and the survival of populations. *Plant Biology* **2**, 379–95.

Borvall, C., Ebenman, B. and Jonsson, T. (2000) Biodiversity lessons, the risk of cascading extinction in model food webs. *Ecology Letters* **3**, 131–6.

Bos, D., Bakker, J.P., de Vries, Y. and van Lieshout, S. (2002) Long-term vegetation changes in experimentally grazed and ungrazed back-barrier marshes in the Wadden Sea. *Applied Vegetation Science* **5**, 45–54.

Bos, D., Loonen, M.J.M.E., Stock, M. *et al.* (2005a) Utilisation of Wadden Sea salt marshes by geese in relation to livestock grazing. *Journal for Nature Conservation*, in press.

Bos, D., Drent, R.H., Ebbinge, B.S. *et al.* (2005b) Capacity of Wadden Sea coastal grasslands for Dark-bellied Geese. *Journal of Wildlife Biology*, in press.

Bos, J.M., Van Geel, B. and Pals, J.P. (1988) Waterland – environmental and economic changes in a Dutch bog area, 1000 AD to 2000 AD. In: *The Cultural Landscape: Past, Present and Future* (eds. H.H. Birks, H.B.J. Birks, P.E. Kaland and D. Moe), pp. 321–31. Cambridge University Press, Cambridge.

Bossuyt, B. and Hermy, M. (2000) Ecological restoration of the understorey layer: evidence from ancient-recent forest ecotones. *Applied Vegetation Science* **3**, 43–50.

Bossuyt, B., Heyn, M. and Hermy, M. (2002) Seed bank and vegetation composition of forest stands of varying age in central Belgium: consequences for regeneration of ancient forest vegetation. *Plant Ecology* **162**, 33–48.

Boström, C. and Bonsdorff, E. (1997) Community structure and spatial variation of benthic invertebrates associated with *Zostera marina* (L.) beds in the northern Baltic Sea. *Journal of Sea Research* **37**, 153–66.

Bottner, P., Coûteaux, M.M. and Vallejo, V.R. (1995) Soil organic matter in Mediterranean-type ecosystems and global climatic changes: a case study: the soils of the Mediterranean Basin. In: *Global Change and Terrestrial Ecosystems* (eds. J.M. Moreno and W.C. Oechel), pp. 306–25. Springer, New York.

Boulton, A.J. (1999) An overview of river health assessment: philosophies, practice, problems and prognosis. *Freshwater Biology* **41**, 469–79.

Boumans, R.M.J., Costanza, R., Farley, J. *et al.* (2002) Modeling the dynamics of the Integrated Earth System and the value of Global Ecosystem Services using the GUMBO model. Special issue: the dynamics and value of ecosystem services: integrating economic and ecological perspectives. *Ecological Economics* **41**, 529–60.

Bourn, N.A.D. and Thomas, J.A. (2002) The challenge of conserving grassland insects at the margins of their range in Europe. *Biological Conservation* **104**, 285–92.

Boyer, M.H.L. and Wheeler, B.D. (1989) Vegetation patterns in spring-fed calcareous fens; calcite precipitation and constraints on fertility. *Journal of Ecology* **77**, 597–609.

Bradshaw, A.D. (1983) The reconstruction of ecosystems. *Journal of Applied Ecology* **20**, 1–17.

Bradshaw, A.D. (1987) Restoration: the acid test for ecology. In: *Restoration Ecology* (eds. W.R. Jordan, III, M.E. Gilpin and J.D. Aber), pp. 53–74. Cambridge University Press, Cambridge.

Bradshaw, A.D. (1997) What do we mean by restoration? In: *Restoration Ecology and Sustainable Development* (eds. K.M. Urbanska, N.R. Webb and P.J. Edwards), pp. 8–14. Cambridge University Press, Cambridge.

Bradshaw, A.D. (2002) Introduction and philosophy. In: *Handbook of Ecological Restoration* (eds. M.R. Perrow and A.J. Davey), pp. 3–9. Cambridge University Press, Cambridge.

Bradshaw, A.D. and Chadwick, M.J. (1980) *The Restoration of Land: The Ecology and Reclamation of Derelict and Degraded Land.* Studies in Ecology, vol. 6. Blackwell Scientific Publications, Oxford.

Bradshaw, A.D. and Hüttl, R.F. (2001) Future minesite restoration involves a broader approach. *Ecological Engineering* **17**, 87–90.

Brady, N.C. and Weil, R.R. (1999) *The Nature and Properties of Soils*, 12th edn. Prentice Hall, London.

Brakke, D.F., Bakker, J.P., Böhmer, J. *et al.* (1994) Group report: physiological and ecological effects of acidification on aquatic biota. In: *Acidification of Freshwaters: Implications for the Future* (eds. C.E.W. Steinberg and R.F. Wright), pp. 275–312. Wiley, Chichester.

Breman, H. and de Wit, C.T. (1983) Rangeland productivity and exploitation in the Sahel. *Science* **221**, 1341–7.

Breshears, D.D., Nyhan, J.W., Heil, C.E. and Wilcox, B.P. (1998) Effects of woody plants on microclimate in a semiarid woodland: soil temperature and evaporation in canopy and intercanopy patches. *International Journal of Plant Science* **159**, 1010–17.

Brett, M.T. (1989) Zooplankton communities and acidification processes (a review). *Water Air Soil Pollution* **44**,
387–414.

Bright, P.W. and Morris, P.A. (1994) Animal translocation for conservation: performance of dormice in relation to release methods, origin and season. *Journal of
Applied Ecology* **31**, 699–708.

Brinson, M.M. (1993) *A Hydrogeomorphic Classification
for Wetlands*. Technical Report WRP-DE-4. US Army
Corps of Engineers, Vicksburg, MS.

Britton, A.J., Pakeman, R.J., Carye, P.D. and Marrs, R.H.
(2001) Impacts of climate, management and nitrogen
deposition on the dynamics of lowland heathland.
Journal of Vegetation Science **12**, 797–806.

Brodin, Y.-W. (1995a) Acidification of lakes and watercourses in a global perspective. In: *Liming of Acidified
Surface Waters: a Swedish Synthesis* (eds. L. Henriksson
and Y.-W. Brodin), pp. 45–62. Springer, Berlin.

Brodin, Y.-W. (1995b) Acidification of Swedish fresh-
waters. In: *Liming of Acidified Surface Waters: A
Swedish Synthesis* (eds. L. Henriksson and Y.-W.
Brodin), pp. 63–80. Springer, Berlin.

Bronstein, J.L. (1994) Conditional outcomes in mutualistic interactions. *Trends in Ecology and Evolution* **9**,
214–17.

Brookes, A. (1984) *Recommendations Bearing on the
Sinuosity of Danish Stream Channels*. Technical
Report no. 6, Freshwater Laboratory, Danish Environ-
mental Protection Agency, Copenhagen.

Brookes, A. (1988) *Channelized Rivers: Perspectives for
Environmental Management*. Wiley, Chichester.

Brooks, J.L. and Dodson, S.I. (1965) Predation, body size,
and the composition of the plankton. *Science* **150**,
28–35.

Brown, J.H. and Kodric-Brown, A. (1977) Turnover rates
in insular biogeography: effect of immigration on
extinction. *Ecology* **58**, 445–9.

Brown, L.F. (2001) *Eco-economy: Building an Economy
for the Earth*. Earth Policy Institute, Washington, DC.

Brown, V.K. and Gange, A.C. (1989) Differential effects
of above- and below-ground insect herbivory during
early plant succession. *Oikos* **54**, 67–76.

Brunberg, A.K. and Boström, B. (1992) Coupling between
benthic biomass of *Microcystis* and phosphorus release
from the sediments of a highly eutrophic lake. *Hydrobiologia* **235**, 375–85.

Bruno, J.F. and Bertness, M.D. (2001) Habitat modification and facilitation in benthic marine communities.
In: *Marine Community Ecology* (eds. M.D. Bertness, S.D.

Gaines and M.E. Hay), pp. 201–18. Sinauer, Sunderland, MA.

Bruno, J.F., Stachowicz, J.J. and Bertness, M.D. (2003)
Inclusion of facilitation into ecological theory. *Trends
in Ecology and Evolution* **18**, 119–25.

Brunsden, D. and Thornes, J.B. (1979) Landscape sensitivity and change. *Transactions of the Institute of British
Geographers New Series* **4**, 463–84.

Buczko, U., Bens, O., Fischer, H. and Hüttl, R.F. (2002)
Water repellency in sandy luvisols under different
forest transformation stages in northeast Germany.
Geoderma **109**, 1–18.

Buijse, A.D., Coop, H., Staras, M. *et al.* (2002) Restoration strategies for river floodplains along large lowland
rivers in Europe. *Freshwater Biology* **47**, 889–907.

Bullock, J.M. (1998) Community translocation in Britain:
setting objectives and measuring consequences. *Biological Conservation* **84**, 199–214.

Bungart, R., Bens, O. and Hüttl, R.F. (2001) Bioenergy pro-
duction and lignite mine reclamation: the need for future
lasting landuse systems. In: *First World Conference on
Biomass for Energy and Industry* (eds. S. Kyritsis, A.A.
Beenackers, P. Helm *et al.*), pp. 593–6. James and James,
London.

Bürger, O. (1995) Prähistorische Landschaftskunde am
Fallbeispiel Pestenacker: pollenanalytische Untersu-
chungen zur Vegetatons- und Siedlungsgeschiehte im
Altmoränengebiet zwischen Lech und Isar. Dissertation,
Ludwig-Maximilians-Universität München, München.

Burkart, A. (1976) Monograph of the genus *Prosopis*.
Journal of the Arnold Arboretum **57**, 219–49, 450–
525.

Burschel, P. and Weber, M. (2001) Wald – Forstwirtschaft
– Holzindustrie: zentrale Größen der Klimapolitik.
Forstarchiv **72**, 75–85.

Bussotti, F., Grossoni, P. and Pantani, F. (1995) The role
of marine salt and surfactants in the decline of
Tyrrhenian coastal vegetation in Italy. *Annales des
Sciences Forestières* **52**, 251–61.

Butaye, J., Jacquemyn, H., Honnay, O. and Hermy, M.
(2002) The species pool concept applied to forests in
a fragmented landscape: dispersal limitation versus
habitat limitation. *Journal of Vegetation Science* **13**,
27–34.

Butzer, K. (1990) The realm of cultural-human ecology:
adaptation and change in historical perspective. In:
The Earth as Transformed by Human Action (eds.
B.L. Turner, III, W.C. Clark, R.W. Kates *et al.*), pp. 685–
701. Cambridge University Press, Cambridge.

Cachón de Mesa, J. (2001) In-stream flows in Spain In: *River Restoration in Europe – Practical Approaches* (eds. H.J. Nijland and M.J.R. Cals), pp. 281–7. Proceedings of the 2000 River Restoration Conference, RIZA, Wageningen.

Cadée, G.C. and Reydon, J.P. (1998) Zouden de Zeegrasslakjes *Rissoa* en *Lacuna* terug kunnen keren in de Waddenzee? *De Levende Natuur* **99**, 68–70.

Cadée, G.C., Boon, J.P., Fischer, C.V., Mensink B.P. and ten Hallers-Tjabbes C.C. (1995) Why the whelk (*Buccinum undatum*) has become extinct in the Dutch Wadden Sea. *Netherlands Journal of Sea Research* **34**, 337–9.

Cairns Jr, J. (1991) The status of the theoretical and applied science of restoration ecology. *Environmental Professional* **13**, 186–94.

Cairns Jr, J. (1993) Ecological restoration: replenishing our national and global ecological capital. In: *Nature Conservation 3: Reconstruction of Fragmented Ecosystems: Global and Regional Perspectives* (eds. D.A. Saunders, R.J. Hobbs and P. Ehrlich), pp. 193–208. Surrey Beatty and Sons, Chipping Norton, Australia.

Cairns, Jr, J. (2000) Setting ecological restoration goals for technical feasibility and scientific validity. *Ecological Engineering* **15**, 171–80.

Cairns, Jr, J. and McCormick, P.V. (1992) Developing an ecosystem-based capability for ecological risk assessments. *Environmental Professional* **14**, 186–96.

Cairns, Jr, J., McCormick, P.V. and Niederlehner, B.R. (1993) A proposed framework for developing indicators of ecosystem health. *Hydrobiologia* **263**, 1–44.

Callaway, R.M. (1995) Positive interactions among plants. *Botanical Review* **61**, 306–49.

Callaway, R.M. and Walker, L.R. (1997) Competition and facilitation: a synthetic approach to interactions in plant communities. *Ecology* **78**, 1958–65.

Callicott, J.B., Crowder, L.B. and Mumford, K. (1999) Current normative concepts in conservation. *Conservation Biology* **13**, 22–35.

Calow, P. (1995) Ecosystem health: a critical analysis of concepts. In: *Evaluating and Monitoring the Health of Large-Scale Ecosystems* (eds. D.J. Rapport, C. Gaudet and P. Calow), pp. 33–41. Springer, Berlin.

Calow, P. (ed.) (1998) *The Encyclopedia of Ecology and Environmental Management*. Blackwell Science, Oxford.

Calow, P. and Petts, G.E. (1992) *The Rivers Handbook Vol. I: Hydrological and Ecological Principles*. Blackwell Scientific Publications, Oxford.

Cannel, M.G.R., Malcolm, D.C. and Robertson, P.A. (1992) *The Ecology of Mixed-Species Stands of Trees: Forest Dynamics*. Oxford University Press, Oxford.

Cappers, R.T.J. (1994) An ecological characterization of plant macro-remains of Heveskesklooster (the Netherlands): a methodological approach. PhD thesis, University of Groningen.

Caraco, N.F., Cole, J.J. and Likens, G.E. (1989) Evidence for sulphate-controlled phosphorus release from sediments of aquatic systems. *Nature* **341**, 156–8.

Carey, A.B. (2003) Restoration of landscape function: reserves or active management? *Forestry* **76**, 221–30.

Carey, A.B. (2004) Restoring ecological function to forested ecosystems and landscapes: active intentional management for multiple values. www.ser.org.

Carlton, J.T. (1999) Molluscan invasions in marine and estuarine communities. *Malacologia* **41**, 439–54.

Carpenter, S.R. and Kitchell, J.F. (1992) Trophic cascade and biomanipulation interface of research and management: a reply to the comment by De Melo *et al.* *Limnology and Oceanography* **37**, 208–13.

Carpenter, S.R. and Kitchell, J.F. (eds.) (1993) *The Trophic Cascade in Lakes*. Cambridge Studies in Ecology. Cambridge University Press, Cambridge.

Carpenter, S.R., Kitchell, J.F. and Hodgson, J.R. (1985) Cascading trophic interactions and lake productivity. *BioScience* **35**, 634–9.

Carpenter, S., Walker, B., Anderies, J.M. and Abel, N. (2001) From metaphor to measurement: resilience of what to what? *Ecosystems* **4**, 765–81.

Carreras, C., Sánchez, J., Reche, P. *et al.* (1997) Primeros resultados de una repoblación mediante siembra con protectores en Vélez-Rubio. *Cuadernos de la SECF (Sociedad Española de Ciencias Forestales)* **4**, 135–9.

Castro, J., Zamora, R., Hódar, J.A. and Gómez, J.M. (2002) The use of shrubs as nurse plants: a new technique for afforestation in Mediterranean mountains. *Restoration Ecology* **10**, 297–305.

CBD (2003) *Handbook of the Convention on Biological Diversity*, 2nd edn. Secretariat of the Convention on Biological Diversity. UNEP, Montreal.

Cellot, B., Mouillot, F. and Henry, C.P. (1998) Flood drift and propagule bank of aquatic macrophytes in a river wetland. *Journal of Vegetation Science* **9**, 631–40.

Cernusca, A., Tappeiner, U., Bahn, M. *et al.* (1996) ECOMONT: Ecological effects of land-use-changes on European terrestrial mountain ecosystems, *Pirineos* **147–8**, 145–72.

Chabrerie, O. (2002) Analyse intégrée de la diversité des communautés végétales dans les pelouses calcicoles de la Basse Vallée de la Seine. PhD thesis, University of Paris-Orsay.

Chambers, J.C. (1997) Restoring alpine ecosystems in the western United States: environmental constraints, disturbance characteristics and restoration success. In: *Restoration Ecology and Sustainable Development* (eds. K.M. Urbanska, N.R. Webb and P.J. Edwards), pp. 161–87. Cambridge University Press, Cambridge.

Chang, E.R., Jefferies, R.L. and Carleton, T.J. (2001) Relationship between vegetation and soil seed banks in an arctic coastal marsh. *Journal of Ecology* 89, 367–84.

Chase, J.M. and Laibold, M. (2002) Spatial scale dictates the productivity-biodiversity relationship. *Nature* 416, 427–30.

Chave, P. (2001) *The EU Framework Directive: An Introduction.* IWA publishing, Cornwall.

Chiarello, N., Hickman, J.C. and Mooney, H.A. (1982) Endomycorrhizal role for interspecific transfer of phosphorus in a community of annual plants. *Science* 217, 941–3.

Chorus, I. (2001) *Cyanotoxins: Occurrence, Causes, Consequences.* Springer, Berlin.

Christensen, N.L., Bartuska, A.M., Brown, J.H. *et al.* (1996) The report of the Ecological Society of America committee on the scientific basis for ecosystem management. *Ecological Applications* 6, 665–91.

Churchill, J.H. (1989) The effect of commercial trawling on sediment resuspension and transport over the Middle Atlantic Bight continental shelf. *Continental Shelf Research* 9, 841–64.

CIPRA (1998) Commission Internationale pour la Protection des Alpes. *1. Alpenreport*, Internationale Alpenschutzkommission CIPRA, Verlag Paul Haupt, Bern.

CIRF (2001) *Manule di Riqualificazione Fluviale: le esperienze pioniere della rinaturalizzazione in Europa.* Centro Italiano per la Riqualificazione Fluviale, Mazzanti Editori, Venice.

Clements, F.E. (1916) *Plant Succession: An Analysis of the Development of Vegetation.* Publication 242. Carnegie Institute of Washington, Washington, DC.

Clewell, A.F. (1999) Restoration of riverine forest at Hall Branch on phosphate-mined land, Florida. *Restoration Ecology* 7, 1–14.

Clewell, A.F. (2000a) Editorial: Restoration of natural capital. *Restoration Ecology* 8, 1.

Clewell, A.F. (2000b) Restoring for natural authenticity. *Ecological Restoration* 18, 216–17.

Clewell, A.F. and Lea, R. (1990) Creation and restoration of forested wetland vegetation in the Southeastern United States. In: *Wetland Creation and Restoration: the Status of the Science* (eds. J.A. Kusler and M.E. Kentula), pp. 199–237. Island Press, Washington, DC.

Codd, G.A. (2000) Cyanobacterial toxins, the perception of water quality, and the prioritisation of eutrophication control. *Ecological Engineering* 16, 51–60.

Colijn, F. and Cadée, G.C. (2003) Is phytoplankton growth in the Wadden Sea light or nitrogen limited? *Journal of Sea Research* 49, 83–94.

Collie, J.S., Hall, S.J., Kaiser M.J. and Poiner, I.R. (2000) A quantitative analysis of fishing impacts on shelf-sea benthos. *Journal of Animal Ecology* 69, 785–98.

Collins, S.L. (1995) The measurement of stability in grasslands. *Trends in Ecology and Evolution* 10, 95–6.

Collins, S.L., Knapp, A.K., Briggs, J.M. *et al.* (1998) Modulation of diversity by grazing and mowing in native tallgrass prairie. *Science* 280, 745–7.

Connell, J.H. (1978) Diversity in tropical rainforests and coral reefs. *Science* 199, 1302–10.

Connell, J.H. and Slatyer, R.O. (1977) Mechanisms of succession in natural communities and their role in community stability and organization. *American Naturalist* 111, 1119–44.

Cook, C.D.K. (1987) Dispersion in aquatic an amphibious vascular plants. In: *Plant Life in Aquatic and Amphibious Habitats* (ed. R.M.M. Crawford), pp. 179–90. Blackwell Scientific Publications, Oxford.

Cook, W.M., Lane, K.T., Foster, B.L. and Holt, R.D. (2002) Island theory, matrix effects and species richness patterns in habitat fragments. *Ecology Letters* 5, 619–23.

Cooke, A.S. and Oldham, R.S. (1995) Establishment of populations of the common frog, *Rana temporaria*, and common toad, *Bufo bufo*, in a newly created reserve following translocation. *Herpetological Journal* 5, 173–80.

Cooke, G.D., Welch, E.B., Peterson, S.A. and Newroth, P.R. (1993) *Restoration and Management of Lakes and Reservoirs*, 2nd edn. Lewis Publishers, Ann Arbor.

Cooper, N.J., Cooper, T. and Burd, F. (2001) 25 years of salt marsh erosion in Essex: implications for coastal defence and nature conservation. *Journal of Coastal Conservation* 7, 31–40.

Coops, H. (2002) Ecology of Charyophytes: an introduction. *Aquatic Botany* 72, 205–8.

Coops, H. and Hosper, S.H. (2002) Water-level management as a tool for the restoration of shallow lakes in the Netherlands. *Lake and Reservoir Management* 18, 292–7.

Copien, J.H. (1942) Untitled. *Zeitschrift für Forst- und Jagdwesen* 74, 43–77, 81.

Correll, D. (1991) Human impact on the functioning of landscape boundaries. In: *Ecotones: The Role of Landscape Boundaries in the Management and Restoration of Changing Environments* (eds M.M. Holland, P.G. Risser and R.J. Naiman), pp. 90–109. Chapman and Hall, New York.

Corsi, F., De Leeuw, J. and Skidmore, A. (2000) Modeling species distribution with GIS. In: *Research Techniques in Animal Ecology: Controversies and Consequences* (eds. L. Boitani and T.K. Fuller), pp. 389–434. Columbia University Press, New York.

Cortina, J. and Vallejo, V.R. (1999) Restoration of Mediterranean Ecosystems. In: *Perspectives in Ecology* (ed. A. Farina), pp. 479–90, Backhuys, Leiden.

Costanza, R. and Daly, H.E. (1992) Natural capital and sustainable development. *Conservation Biology* 6, 37–46.

Costanza, R., Norton, B.G. and Haskell, B.D. (1992) *Ecosystem Health: New Goals for Environmental Management*. Island Press, Washington, DC.

Costanza, R., d'Arge, R., de Groot, R. *et al.* (1997) The value of the world's ecosystem services and natural capital. *Nature* 387, 253–60.

Coulson, S.J., Bullock, J.M., Stevenson, M.J. and Pywell, R.F. (2001) Colonization of grassland by sown species: dispersal versus microsite limitation in responses to management. *Journal of Applied Ecology* 38, 204–16.

Crabbé, P., Holland, A., Ryszkowski, L. and Westra, L. (eds.) (2000) *Implementing Ecological Integrity: Restoring Regional and Global Environmental and Human Health*. Kluwer Academic Publishers, Dordrecht.

Crawley, M.J. (1983) *Herbivory: The Dynamics of Animal-Plant Interactions*. Blackwell Scientific Publications, Oxford.

Crawley, M.J. (1987) What makes a community invasible? In: *Colonization, Succession and Stability* (eds. A.J. Gray, M.J. Crawley and P.J. Edwards), pp. 429–51. Blackwell Scientific Publications, Oxford.

Crosby, A. (1986) *Ecological Imperialism. The Biological Expansion of Europe 900–1900*. Cambridge University Press, Cambridge.

Cross, J.R. (1975) Biological flora of the British Isles: *Rhododendron ponticum. Journal of Ecology* 63, 345–64.

Crutzen, P.J. (2002) Geology of mankind. *Nature* 415, 23.

Crutzen, P.J. and Stoermer, E.F. (2000) The 'Anthropocene'. *International Geosphere Biosphere Programme Newsletter* 41, 17–18.

Cui, M. and Caldwell, M.M. (1996a) Facilitation of plant phosphate acquisition by arbuscular mycorrhizas from enriched soil patches. I. Roots and hyphae exploiting the same soil volume. *New Phytologist* 133, 453–60.

Cui, M. and Caldwell, M.M. (1996b) Facilitation of plant phosphate acquisition by arbuscular mycorrhizas from enriched soil patches. II. Hyphae exploiting root-free soil. *New Phytologist* 133, 461–7.

Cullen, P. and Forsberg, C. (1988) Experience with reducing point source of phosphorus to lakes. *Hydrobiologia* 170, 321–36.

Cunningham, S. (2002) *The Restoration Economy*. Berrett-Koehler Publishers, San Francisco.

Cyrulnik, B. (2001) *Les villains petits canards*. Odile Jacob, Paris.

Dabbert, S., Dubgaard, A., Slangen, L. and Whitby, M. (eds.) (1998) *The Economics of Landscape and Wildlife Conservation*. CAB International, Wallingford.

Daily, G.C. (ed.) (1997) *Nature's Services. Societal Dependance on Natural Ecosystems*. Island Press, Washington, DC.

Daily, G.C. and Ellison, K. (2002) *The New Economy of Nature: The Quest to Make Conservation Profitable*. Island Press, Washington, DC.

Dale, V.H. and Beyeler, S.C. (2001) Challenges in the development and use of ecological indicators. *Ecological Indicators* 1, 3–10.

Daly, H.E. and Farley, J. (2004) *Ecological Economics: Principles and Applications*. Island Press, Washington, DC.

Dalziel, T.R.K., Wilson, E.J. and Proctor, M.V. (1994) The effectiveness of catchment liming in restoring acid waters at Loch-Fleet, Galloway, Scotland. *Forest Ecology and Management* 68, 107–17.

Danielson, B.J. and Gaines, M.S. (1987) The influences of conspecific and heterospecific residents on colonization. *Ecology* 68, 1778–84.

D'Antonio, C. and Meyerson, L.A. (2002) Exotic plant species as problems and solutions in ecological restoration: a synthesis. *Restoration Ecology* 10, 703–13.

D'Antonio, C.M., Hughes, R.F. and Vitousek, P.M. (2001) Factors influencing dynamics of two invasive C4-grasses in seasonally dry Hawaiian woodlands. *Ecology* 82, 89–104.

Danvid, M. and Nilsson, C. (1997) Seed floating ability and distribution of alpine plants along a northern Swedish river. *Journal of Vegetation Science* 8, 271–6.

Dasgupta, P., Levin, S. and Lubchenco, J. (2000) Economic pathways to ecological sustainability. *BioScience* 50, 339–45.

Davis, M.A., Grime, J.P. and Thompson, K. (2000) Fluctuating resources in plant communities: a general theory of invasibility. *Journal of Ecology* **88**, 528–34.

Davis, M.B., Woods, K.D., Webb, S.L. and Futyma, R.P. (1986) Dispersal vs. climate: expansion of *Fagus* and *Tsuga* into the upper Great Lakes region. *Vegetatio* **67**, 93–103.

Dayton, P.K., Thrush, S.F., Agardy, M.T. and Hofman, R.J. (1995) Environmental effects of marine fishing. *Aquatic Conservation: Marine and Freshwater Ecosystems* **5**, 205–32.

DeAngelis, D.L. (1992) *Dynamics of Nutrient Cycling and Food Webs.* Chapman and Hall, London.

De Bolós, O. (1967) Comunidades vegetales de las comarcas próximas al litoral situadas entre los ríos Llibregat y Segura. *Memorias de la Real Academia de Ciencias y Artes de Barcelona* **38**, 1269.

de Deyn, G.B., Raaijmakers, C.E., Zoomer, R.H. *et al.* (2003) Soil invertebrate fauna enhances grassland succession and diversity. *Nature* **422**, 711–13.

DEFRA (2002) *Directing the Flow: priorities for future water policy.* DEFRA publication PB 7510, Department for Environment, Food and Rural Affairs, London.

de Graaf, M.C.C., Verbeek, P.J.M., Bobbink, R. and Roelofs, J.G.M. (1998a) Restoration of species-rich dry heaths: the importance of appropriate soil conditions. *Acta Botanica Neerlandica* **47**, 89–111.

de Graaf, M.C.C., Bobbink, R., Roelofs, J.G.M. and Verbeek, P.J.M. (1998b) Differential effects of ammonium and nitrate on three heathland species. *Plant Ecology* **135**, 185–96.

de Groot, R.S., Wilson, M. and Boumans, R.M.J. (2002) A typology for the classification, description and valuation of ecosystem functions, goods and services. *Ecological Economics* **41**, 393–408.

de Jonge, V.N., van den Bergs, J. and de Jong, D.J. (1997) *Zeegras in de Waddenzee, een toekomstperspectief: beheersaanbevelingen voor het herstel van groot en klein zeegras (Zostera marina L. en Zostera noltii Hornem).* Rapport RIKZ-97.016. Rijkswaterstaat, Haren.

Delcourt, P.A. and Delcourt, H.R. (1983) Late-Quaternary vegetational dynamics and community stability reconsidered. *Quaternary Research* **19**, 265–71.

Delcourt, H.R. and Delcourt, P.A. (1991) *Quaternary Ecology: a Palaeological Perspective.* Chapman and Hall, London.

Delcourt, H.R., Delcourt, P.A. and Webb, III, T. (1983) Dynamic plant ecology: the spectrum of vegetation change in time and space. *Quaternary Science Reviews* **1**, 153–75.

de Leeuw, J., de Munck, W., Olff, H. and Bakker, J.P. (1993) Does zonation reflect the succession of salt marsh vegetation? A comparison of an estuarine and a coastal bar island marsh in the Netherlands. *Acta Botanica Neerlandica* **42**, 435–45.

de Leeuw, J., de Jong, D.J., Apon. L.P. *et al.* (1994) The response of salt marsh vegetation to tidal reduction caused by the Oosterschelde storm surge-barrier. *Hydrobiologia* **282/283**, 335–53.

De Luis, M., García-Cano, M.F., Cortina, J. *et al.* (2001) Climatic trends, disturbances and short-term vegetation dynamics in a Mediterranean shrubland. *Forest Ecology and Management* **147**, 25–37.

de Meester, L., Gomez, A., Okamura, B. and Schucik, K. (2002) The monopolization hypothesis on the dispersal-gene flow paradox in aquatic organisms. *Acta Oecologica* **23**, 121–35.

De Melo, R., France, R. and McQueen, D.J. (1992) Biomanipulation hit or myth? *Limnology and Oceanography* **37**, 192–207.

den Boer, P.J. (1968) Spreading of risk and stabilization of animal numbers. *Acta Biotheoretica* **18**, 165–94.

den Boer. P.J. (1990) The survival value of dispersal in terrestrial arthropods. *Biological Conservation* **54**, 175–92.

Denton, J.S., Hitchings, S.P., Beebee, T.J.C. and Gent, A. (1997) A recovery program for the natterjack toad (*Bufo calamita*) in Britain. *Conservation Biology* **11**, 1329–38.

Derrickson, S.R. and Carpenter J.W. (1983) Techniques for reintroducing cranes to the wild. *Annual Proceedings of the American Association of Zoo Veterinarians* **1983**, 148–52.

de Ruiter, P.C., Neutel, A.M. and Moore, J.C. (1995) Energetics, patterns of interaction strengths, and stability in real ecosystems. *Science* **269**, 1257–60.

Diamond, J.M. (1975) Assembly of species communities. In: *Ecology and Evolution of Communities* (eds. M.L. Cody and J.M. Diamond), pp. 342–444. Harvard University Press, Cambridge, MA.

Dicke, M. and Vet, L.E.M. (1999) Plant-carnivore interactions: evolutionary and ecological consequences for plant, herbivore and carnivore. In: *Herbivores: Between Plants and Predators* (eds. H. Olff, V.K. Brown and R.H. Drent), pp. 483–520. Blackwell Science, Oxford.

Dickman, C.R. (1992) Commensal and mutualistic interactions among terrestrial vertebrates. *Trends in Ecology and Evolution* **7**, 194–7.

Dickson, W. and Brodin, Y.-W. (1995) Strategies and methods for freshwater liming. In: *Liming of Acidified*

Surface Waters: A Swedish Synthesis (eds. L. Henriksson and Y.-W. Brodin), pp. 81–116. Springer, Berlin.

Diemer, M. and Prock, S. (1993) Estimates of alpine seed bank size in two central European and one Scandinavian subarctic plant communities. *Arctic and Alpine Research* **25**, 194–200.

Dierssen, K. (1996) *Vegetation Nordeuropas.* Ulmer, Stuttgart.

Dierssen, K. and Dierssen, B. (2001) *Moore: Ökosysteme Mitteleuropas aus geobotanischer Sicht.* Ulmer, Stuttgart.

Dieter, B.E., Wion, D.A. and McConnaughey, R.A. (2003) *Mobile Fishing Gear Effects on Benthic Habitats: A Bibliography.* National Oceanic and Atmospheric Administration, Department of Fisheries, Seattle.

Dijk, E., Willems, J.H. and van Andel J. (1997) Nutrient responses as a key factor to the ecology of orchid species. *Acta Botanica Neerlandica* **46**, 339–63.

Dijkema, K.S. (1984) *Salt Marshes in Europe.* Council of Europe, Strasbourg.

Dijkema, K.S. (1987) Changes in salt-marsh area in the Netherlands Wadden Sea after 1600. In: *Vegetation Between Land and Sea* (eds. A.H.L. Huiskes, C.W.P.M. Blom and J. Rozema), pp. 42–9. Junk, Dordrecht.

Dijkema, K.S. (1990) Salt and brackish marshes around the Baltic Sea and adjacent parts of the North Sea; their development and management. *Biological Conservation* **51**, 191–209.

Dijkema, K.S. (1997) Impact prognosis for salt marshes from subsidence by gas extraction in the Wadden Sea. *Journal of Coastal Research* **13**, 1294–304.

Dionisio Pires, L.M., Jonker, R.R., van Donk, E. and Laarbroek, H.J. (2004) Selective grazing by adults and larvae of zebra mussel (*Dreissena polymorpha*): application of flowcytometry in natural seston. *Freshwater Biology* **49**, 116–26.

Dobson, A. and Crawley, M. (1994) Pathogens and the structure of plant communities. *Trends in Ecology and Evolution* **9**, 393–8.

Dodd, C.K. and Seigel R.A. (1991) Relocation, repatriation, and translocation of amphibians and reptiles – are they conservation strategies that work? *Herpetologica* **47**, 336–50.

Domesday Book (1086) or *The Great Survey of England of William the Conqueror.* Facsimile of the part relating to Nottinghamshire (1862). Ordnance Survey Office, Southampton.

Donker, A. (1999) Pitrus, een verrassend goed reptielbiotoop. *De Levende Natuur* **100**, 222–3.

Dorland, E., Bobbink, R., Messelink, J.H. and Verhoeven, J.T.M. (2003) Soil ammonium accumulation after sod cutting hampers the restoration of degraded wet heathland. *Journal of Applied Ecology* **40**, 804–14.

Downs, P. and Thorne, C. (1998) Design principles and suitability testing for rehabilitation in a flood defence channel: the River Idle, Nottinghamshire. *UK Aquatic Conservation: Marine and Freshwater Ecosystems* **8**, 17–38.

Drenner, R.W. and Hambright, K.D. (2002) Piscivores, trophic cascades, and lake management. *Scientific World* **2**, 284–307.

Drent, R.H. and Prins, H.H.T. (1987) The herbivore as prisoner of its food supply. In: *Disturbance in Grasslands: Causes, Effects and Processes* (eds. J. van Andel, J.P. Bakker and R.W. Snaydon), pp. 131–48. Junk, Dordrecht.

Drent, R.H. and van der Wal, R. (1999) Cyclic grazing in vertebrates and the manipulation of the food resource. In: *Herbivores: Between Plants and Predators* (eds. H. Olff, V.K. Brown and R.H. Drent), pp. 271–99. Blackwell Science, Oxford.

Drinkwaard, A.C. (1999) Introductions and developments of oysters in the North Sea area: a review. *Helgoländer Meeresuntersuchungen* **52**, 301–8.

Düker, C. (2003) Untersuchungen zur Enchytraeidenfauna (Oligochaeta, Annelida) ausgewählter Altersstadien forstlich rekultivierter Kippenstandorte im Lausitzer Braunkohlerevier. *Cottbuser Schriften* **23**.

Düker, C., Keplin, B. and Hüttl, R.F. (1999) Development of Enchytraeid communities in reclaimed lignite mine spoil. *Newsletter on Enchytraeidae* **6**, 77–89.

Dukes, J.S. and Mooney, H.A. (1999) Does global change increase the success of biological invaders? *Trends in Ecology and Evolution* **14**, 135–9.

EC (1991) European Community Convention on the Protection of the Alps (Alpine Convention). *Official Journal of the European Communities* **34**, 1–8.

EC (1992a) Habitats Directive 92/43/EEC. European Communities Council Directive of 21 May 1992 on the conservation of natural habitats and of wild fauna and flora. *Official Journal of the European Communities* **L206**, 7–50.

EC (1992b) Council Regulation (EEC) No 2080/92 of 30 June 1992 instituting a Community aid scheme for forestry measures in agriculture. *Official Journal of the European Communities* **L215**, 96–9.

EC (1992c) *Treaty on the European Union: Treaty of Maastricht.* 7 February 1992. European Communities, Maastricht.

EC (2000) Directive 2000/60/EC of the European Parliament and of the Council of 23 October 2000, establishing a framework for the Community action in the field of water policy. *Official Journal of the European Communities* L327, 1–73.

Edelvang, K. (1997) Tidal variation in the settling diameters of suspended matter on a tidal mud flat. *Helgoländer Meeresuntersuchungen* 51, 269–79.

Edmondson, W.T. (1991) *The Use of Ecology: Lake Washington and Beyond*. University of Washington Press, Seattle.

Edmondson, W.T. (1994) Sixty years of Lake Washington: a curriculum vitae. *Lake and Reservoir Management* 10, 75–84.

Edmondson, W.T. and Lehman, J.T. (1981) The effects of changes in nutrient income on the condition of Lake Washington. *Limnology and Oceanography* 26, 1–29.

Egan, D. and Howell, E.A. (eds.) (2001) *The Historical Ecology Handbook: A Restorationist's Guide to Reference Ecosystems*. Island Press, Washington, DC.

Egler, F.E. (1954) Vegetation science concepts. I. Initial floristic composition, a factor in old field vegetation development. *Vegetatio* 4, 412–17.

Ehrenfeld, J.G. (2000) Defining the limits of restoration: the need for realistic goals. *Restoration Ecology* 8, 2–9.

Ehrenfeld, J.G. and Toth, L.A. (1997) Restoration ecology and the ecosystem perspective. *Restoration Ecology* 5, 307–17.

Ehrlich, P.R. and Ehrlich, A.H. (1981) *Extinction: The Causes and Consequences of the Disappearance of Rare Species*. Random House, New York.

Ehrlich, P.R. and Ehrlich, A.H. (1992) The value of biodiversity. *Ambio* 21, 219–26.

Eisma, D., de Boer, P.L. and Cadée, G.C. (1998) *Intertidal Deposits: River Mouths, Tidal Flats, and Coastal Lagoons*. CRC Press, Boca Raton.

Elgersma, A.-M. and Dhillion, S.S. (2002) Soil chemistry and climatic regions: tools for restoration of forest communities. In: *Proceedings of the IUFRO Conference on Restoration of Boreal and Temperate Forests* (eds. E.S. Gardiner and L.J. Breland). Report No. 11, pp. 181–2. Danish Centre for Forest, Landscape and Planning, Vejle.

Ellenberg, H. (1986) *Vegetation Mitteleuropas mit den Alpen in ökologischer Sicht*. Ulmer, Stuttgart.

Ellenberg, H. (1996) *Vegetation Mitteleuropas mit den Alpen in ökologischer, dynamischer und historischer Sicht*, 5th edn. Ulmer, Stuttgart.

Ellenberg, H., Weber, H.E., Düll, R. *et al.* (1991) *Zeigerwerte von Pflanzen in Mitteleuropa*. Verlag Erich Goltze, Göttingen.

Elliot, R. (1997) *Faking Nature: The Ethics of Environmental Restoration*. Routledge Press, London.

Ellstrand, N.C. (1992) Gene flow by pollen: implications for plant conservation genetics. *Oikos* 63, 77–86.

Elton, C.S. (1958) *The Ecology of Invasions by Animals and Plants*. Methuen, London.

EMEP (1999) Flächendeckende Modellierung der Gesamtdeposition von oxidiertem und reduziertem Stickstoff sowie oxidiertem Schwefel in Europa [Cooperative Programme for Monitoring and Evaluation of the Long-Range Transmission of Air Pollutants in Europe]. In: *Daten Zur Umwelt*, 2000, pp. 178–82. Umweltbundesamt, Berlin.

Emerson, C.W. (1989) Wind stress limitation of benthic secondary production in shallow, soft-sediment communities. *Marine Ecology: Progress Series* 53, 65–77.

Ens, B.J. and Alting, D. (1996) The effect of an experimentally created mussel bed on bird densities and food intake of the oystercatcher *Haematopus ostralegus*. *Ardea* 84A, 493–508.

Erchinger, H.F., Coldewey, H.-G., Frank, U. *et al.* (1994) *Erosionsfestigkeit von Hellern*. Staatliches Amt für Insel- und Küstenschutz, Norden.

Eriksson, M.O.G., Henrikson, L., Nilsson, I.-B. *et al.* (1980) Predator-prey relations important for biotic changes in acidified lakes. *Ambio* 9, 248–9.

Eriksson, O. (1996) Regional dynamics of plants: a review of evidence for remnant, source-sink and metapopulations. *Oikos* 77, 248–58.

Ernest, S.K.M. and Brown, J.H. (2001) Homeostasis and compensation: the role of species and resources in ecosystem stability. *Ecology* 82, 2118–32.

Ernst, W.H.O. (1998) Invasion, dispersal and ecology of the South African neophyte *Senecio inaequidens* in The Netherlands: from wool alien to railway and road alien. *Acta Botanica Neerlandica* 47, 131–51.

Esselink, P. (2000) Nature management of coastal salt marshes: interactions between anthropogenic influences and natural dynamics. PhD thesis, University of Groningen.

Esselink, P., Dijkema, K.S., Reents, S. and Hageman, G. (1998) Vertical accretion and profile changes in man-made tidal marshes in the Dollard Estuary in the Netherlands. *Journal of Coastal Research* 14, 570–82.

Esselink, P., Fresco, L.F.M. and Dijkema, K.S. (2002) Vegetation change in a man-made salt marsh affected by a reduction in both grazing and drainage. *Applied Vegetation Science* 5, 17–32.

Essink, K. (1986) On the occurrence of the American jack-knife clam *Ensis directus* (Conrad, 1843) in N.W. Europe (Bivalvia, Cultellidae). *Basteria* **50**, 33–4.

Essink, K. and Esselink, P. (eds.) (1998) *Het Eems-Dollard estuarium: interacties tussen menselijke beinvloeding en natuurlijke dynamiek.* Report RIKZ-98.020, Haren.

Essink, K., Beukema, J.J., Madsen, P.B. *et al.* (1998) Long-term development of biomass of intertidal macrozoobenthos in different parts of the Wadden Sea. Governed by nutrient loads? *Senckenbergiana maritima* **29**, 25–35.

Etienne, M., Aronson, J. and Le Floc'h, E. (1998) Abandoned lands and land use conflicts in southern France: piloting ecosystem trajectories and redesigning outmoded landscapes in the 21st century. In: *Landscape Degradation and Biodiversity in Mediterranean-type Ecosystems* (eds. P.W. Rundel, G. Montenegro and F. Jaksic), pp. 127–40, Ecological Studies Series, vol. 136. Springer, Berlin.

Everts, F.H. and de Vries, N.P.J. (1991) *De Vegetatieontwikkeling van Beekdalsystemen: een landschapsoecologische analyse van enkele Drentse beekdalen.* Historische Uitgeverij, Groningen.

Eysink, W.D., Dijkema, K.S., Van Dobben, H.F. *et al.* (2000) *Monitoring effecten bodemdaling op Ameland-Oost. Evaluatie na 13 jaar gaswinning.* Begeleidingscommissie Monitoring Bodemdaling Ameland, Assen.

Faber, A. (1937) *Erläuterungen zum pflanzensoziologischen Kartenblatt des mittleren Neckar- und des Ammertalgebietes* (eds. Württemberg Forstdirektion und Württemberg Naturaliensammlung), Stuttgart.

Fagan, W.F. (1997) Omnivory as a stabilizing feature of natural communities. *American Naturalist* **150**, 554–67.

Fahrig, L. (2003) Effects of habitat fragmentation on biodiversity. *Annual Review of Ecology and Systematics* **34**, 487–515.

Falk, D.A., Millar, C.I. and Olwell, M. (eds.) (1996) *Restoring Diversity: Strategies for Reintroduction of Endangered Plants.* Island Press, Washington, DC.

Fanelli, G., Piraina, S., Belmonte, G. *et al.* (1994) Human predation along Apulian rocky coasts (SE Italy): desertification caused by *Lithophaga lithophaga* (Mollusca) fisheries. *Marine Ecology: Progress Series* **110**, 1–8.

FAO (2001) *Global Forest Resources Assessment 2001,* main report. Food and Agriculture Organization of the United Nations, Rome.

Farina, A. (2000) The cultural landscape as a model for the integration of ecology and economics. *BioScience* **50**, 313–20.

Felinks, B., Pilarski, M. and Wiegleb, G. (1998) Vegetation survey in the former brown coal mining area of Eastern Germany by integrating remote sensing and ground-based methods. *Applied Vegetation Science* **1**, 233–40.

Fell, P.E., Warren, R.S. and Niering, W.A. (2000) Restoration of salt and brackish tidelands in southern New England. In: *Concepts and Controversies in Tidal Marsh Ecology* (eds. M.P. Weinstein and D.A. Kreeger), pp. 845–58. Kluwer Academic Publishers, Dordrecht.

Fettweis, U., Bens, O. and Hüttl, R.F. (2005) Soil organic matter accumulation and humus characteristics in reclaimed mine sites afforested with *Pinus* spec.: information from carbon pool differentiation and ^{13}C CPMAS NMR spectroscopy. *Geoderma*, in press.

Finlay, R.D., Ek, H., Odham, G. and Söderström, B. (1988) Mycelial uptake, translocation and assimilation of nitrogen from ^{15}N-labelled ammonium by *Pinus sylvestris* plants infected with four different ectomycorrhizal fungi. *New Phytologist* **110**, 59–66.

Fischer, A. (1993) Zehnjährige vegetationskundliche Dauerbeobachtungen stadtnaher Waldbestände. *Forstwissenschaftliches Centralblatt* **112**, 141–58.

Fischer, A. (1997) Vegetation dynamics in European beech forests. *Annali di Botanica* **55**, 59–76.

Fischer, A. (ed.) (1998) *Die Entwicklung von Wald-Biozönosen nach Sturmwurf.* Ecomed-Verlagsgesellschaft, Landsberg.

Fischer, A. (1999) Floristical changes in Central European forest ecosystems during the past decades as an expression of changing site conditions. In: *Causes and Consequences of Accelerating Tree Growth in Europe* (eds. T. Karjalainen, H. Spiecker and O. Laroussinie). Proceedings of the European Forest Institute, vol. 27, pp. 53–64. European Forest Institute, Joensuu.

Fischer, A., Lindner, M., Abs, C. and Lasch, P. (2002) Vegetation dynamics in Central European forest ecosystems (near natural as well as managed) after storm events. *Folia Geobotanica* **37**, 17–32.

Fischer, H. (1998) Acker-Erstaufforstungen. Bestandesbegründung, Wachstum und Ökologie an Fallbeispielen. PhD thesis, University of Göttingen.

Fischer, H. (1999) Begleitwuchsregulierung mittels Hilfspflanzendecken: mögliche Negativeffekte einer waldbaulichen Konzeption für Acker-Erstaufforstungen. *Forstarchiv* **70**, 207–18.

Fischer, H. (2000) Zur Präsenz der Gemeinen Quecke (*Agropyron repens* (L.) P.B.) auf Ackeraufforstungsflächen. *Forstarchiv* **71**, 199–204.

Fischer, H., Bens, O. and Hüttl, R.F. (2002) Veränderung von Humusform, -vorrat und -verteilung im Zuge von Waldumbaumaßnahmen im Nordwestdeutschen Tiefland. *Forstwissenschaftliches Centralblatt* 121, 322–34.

FISRWG (1998) *Stream Corridor Restoration: Principles, Processes, and Practices.* The Federal Interagency Stream Restoration Working Group, Washington, DC.

Floret, C., Le Floc'h, E., Romane, F. and Pontanier, R. (1981) Dynamique de systèmes écologiques de la zone aride. *Acta Oecologica* 2, 195–214.

Florineth, F. (1992) Hochlagenbegrünung in Südtirol. *Rasen-Turf-Gazon* 23, 74–80.

Florineth, F. (2000) Erosionsschutz im Gebirge: neue Begrünungsmethoden. *Ingenieurbiologie: Mitteilungsblatt* 2, 62–71.

Fonseca, M.S. and Fisher, J.S. (1986) A comparison of canopy friction and sediment movement between four species of seagrass with reference to their ecology and restoration. *Marine Ecology: Progress Series* 29, 15–22.

Foreman, D., Davis, J., Johns, D. *et al.* (1995) The Wildland Mission Statement. *Wild Earth* 1, 3–4.

Forsberg, C. (1987) Evaluation of lake Restoration in Sweden. *Schweizerische Zeitschrift für Hydrologie* 49, 260–74.

Francis, C.F. (1990) Variaciones sucesionales y estacionales de vegetación en campos abandonados de la provincia de Murcia, España. *Ecología* 4, 35–47.

Francis, R. and Read, D.J. (1994) The contribution of mycorrhizal fungi to the determination of plant community structure. *Plant and Soil* 159, 11–25.

Freckleton, R.P. and Watkinson, A.R. (2002) Large-scale dynamics of plants: metapopulations, regional assemblages and patchy populations. *Journal of Ecology* 90, 419–34.

Fretwell, S.D. (1987) Food chain dynamics: the central theory of ecology? *Oikos* 50, 291–301.

Friedel, M.H. (1991) Range condition assessment and the concept of thresholds: a viewpoint. *Journal of Range Management* 44, 422–6.

Friedel, M.H., Bastin, G.N. and Griffin, G.F. (1988) Range assessment and monitoring of arid lands: the derivation of functional groups to simplify vegetation data. *Journal of Environmental Management* 27, 85–97.

Fritts, S.H. and Carbyn, L.N. (1995) Population viability, nature reserves, and the outlook for gray wolf conservation in North America. *Restoration Ecology* 3, 26–38.

Fritts, S.H., Bangs, E.E., Fontaine, J.A. *et al.* (1997) Planning and implementing a reintroduction of wolves in Yellowstone National Park and Central Idaho. *Restoration Ecology* 5, 7–27.

Gadgil, M. (1995) Prudence and profligacy: a human ecological perspective. In: *The Economics and Ecology of Biodiversity Decline* (ed. T.M. Swanson), pp. 99–110. Cambridge University Press, Cambridge.

Galatowitsch, S.M. and van der Valk, A.G. (1996) The vegetation of restored and natural prairie wetlands. *Ecological Applications* 6, 102–12.

Gange, A. (2000) Arbuscular mycorrhizal fungi, Collembola and plant growth. *Trends in Ecology and Evolution* 15, 369–72.

García, A. (1992) Conserving the species-rich meadows of Europe. *Agriculture, Ecosystems and Environment* 40, 219–32.

Gardner, R.H., O'Neill, R.V., Turner, M.G. and Dale, V.H. (1989) Quantifying scale-dependent effects of animal movement with simple percolation models. *Landscape Ecology* 4, 217–27.

Gaston, K.J. (1996a) Species richness: measure and measurement. In: *Biodiversity: A Biology of Numbers and Difference* (ed. K.J. Gaston), pp. 77–113. Blackwell Science, Oxford.

Gaston, K.J. (1996b) Species-range-size distributions: patterns, mechanisms and implications. *Trends in Ecology and Evolution* 11, 197–201.

Gaston, K.J. (1997) What is rarity? In: *The Biology of Rarity* (eds. W.E. Kunin and K.J. Gaston), pp. 30–47. Chapman and Hall, London.

Geist, C. and Galatowitsch, S.M. (1999) Reciprocal model for meeting ecological and human needs in restoration projects. *Conservation Biology* 13, 970–79.

Gerke, H.H., Hangen, E., Schaaf, W. and Hüttl, R.F. (2001) Spatial variability of water repellency in a lignitic mine soil afforested with *Pinus nigra*. *Geoderma* 102, 255–74.

Gerken, B. and Meyer, C. (eds.) (1996) Wo lebten Pflanzen und Tiere in der Naturlandschaft und der frühen Kulturlandschaft Europas? *Natur- und Kulturlandschaft* 1, 1–205.

Giller, P.S. and O'Donavan, G. (2002) Biodiversity and ecosystem function: do species matter? *Biology and Environment: Proceedings of the Royal Irish Academy* 102B, 129–39.

Girel, J. (1994) Old distribution procedure of both water and matter fluxes in floodplains of western Europe: impact on present vegetation. *Environmental Management* 18, 203–21.

Gitay, H. and Noble, I.R. (1997) What are functional types and how should we seek them? In: *Plant Functional*

Types (eds. T.M. Smith, H.H. Shugart and F.I. Woodward), pp. 3–19. Cambridge University Press, Cambridge.

Gitay, H., Suárez, A., Watson, H.T. and Dokken D.J. (eds.) (2002) *Climate Change and Biodiversity*. Technical Paper V. IPCC, Geneva.

Glaser, P.H., Janssens, J.A. and Siegel, D.I. (1990) The response of vegetation to chemical and hydrological gradients in the lost river peatland, northern Minnesota. *Journal of Ecology* 78, 1021–48.

Glatzel, G. (1999) Historic forest use and possible implications to recently accelerated tree growth in central Europe. In: *Causes and Consequences of Accelerating Tree Growth in Europe* (eds. T. Karjalainen, H. Spiecker and O. Laroussinie). Proceedings of the European Forest Institute, vol. 27, pp. 75–86. European Forest Institute, Joensuu.

Gleason, M.L., Elmer, D.A., Pien, N.C. and Fisher, J.S. (1979) Effects of stem density upon sediment retention by salt marsh cord grass *Spartina alterniflora* Loisel. *Estuaries* 2, 271–3.

Golldack, J., Münzenberger, B. and Hüttl, R.F. (2000) Mykorrhizierung der Kiefer (*Pinus sylvestris* L.) auf forstlich rekultivierten Kippenstandorten des Lausitzer Braunkohlereviers. In: *Rekultivierung in Bergbaufolgelandschaften* (eds. G. Broll, W. Dunger, B. Keplin and W. Topp), pp. 131–46. Springer, Berlin.

Golley, F.B. (1995) Reaching a landmark. *Landscape Ecology* 10, 3–4.

Gómez, J.M., Hódar, J.A., Zamora, R. *et al.* (2001) Ungulate damage on Scots pines in Mediterranean environments. Effects of association with shrubs. *Canadian Journal of Botany* 79, 739–46.

Gondard, H., Jauffret, S., Aronson, J. and Lavorel, S. (2003) Plant functional types: a promising tool for management and restoration of degraded lands. *Applied Vegetation Science* 6, 223–234.

González-Bernáldez, F. (1992) La frutalización del paisaje mediterráneo. In: *Paisaje mediterráneo*, pp. 136–41. Electa, Milan.

Gordon, D.R. (1998) Effects of invasive, non-indigenous plant species on ecosystem processes: lessons from Florida. *Ecological Applications* 8, 975–89.

Gossow, H. and Honsig-Erlenburg, P. (1986) Management problems with reintroduced lynx in Austria. In: *Cats of the World: Biology, Conservation and Management* (eds. S.D. Miller and D.D. Everett), pp. 77–83. National Wildlife Federation, Washington, DC.

Grabherr, G. and Hohengartner, H. (1989) Die 'Junggärtnermethode': eine neue Methode zur Renaturierung hochalpiner Rohbodenflächen mit autochthonem Pflanzgut. *Die Bodenkultur* 40, 85–94.

Grantz, D.A., Vaughn, D.L., Farber, R.J. *et al.* (1998) Transplanting native plants to revegetate abandoned farmland in the Western Mojave desert. *Journal of Environmental Quality* 27, 960–7.

Grashof-Bokdam, C. (1997) Forest species in an agricultural landscape in the Netherlands: effects of habitat fragmentation. *Journal of Vegetation Science* 8, 21–8.

Gray, D.H. and Sotir, R.B. (1996) *Biotechnical and Soil Bioengineering Slope Stabilization*. Wiley, New York.

Gregory, K.J. (1997) *Fluvial Geomorphology of Great Britain*. Chapman and Hall, London.

Gregory, K.J. and Lewin, J. (1987) Conclusion palaeohydrological synthesis and application. In: *Palaeohydrology in Practice* (eds. K.J. Gregory, J. Lewin and J.B. Thornes). Wiley, Chichester.

Griffith, B., Scott, J.M., Carpenter, J.W. and Reed, C. (1989) Translocation as a species conservation tool: status and strategy. *Science* 245, 477–80.

Grime, J.P. (1973) Competitive exclusion in herbaceous vegetation. *Nature* 242, 344–7.

Grime, J.P. (1979) *Plant Strategies and Vegetation Processes*, 1st edn. Wiley, Chichester.

Grime, J.P. (1998) Benefits of plant diversity to ecosystems: immediate, filter and founder effects. *Journal of Ecology* 86, 902–10.

Grime, J.P. (2001) *Plant Strategies, Vegetation Processes, and Ecosystem Properties*, 2nd edn. Wiley, Chichester.

Grime, J.P., Mackey, J.M., Hillier, S.H. and Read, D.J. (1987) Floristic diversity in a model system using experimental microcosms. *Nature* 328, 420–2.

Grimm, V. and Wissel, C. (1997) Babel, or the ecological stability discussions: an inventory and analysis of terminology and a guide for avoiding confusion. *Oecologia* 109, 323–34.

Gritten, R.H. (1995) *Rhododendron ponticum* and some other invasive plants in the Snowdonia National Park. In: *Plant Invasions: General Aspects and Special Problems* (eds. P. Pyšek, K. Prach, M. Rejmánek and M. Wade), pp. 213–19. SPB Academic Publishing, Amsterdam.

Groombridge, B. and Jenkins, M.D. (2000) *Global Biodiversity: Earth's Living Resources in the 21st Century*. World Conservation Press, Cambridge.

Grootjans, A.P. and van Diggelen, R. (1995) Assessing restoration perspectives of degraded fens. In: *Restoration of Temperate Wetlands* (eds. B.D. Wheeler, S.S. Shaw, W.J. Fojt and R.A. Robertson), pp. 73–90. Wiley, Chichester.

Grootjans, A.P., Schipper, P.C. and Van der Windt, H.J. (1985) Influence of drainage on N-mineralization and vegetation response in wet meadows. I: Calthion palustris stands. *Acta Oecologica* 6, 403–17.

Grootjans, A.P., Allersma, G. and Kik, C. (1987) Hybridization of the habitat in disturbed hay meadows. In: *Disturbance in Grasslands: Causes, Effects and Processes* (eds. J. van Andel, J.P. Bakker and R.W. Snaydon), pp. 67–77. Junk Publishers, Dordrecht.

Grootjans, A.P., van Diggelen, R., Wassen, M.J. and Wiersinga, W.A. (1988) The effects of drainage on groundwater quality and plant species distribution in stream valley meadows. *Vegetatio* 75, 37–48.

Grootjans, A.P., van Wirdum, G., Kemmers, R.H. and van Diggelen, R. (1996) Ecohydrology in The Netherlands: principles of an application-driven interdiscipline. *Acta Botanica Neerlandica* 45, 491–516.

Grootjans, A.P., Ernst, W.H.O. and Stuyfzand, P.J. (1998) European dune slacks: strong interactions of biology, pedogenesis and hydrology. *Trends in Ecology and Evolution* 13, 96–100.

Grootjans, A.P., Bakker, J.P., Jansen, A.J.M. and Kemmers, R.M. (2002) Restoration of brook valley meadows in The Netherlands. *Hydrobiologia* 478, 149–70.

Grootjans, A.P., Adema, E.B., Baaijens, G.J., Rappoldt, K. and Verschoor, A. (2003) Mechanisms behind restoration of small bog ecosystems in a cover sand landscape. *Archiv für Naturschutz und Landschaftsforschung* August 2003, 43–48.

Grosholz, E. (2002) Ecological and evolutionary consequences of coastal invasions. *Trends in Ecology and Evolution* 17, 22–7.

Grove, A.T. and Rackham, O. (2000) *The Nature of Mediterranean Europe: An Ecological History.* Yale University Press, New Haven, CT.

Grover, J. (1994) Assembly rules for communities of nutrient-limited plants and specialist herbivores. *American Naturalist* 143, 258–82.

Grubb, P.J. (1985) Plant populations in relation to habitat, disturbance and competition: problems of generalization. In: *The Population Structure of Vegetation* (ed. J. White), pp. 595–621. Junk Publishers, Dordrecht.

Grubb, P.J. (1998) A reassessment of strategies of plants which cope with shortages of resources. *Perspectives in Plant Ecology, Evolution and Systematics* 1, 3–31.

Grünewald, U. and Nixdorf, B. (1995) Erfassung und Prognose der Gewässergüte der Lausitzer Restseen. In: *Fachtagung Rezente Flutungsprobleme mitteldeutscher und Lausitzer Tagebaurestlöcher.* Proceedings of the Dresner Grundwasserforschungszentrum e.V. *Coswig* 8, 159–79.

Grünig, A. (ed.) (1994) *Mires and Man: Mire Conservation in a Densely Populated Country: The Swiss Experience.* Swiss Federal Institute for Forest, Snow and Landscape Research, Birmensdorf.

Gulati, R.D. (1989) Concept of stress and recovery in aquatic ecosystems. In: *Ecological Assessment of Environmental Degradation, Pollution and Recovery* (ed. O. Ravera), pp. 81–119. Elsevier, Amsterdam.

Gulati, R.D. (1990a) Zooplankton structure in the Loosdrecht lakes in relation to trophic status and recent restoration measures. *Hydrobiologia* 191, 173–88.

Gulati, R.D. (1990b) Structural and grazing responses of zooplankton community to biomanipulation of some Dutch water bodies. *Hydrobiologia* 200/201, 99–118.

Gulati, R.D. (1995) Manipulation of fish population for lake recovery from eutrophication in the temperate region. In: *Biomanipulation in Lakes and Reservoirs Management* (eds. R. De Bernardi and G. Giussani). Guidelines of Lake Management, vol. 7, pp. 53–79. International Lake Environment Committee, Shiga.

Gulati, R.D. and van Donk, E. (2002) Lakes in the Netherlands, their origin, eutrophication and restoration: state-of-the-art-review. *Hydrobiologia* 478, 73–106.

Gulati, R.D., Lammens, E.H.R.R., Meijer, M.-L. and van Donk, E. (eds.) (1990) Biomanipulation, tool for water management. *Hydrobiologia* 200/201.

Gulati, R.D., Ooms-Wilms, A.L., van Tongeren, O.F.R. *et al.* (1992) The dynamics and role of limnetic zooplankton in Loosdrecht Lakes (the Netherlands). *Hydrobiologia* 233, 69–86.

Gunderson, L.H. (2000) Ecological resilience: in theory and application. *Annual Review of Ecology and Systematics* 31, 425–39.

Gunderson, L.H. and Holling, C.S. (eds.) (2002) *Panarchy: Understanding Transformations in Human and Natural Systems.* Island Press, Washington, DC.

Gurd, D.B., Nudds, T.D. and Rivard, D.H. (2001) Conservation of mammals in eastern North American wildlife reserves: how small is too small? *Conservation Biology* 15, 1355–63.

Hadri, H. and Tschinkel, H. (1975) *La régénération de Pinus halepensis après coupe rase et sous peuplement* (note de recherche). Institut National de Recherches Forestières, Tunisie.

Häge, K., Drebenstedt, C. and Angelov, E. (1996) Landscaping and ecology in the lignite mining area of Maritza-East, Bulgaria. *Water, Air and Soil Pollution* 91, 135–44.

Hairston, N.G., Smith, F.E. and Slobodkin, L.E. (1960) Community structure, population control, and competition. *American Naturalist* **94**, 421–5.

Halahan, R. (2000) *Species Alert! Natura 2000: a Last Chance for European Biodiversity.* World Wildlife Fund UK.

Hall, D.J., Threlkeld, S.T., Burns, C.W. and Crowley P.H. (1976) The size-efficiency hypothesis and the size structure of zooplankton communities. *Annual Review of Ecology and Systematics* **7**, 177–208.

Hall, M. (1998) Ideas from overseas: American preservation and Italian restoration. *The George Wright Forum: A Journal of Cultural and Natural Parks and Reserves* **15**, 24–9.

Hall, M. (2000) Comparing damages: American and Italian concepts of degradation. In: *Methods and Approaches in Forest History* (eds. M. Agnoletti and S. Anderson), pp. 145–52. CAB International, Wallingford.

Hall, S.J. (1994) Physical disturbance and marine benthic communities: life in unconsolidated sediments. *Oceanography and Marine Biology: An Annual Review* **32**, 179–239.

Hall, S.J. and Harding, M.J.C. (1997) Physical disturbance and marine benthic communities: the effects of mechanical harvesting of cockles on non-target benthic infauna. *Journal of Applied Ecology* **34**, 497–517.

Hall, S.J., Basford, D.J. and Robertson, M.R. (1991) The impact of hydraulic dredging for razor clams *Ensis* sp. on an infaunal community. *Netherlands Journal of Sea Research* **27**, 119–25.

Hammill, K.A., Bradstock, R.A. and Allaway, W.G. (1998) Post-fire seed dispersal and species re-establishment in proteaceous heath. *Australian Journal of Botany* **46**, 407–19.

Handa, I.T. and Jefferies, R.L. (2000) Assisted revegetation trials in degraded salt marshes. *Journal of Applied Ecology* **37**, 944–58.

Handel, S.N. (1997) The role of plant-animal mutualism in the design and restoration of natural communities. In: *Restoration Ecology and Sustainable Development* (eds. K. Urbanska, N.R. Webb and P.J. Edwards), pp. 111–32. Cambridge University Press, Cambridge.

Hangen, E. (2003) Präferenzieller Fluss in einem heterogenen aufgeforsteten Kippboden. *Cottbuser Schriften* **19**.

Hansen, H.O. (ed.) (1996) *River Restoration: Danish Experiences and Examples.* National Environmental Research Institute, Silkeborg.

Hansen, J., Reitzel, K., Jensen, H.S. and Andersen, F.O. (2003) Effect of aluminium, iron, oxygen, and nitrate additions on phosphorus release from the sediment of a Danish softwater lake. *Hydrobiologia* **492**, 139–49.

Hanski, I. (1997) Metapopulation dynamics: from concepts and observations to predictive models. In: *Metapopulation Biology: Ecology, Genetics and Evolution* (eds. I. Hanski and M.E. Gilpin), pp. 69–91. Academic Press, San Diego.

Hanski, I. (1999) *Metapopulation Ecology.* Oxford University Press, Oxford.

Hanski, I. and Gilpin, M.E. (1991) Metapopulation dynamics: brief history and conceptual domain. *Biological Journal of the Linnean Society* **42**, 3–16.

Hanski, I. and Simberloff, D. (1997) The metapopulation approach, its history, conceptual domain, and application to conservation. In: *Metapopulation Biology: Ecology, Genetics and Evolution* (eds. I. Hanski and M.E. Gilpin), pp. 5–26. Academic Press, San Diego.

Hanski, I., Moilanen, A. and Gyllenberg, M. (1996) Minimum viable metapopulation size. *American Naturalist* **147**, 527–41.

Hanson, M.T. and Willison, J.M. (1983) The 1978 relocation of tule elk at Fort Hunter Liggett – reasons for its failure. *Cal-Neva Wildlife: Transactions* 43–9.

Hansson, L.A., Annadotter, H., Bergmann, E. *et al.* (1998) Biomanipulation as an application of food-chain theory: constraints, synthesis, and recommendations for temperate lakes. *Ecosystems* **1**, 558–74.

Hardin, G. (1985) Human ecology: the subversive, conservative science. *American Zoologist* **25**, 469–76.

Hardin, G. (1991) Paramount positions in ecological economics. In: *Ecological Economics: The Science and Management of Sustainability* (ed. R. Costanza), pp. 47–57. Columbia University Press, New York.

Hardin, G. (1993) *Living within Limits: Ecology, Economics and Population Taboos.* Oxford University Press, New York.

Harding, M. (1993) Redgrave and Lopham fens, East Anglia, England: a case study of change in flora and fauna due to groundwater abstraction. *Biological Conservation* **66**, 35–45.

Harms, B., Knaapen, J.P. and Rademakers, J.G. (1993) Landscape planning for nature restoration: comparing regional scenarios. In: *Landscape Ecology of a Stressed Environment* (eds. C.C. Vos and P. Opdam), pp. 197–218. Chapman and Hall, London.

Harper, D., Brierley, B., Ferguson, A.J.D. and Phillips, G. (eds.) (1999) The ecological basis of lake and reservoir management. *Hydrobiologia* **395/396**.

Harper, J.L. (1977) *Population Biology of Plants.* Academic Press, London.

Harper, J.L. (1987) The heuristic value of restoration. In: *Restoration Ecology* (eds. W.R. Jordan, III, M.E. Gilpin and J.D. Aber), pp. 35–45. Cambridge University Press, Cambridge.

Harris, J.A. and Hobbs, R.J. (2001) Clinical practice for ecosystem health: the role of ecological restoration. *Ecosystem Health* 7, 195–202.

Harris, J.A., Birch, P. and Palmer, J. (1996) *Land Restoration and Reclamation: Principles and Practice.* Longman Higher Education, Harlow.

Hartman, M. (1999) Species dependent root decomposition in rewetted fen soils. *Plant and Soil* 213, 93–8.

Hasenauer, H. (2003) Glossary of terms and definitions relevant for conversion. In: *Norway Spruce Conversion: Options and Consequences* (eds. H. Spiecker, J. Hansen, E. Klimo *et al.*) European Forest Institute Research Report. Brill Academic Publishers, Leiden.

Haskell, J.P., Ritchie, M.E. and Olff, H. (2002) Fractal geometry predicts varying body size scaling relationships for mammals and bird home ranges. *Nature* 418, 527–30.

Hassler, D., Hassler, M. and Glaser, H.H. (1995) *Wässerwiesen. Geschichte, Technik und Ökologie der bewässerte Wiesen, Bache und Graben in Kraichgau, Hardt und Bruhrain.* LNV Verlag Regiokultur, Karlsruhe.

Hawken, P., Lovins, A. and Lovins, L.H. (1999) *Natural Capitalism: Creating the Next Industrial Revolution.* Little, Brown and Company, Boston.

Heal, G. (2000) *Nature and the Marketplace.* Island Press, Washington, DC.

Heaney, S.I., Correy, J.E. and Lishman, J.P. (1992) Changes of water quality and sediment phosphorus of a small productive lake following decreased phosphorus loading. In: *Eutrophication: Research and Applications to Water Supply* (eds. D.W. Sutcliffe and J.G. Jones), pp. 119–31. Freshwater Biological Association, Ambleside.

Hedrick, P.W. and Gilpin, M.E. (1997) Genetic effective size of a metapopulation. In: *Metapopulation Biology: Ecology, Genetics and Evolution* (eds. I. Hanski and M.E. Gilpin), pp. 166–81. Academic Press, San Diego.

Heer, C. and Körner, C. (2002) High elevation pioneer plants are sensitive to mineral nutrient addition. *Basic and Applied Ecology* 3, 39–47.

Heinsdorf, D.L. (1992) *Untersuchungen zur Düngerbedürftigkeit von Forstkulturen auf Kipprohböden der Niederlausitz.* TU Dresden, Tharandt.

Heinsdorf, D.L. (1996) Development of forest stands in the Lusatian lignite mining district after mineral fertilization adapted to site and tree species. *Water, Air and Soil Pollution* 91, 33–42.

Heip, C. (1995) Eutrophication and zoobenthos dynamics. *Ophelia* 41, 113–36.

Helliwell, J. (1969) A structural model of the foreign exchange market. *Canadian Journal of Economics* 2, 90–105.

Hem, J.D. (1959) *Study and Interpretation of the Chemical Characteristics of Natural Waters.* Water Supply Paper, 2254, U.S. Geological Survey, Washington, DC.

Hengeveld, R. (1999) Modelling of the impact of biological invasions. In: *Invasive Species and Biodiversity Management* (eds. O.T. Sandlund, P.J. Schei and Å. Viken), pp. 127–38. Kluwer Academic Publishers, Dordrecht.

Henriksen, A. and Hindar A. (1993) Counter measures in water: is it possible to calculate the amount of lime needed in Norway? In: *Liming of Lakes and Rivers* (ed. A.J. Romundstad), pp. 162–70. Directorate for Nature Management, Trondheim.

Henriksen, A., Lien, L., Rosseland, B.O. *et al.* (1989) Lake acidification in Norway: present and predicted fish status. *Ambio* 18, 314–21.

Henrikson, L. and Brodin, Y.-W, (1995), Liming of surface water in Sweden: a synthesis. In: *Liming of Acidified Surface Water: A Swedish Synthesis* (eds. L. Henrikson and Y.-W. Brodin), pp. 1–44. Springer, Berlin.

Heter, E.W. (1950) Transplanting beavers by airplane and parachute. *Journal of Wildlife Management* 14, 143–7.

Heuson, R. (1929) *Praktischen Kulturvorschläge für Kippen, Bruchfelder, Dünen und Ödländerreien.* Neudamm, Berlin.

Higgs, E.S. (1997) What is good ecological restoration? *Conservation Biology* 11, 338–48.

Higgs, E. (2003) *Nature by Design: People, Natural Process, and Ecological Restoration.* MIT Press, Cambridge, MA.

Hillel, D. (1992) *Out of the Earth. Civilization and the Life of Soil.* University of California Press, Berkeley.

Hindar A., Henriksen, A., Sandoy, A. and Romundstad, A.J. (1998) Critical load concept to restoration goals for liming acidified Norwegian waters. *Restoration Ecology* 6, 353–63.

Hobbs, R.J. (2002) The ecological context: a landscape perspective. In: *Handbook of Ecological Restoration* (eds. M. Perrow and A. Davy), vol. 1, pp. 24–45. Cambridge University Press, Cambridge.

Hobbs, R.J. and Hopkins, A.J.M. (1990) From frontier to fragments: European impact on Australia's vegetation. *Proceedings of the Ecological Society of Australia* 16, 93–114.

Hobbs, R.J. and Saunders, D.A. (eds.) (1992) *Reintegrating Fragmented Landscapes: Towards Sustainable Production and Nature Conservation.* Springer, New York.

Hobbs, R.J. and Norton, D.A. (1996) Towards a conceptual framework for restoration ecology. *Restoration Ecology* 4, 93–110.

Hobbs, R.J. and Mooney, H.A. (1998) Broadening the extinction debate: population deletions and additions in California and Western Australia. *Conservation Biology* 12, 271–83.

Hobbs, R.J. and Morton, S.R. (1999) Moving from descriptive to predictive ecology. *Agroforestry Systems* 45, 43–55.

Hobbs, R.J. and Harris, J.A. (2001) Restoration ecology: repairing the Earth's damaged ecosystems in the new millennium. *Restoration Ecology* 9, 239–46.

Hoelzel, A.R., Halley, J., O'Brien, S.J. *et al.* (1993) Elephant seal genetic variation and the use of simulation models to investigate historical population bottlenecks. *Journal of Heredity* 84, 443–9.

Hoffmann, C.C. (1998) *Nutrient Retention in Wet Meadows and Fens.* PhD Thesis, National Environmental Research Institute, Silkeborg.

Holaus, K. and Partl, C. (1996) Verbesserung und Erhaltung der Hochlagenvegetation durch Düngungsmaßnahmen. Sonderdruck *Der Alm- und Bergbauer*, Innsbruck.

Holland, M.M., Risser, P.G. and Naiman, R.J. (eds.) (1991) *Ecotones: The Role of Landscape Boundaries in the Management and Restoration of Changing Environments.* Chapman and Hall, New York.

Hollick, M. (1981) The role of quantitative decision making methods in environmental impact assessment. *Journal of Environmental Management* 12, 65–78.

Holling, C.S. (1973) Resilience and stability of ecological systems. *Annual Review of Ecology and Systematics* 4, 1–24.

Holling, C.S. (2000) Theories for sustainable futures. *Conservation Ecology* 4, 7.

Holling, C.S. (2001) Understanding the complexity of economic, ecological, and social systems. *Ecosystems* 4, 390–405.

Holt, R.D. (1997) From metapopulation dynamics to community structure: some consequences of spatial heterogeneity. In: *Metapopulation Biology: Ecology,*

Genetics and Evolution (eds. I. Hanski and M.E. Gilpin), pp. 149–64. Academic Press, San Diego.

Horn, H.S. (1976) Succession. In: *Theoretical Ecology: Principles and Applications* (ed. R.M. May), pp. 187–204, Blackwell Scientific Publications, Oxford.

Horppila, J., Peltonen, H., Malinen, T. *et al.* (1998) Top-down or bottom-up effects by fish: issues of concern in biomanipulation of lakes. *Restoration Ecology* 6, 20–8.

Hosper, S.H. (1984) Restoration of Lake Veluwe, The Netherlands, by reduction of phosphorus loading and flushing. *Water Science and Technology* 17, 757–68.

Hosper, S.H. (1997) Clearing lakes: an ecosystem approach to the restoration and management of shallow lakes in the Netherlands. PhD thesis, Wageningen University.

Hosper, S.H. (1998) Stable states, buffers and switches: an ecosystem approach to the restoration and management of shallow lakes in The Netherlands. *Water Science and Technology* 37, 151–64.

Hosper, H. and Meijer, M.-L. (1986) Control of phosphorus loading and flushing as restoration methods for Lake Veluwe, The Netherlands. *Hydrobiological Bulletin* 20, 183–94.

Howe, H.F. and Smallwood, J. (1982) Ecology of seed dispersal. *Annual Review of Ecology and Systematics* 13, 201–28.

Howells, O. and Edwards-Jones, G. (1997) A feasibility study of reintroducing wild boar *Sus scrofa* to Scotland: are existing woodlands large enough to support minimum viable populations. *Biological Conservation* 81, 77–89.

Hsü, K.J. (1986) *The Great Dying.* Harcourt Brace Jovanovich, San Diego.

Hubbell, S.P. (2001) *The Unified Neutral Theory of Biodiversity and Biogeography.* Princeton University Press, Princeton.

Hudson, P. and Greenman, J. (1998) Competition mediated by parasites: biological and theoretical progress. *Trends in Ecology and Evolution* 13, 387–90.

Hughes, J.D. (1982) Deforestation, erosion and forest management in ancient Greece and Rome. *Journal of Forest History* 26, 60–75.

Huisman, J. and Weissing, F.J. (1999) Biodiversity of plankton by species oscillations and chaos. *Nature* 402, 407–10.

Hutchings, J.A. (2000) Collapse and recovery of marine fishes. *Nature* 406, 882–5.

Hutchings, M.J. and Stewart, A.J.A. (2002) Calcareous grasslands. In: *Handbook of Ecological Restoration,* vol. 2, *Restoration in Practice* (eds. M.R. Perrow and

A.J. Davy), pp. 419–42. Cambridge University Press, Cambridge.

Hüttl, R.F. (1998) Ecology of post strip-mining landscapes in Lusatia, Germany. *Environmental Science and Policy* 1, 129–35.

Hüttl, R.F. and Bradshaw, A.D. (2001) Open-cast mining and water. *Journal of Geochemical Exploration* 73, 61–2.

Hüttl, R.F. and Weber, E. (2001) Forest ecosystem development in post-mining landscapes: a case study of Lusatian lignite district. *Naturwissenschaften* 88, 322–9.

ICONA (1989) *Técnicas de reforestación en los países Mediterráneos*. Instituto Nacional para la Conservación de la Naturaleza, Madrid.

ICPR (1998) *Bestandsaufnahme der ökologisch wertvollen Gebiete am Rhein und erste Schritte auf dem Weg zum Biotopverbund*. Report. International Committee for the Protection of the Rhine, Koblenz.

IDF (2001) *Evaluation du système d'aide communautaire pour les mesures forestières en agriculture du Reglement 2080/92*. Final report. Institut pour le Developpement Forestier, Paris.

Ilnicki, P. (ed.) (2002) *Restoration of Carbon Sequestrating Capacity and Biodiversity in Abandoned Grassland on Peatland in Poland*. Agricultural University Poznan, Poznan.

IPCC (2001) *Summary for Policymakers*. A report of Working Group I of the Intergovernmental Panel on Climate Change, Geneva.

IUCN (1995) *IUCN/SSC Guidelines for Re-introductions*. Species Survival Commission of the International Union for the Conservation of Nature. International Union for the Conservation of Nature, Gland.

Iversen, J. (1941) Landnam i Danmarks Stenalder. En pollenanalytisk undersogelse over det forste landbrugs indvirkning paa vegetationsudviklingen. *Danmarks Geologiske Undersogelse* 2, 7–68.

Jackson, L.L., Lopoukhine, N. and Hillyard, D. (1995) Ecological restoration: a definition and comments. *Restoration Ecology* 3, 71–5.

James, D., Fooks, L.G. and Preston, J.R. (1983) Success of wild-trapped compared to captivity-raised birds in restoring Wild Turkey populations to northwestern Arkansas. *Proceedings of the Arkansas Academy of Science* 37, 38–41.

Janse, J.H., Aldenberg, T. and Kramer, P.R.G. (1992) A mathematical model of the phosphorus cycle in Lake Loosdrecht and simulation of additional measures. *Hydrobiologia* 233, 119–36.

Jansen, A.J.M. and Roelofs, J.G.M. (1996) Restoration of Cirsio-Molinietum wet meadows by sod cutting. *Ecological Engineering* 7, 279–98.

Jansen, A.J.M., Grootjans, A.P. and Jalink, M. (2000) Hydrology of Dutch Cirsio-Molinietum meadows: prospects for restoration. *Applied Vegetation Science* 3, 51–64.

Jansen, A.J.M., Fresco, L.F.M., Grootjans, A.P. and Jalink, M.H. (2004) Effects of restoration measures on plant communities of wet heathland ecosystems. *Applied Vegetation Science* 7, 243–52.

Janssens, F., Peeters, A.A., Tallowin, J.R.B. *et al.* (1998) Relationship between soil chemical factors and grassland diversity. *Plant and Soil* 202, 69–78.

Janzen, D. (1998) Gardenification of wildland nature and the human footprint. *Science* 279, 1312–13.

Janzen, D.H. (2002) Tropical dry forest restoration: area de conservación Guanacaste, northwestern Costa Rica. In: *Handbook of Ecological Restoration*, vol. 2. *Restoration in Practice* (eds. M.R. Perrow and A.J. Davy), pp. 559–84. Cambridge University Press, Cambridge.

Jefferies, R.L. (1999) Herbivores, nutrients and trophic cascades in terrestrial environments. In: *Herbivores: Between Plants and Predators* (eds. H. Olff, V.K. Brown and R.H. Drent), pp. 301–30. Blackwell Science, Oxford.

Jefferies, R.L. and Rockwell, R.F. (2002) Foraging geese, vegetation loss and soil degradation in an Arctic salt marsh. *Applied Vegetation Science* 5, 7–16.

Jentsch, A. and Beyschlag, W. (2003) Vegetation ecology of dry acidic grasslands in the lowland area of Central Europe. *Flora* 198, 3–25.

Jepma, C.J. and Munasinghe, M. (1998) *Climate Change Policy: Facts, Issues and Analyses*. Cambridge University Press, Cambridge.

Jeppesen, E. (1998) The ecology of shallow lakes: trophic interactions in the pelagial. DSc thesis, University of Copenhagen.

Jeppesen, E., Kristensen, P., Jensen, J.P. *et al.* (1991) Recovery resilience following a reduction in external phosphorus loading of shallow, eutrophic Danish lakes: duration, regulating factors and methods for overcoming resilience. *Memorie dell'Istituto Italiano di Idrobiologia* 48, 127–48.

Jeppesen, E., Jensen, J.P., Søndergaard, M. and Lauridsen, T. (1999) Trophic dynamics in turbid and clear-water lakes with special emphasis on the role of zooplankton for water clarity. *Hydrobiologia* 408/409, 217–31.

Jeschke, L. (1987) Vegetationsdynamik des Salzgraslandes in Bereich der Ostseeküste der DDR unter dem Einfluss des Menschen. *Hercynia N.F.* 24, 321–8.

Jeschke, L. and Paulson, Ch. (2001) Revitalisierung von Kesselmooren in Serrahner Wald. In: *Landschaftsökologische Moorkunde* (eds M. Succow and H. Joosten), pp. 523–8. Schweizerbart'sche Verlagsbuchhandlung, Stuttgart.

Jochimsen, M.E.A. (1991) Begrünung von Bergehalden auf der Grundlage der natürlichen Sukzession. In: *Bergehalden im Ruhrgebiet. Beanspruchung und Veränderung eines industriellen Ballungsraumes* (eds. H. Wiggering and M. Kerth), pp. 189–94. Vieweg, Braunschweig/Wiesbaden.

Jochimsen, M.E.A. (1996) Reclamation of colliery mine spoil founded on natural succession. *Water, Air and Soil Pollution* 91, 99–108.

Joffre, R. and Rambal, S. (1993) How tree cover influences the water balance of Mediterranean rangelands. *Ecology* 74, 570–82.

Joffre, R., Vacher, J., De los Llanos, C. and Long, G. (1988) The dehesa: an agrosilvopastoral system of the Mediterranean region with special reference to the Sierra Morena area of Spain. *Agroforestry Systems* 6, 71–96.

Joffre, R., Rambal, S. and Ratte, J.P. (1999) The dehesa system of southern Spain and Portugal as a natural ecosystem mimic. *Agroforestry Systems* 45, 57–79.

Johansson, M.E., Nilsson, C. and Nilsson, E. (1996) Do rivers function as corridors for plant dispersal? *Journal of Vegetation Science* 7, 593–8.

Johnson, J.E., Pardew, M.G. and Lyttle, M.M. (1993) Predator recognition and avoidance by larval razorback sucker and northern hog sucker. *Transactions of the American Fisheries Society* 122, 1139–45.

Johnson, K.J., Vogt, K.A., Clark, H.J. *et al.* (1996) Biodiversity and the productivity and stability of ecosystems. *Trends in Ecology and Evolution* 11, 372–7.

Johnson, S.D. and Steiner, K.E. (2000) Generalization versus specialization in plant pollination systems. *Trends in Ecology and Evolution* 15, 140–3.

Jones, C.G., Heck, W., Lewis, R.E. *et al.* (1995) The restoration of the Mauritius Kestrel *Falco punctatus* population. *Ibis* 137 (suppl. 1), 173–80.

Jones, N. (2003) South Aral 'gone in 15 years'. *New Scientist* 179, 9.

Jongejans, E. and Schippers, P. (1999) Modeling seed dispersal by wind in herbaceous species. *Oikos* 87, 362–72.

Jongejans, E. and Telenius, A. (2001) Field experiments on seed dispersal by wind in ten umbelliferous species (Apiaceae). *Plant Ecology* 152, 67–78.

Joosten, H. and Clarke, D. (2002) *Wise Use of Peatlands.* International Mire Conservation Group, International Peat Society, Jyväskylä.

Jurdant, M., Bélair, J.L., Gerardin, V. and Ducroc, J.P. (1977) *L'Inventaire du Capital-Nature. Méthode de Classification et de Cartographie Écologique du Territoire.* Service des Études Écologiques Régionales, Direction Régionale des Terres, Pêches et Environnement, Québec.

Kahmen, S., Poschlod, P. and Schreiber, K.F. (2002) Conservation management of calcareous grasslands: changes in plant species composition and the response of plant functional traits during 24 years. *Biological Conservation* 104, 319–28.

Kaiser, M.J., Ramsay, K., Richardson, C.A. *et al.* (2000) Chronic fishing disturbance has changed shelf sea benthic community structure. *Journal of Animal Ecology* 69, 494–503.

Kalchreuter, H. and Wagner, W. (1982) Preliminary results of reintroduction programs of Blackgrouse (*Lyrurus tetrix*) and Capercaillie (*Tetrao urogallus*) in southern Germany (Frg). In: *Proceedings of the Second International Symposium on Grouse*, March 1981 (ed. T.W.I. Lovel), pp. 202–3. World Pheasant Association, Exning.

Kamermans, P. and Smaal, A.C. (2002) Mussel culture and cockle fisheries in the Netherlands: finding a balance between economy and ecology. *Journal of Shellfish Research* 21, 509–17.

Kasprzak, P., Benndorf, J., Mehner, T. and Koschel, R. (2002) Biomanipulation of lake ecosystems: an introduction. *Freshwater Biology* 47, 2277–81.

Kates, R.W., Clark, W.C., Corell, R. *et al.* (2001) Sustainability science. *Science* 292, 641–2.

Katz, E. (1992) The big lie: human restoration of nature. *Research in Philosophy and Technology* 12, 231–41.

Katzur, J. and Haubold-Rosar, M. (1996) Amelioration and reforestation of sulfurous mine soils in Lusatia (Eastern Germany). *Water, Air and Soil Pollution* 91, 17–32.

Katzur, J., Fischer, K., Böcker, L. *et al.* (2001) *Afforestation of Industrial Dumps Using Selected Soil Improvement Agents – New Frontiers in Reclamation: Facts & Procedures in Extractive Industry.* Milos 2001, Greece.

Kay, J.J. and Regier, H.A. (2000) Uncertainty, complexity, and ecological integrity: insights from an ecosystem approach. In: *Implementing Ecological Integrity* (eds. P. Crabbé *et al.*), pp. 121–56. Kluwer Academic Publishers, Dordrecht.

Kearns, C.A., Inouye, D.W. and Waser, N.M. (1998) Endangered mutualisms: the conservation of plant-pollinator interactions. *Annual Review of Ecology and Systematics* 29, 83–112.

Keller, M., Kollmann, J. and Edwards, P.J. (2000) Genetic introgression from distant provenances reduces fitness in local weed populations. *Journal of Applied Ecology* 37, 647–59.

Kelty, M.J., Larson, B.C. and Oliver, C.D. (1992) *The Ecology and Silviculture of mixed-species Forests.* Kluwer Academic Publishers, Dordrecht.

Kenk, G. and Guehne, S. (2001) Management of transformation in Central Europe. *Forest Ecology and Management* 151, 107–19.

Kentula, M.E. (1997) A step toward a landscape approach in riparian restoration. *Restoration Ecology* 5 (4S), 2–3.

Kerfoot, W.C. and DeAngelis, D.L. (1989) Scale-dependent dynamics: zooplankton and the stability of freshwater food webs. *Trends in Ecology and Evolution* 4, 167–71.

Kershner, J.L. (1997) Setting riparian/aquatic restoration objectives within a watershed context. *Restoration Ecology* 5 (suppl.), 15–24.

Keuning, J.A. (1994) Botanische samenstelling grasmat op de stikstofbedrijven sterk veranderd van 1949–1991. *Meststoffen* 99, 44–52.

Kiers, E.T., Lovelock, C.E., Krueger, E.L. and Herre, E.A. (2000) Differential effects of tropical arbuscular mycorrhizal fungal inocula on root colonization and tree seedling growth: implications for tropical forest diversity. *Ecology Letters* 3, 106–13.

Kinzig, A.P., Pacala, S.W. and Tilman, D. (eds.) (2001) *The Functional Consequences of Biodiversity: Empirical Progress and Theoretical Extensions.* Princeton University Press, Princeton.

Kleijn, D., Berendse, F., Smit, R. and Gillissen, N. (2001) Agri-environment schemes do not effectively protect biodiversity in Dutch agricultural landscapes. *Nature* 413, 723–5.

Kleinman, D.L. (2000) Democratizations of science and technology. In: *Science, Technology, and Democracy* (ed. D.L. Kleinman), pp. 139–65. SUNY Press, Albany, NY.

Kleyer, M., Kratz, R., Lutze, G. and Schröder, B. (2000) Habitatmodelle für Tierarten: Entwicklung, Methoden und Perspektiven für die Anwendung. *Zeitschrift für Ökologie und Naturschutz* 8, 177–94.

Klijn, F. and Dijkman, J. (2001) *Room for the Rhine in the Netherlands*, Directorate – General of Public Works and Water Management, The Hague.

Klitgaard, O.L. (2002) Problems and plans with restocking in the Danish forest district Aabenraa. In: *Restocking of Storm-felled Forests: New Approaches* (ed. A. Brunner). *Skov og Landskab Reports* 12, 53–60.

Klötzli, F. (1987) Disturbance in transplanted grasslands and wetlands. In: *Disturbance in Grasslands: Causes, Effects and Processes* (eds. J. van Andel, J.P. Bakker and R.W. Snaydon), pp. 79–96. Junk Publishers, Dordrecht.

Klötzli, F. and Walther, G.-R. (1999) Recent vegetation shifts in Switzerland. In: *Recent Shifts in Vegetation Boundaries of Deciduous Forests, Especially Due to General Global Warming* (eds. F. Klötzli and G.-R. Walther), pp. 317–31. Birkhäuser, Basel.

Klug, B., Scharfetter-Lehrl, G. and Scharfetter, E. (2002) Effects of trampling on vegetation above the timberline in the eastern Alps, Austria. *Arctic, Antarctic and Alpine Research* 34, 377–88.

Knapp, A.K., Briggs, J.M., Hartnett, D.C. *et al.* (eds.) (1998) *Grassland Dynamics: Long-term Ecological Research in Tallgrass Prairie.* Oxford University Press, Oxford.

Knapp, A.K., Blair, J.M., Briggs, J.M. *et al.* (1999) The keystone role of bison in North American tallgrass prairie. *BioScience* 49, 39–50.

Knevel, I.C., Bekker, R.M., Bakker, J.P. and Kleyer, M. (2003) Life-history traits of the Northwest European flora: the LEDA database. *Journal of Vegetation Science* 14, 611–14.

Knighton, D. (1994) *Fluvial Forms and Process.* Arnold, Baltimore.

Knoop, W.T. and Walker, B.H. (1985) Interactions of woody and herbaceous vegetation in a southern African savanna. *Journal of Ecology* 73, 235–53.

Knörzer, K. (1996) Pflanzentransport im Rhein zur Römerzeit, im Mittelalter und heute. *Decheniana (Bonn)* 149, 81–123.

Koeman, J.H., Oskamp, A.A.G., Veen, J. *et al.* (1967) Insecticides as a factor in the mortality of the sandwich tern (*Sterna sandvicensis*). A preliminary communication. *Mededelingen Rijksfaculteit Landbouwwetenschappen, Gent* 32, 841–54.

Koerselman, W. and Verhoeven, J.T.A. (1995) Eutrophication of fen ecosystems: external and internal nutrient sources and restoration strategies. In: *Restoration of Temperate Wetlands* (eds. B.D. Wheeler, S.C. Shaw, W.J. Foyt, and R.A. Robertson), pp. 91–112. Wiley, Chichester.

Koerselman, W., Bakker, S.A. and Blom, M. (1990) Nitrogen, phosphorus and potassium budgets for two small fens surrounded by heavy fertilized pastures. *Journal of Ecology* 78, 428–42.

Kohlmaier, G.H., Weber, M. and Houghton, R.A. (eds.) (1998) *Carbon Dioxide Mitigation in Forestry and Wood Industry.* Springer, Berlin.

Kolar, C.S. and Lodge, D.M. (2001) Progress in invasion biology: predicting invaders. *Trends in Ecology and Evolution* 16, 199–204.

Kolk, A. and Bungart, R. (2000) Bodenmikrobiologische Untersuchungen an forstlich rekultivierten Kippenflächen im Lausitzer Braunkohlenrevier. In: *Rekultivierung in Bergbaufolgelandschaften: Bodenorganismen, bodenökologische Prozesse und Standortentwicklung* (eds. G. Broll, W. Dunger, B. Keplin and W. Topp), *Geowissenschaften und Umwelt*, vol. 4, Springer, Berlin.

Komers, P.E. and Curman, G.P. (2000) The effect of demographic characteristics on the success of ungulate re-introductions. *Biological Conservation* 93, 187–93.

Komor, S.C. (1994) Geochemistry and hydrology of a calcareous fen within the Savage Fen wetlands complex, Minnesota, USA. *Geochimica et Cosmochimica Acta* 58, 3353–67.

König, A., Mössmer, R. and Bäumler, A. (1995) Waldbauliche Dokumentation der flächigen Sturmschäden des Frühjahrs 1990 in Bayern und meteorologische Situation zur Schadenszeit. *Berichte aus der Bayerischen Landesanstalt für Wald und Forstwirtschaft (LWF)* 2, Freising.

Kooijman, A.M. and Bakker, C. (1995) Species replacement in the bryophyte layer in mires; the role of water type, nutrient supply and interspecific interactions. *Journal of Ecology* 83, 1–8.

Koonce, J.F., Busch, W.D. and Czapla, T. (1996) Restoration of Lake Erie: contribution of water quality and natural resource management. *Canadian Journal of Fisheries and Aquatic Sciences* 53 (suppl. 1), 105–12.

Körner, C. (1993) Scaling from species to vegetation: the usefulness of functional groups. In: *Biodiversity and Ecosystem Function* (eds. E.D. Schulze and H.A. Mooney), pp. 117–42. Ecological Studies Series, vol. 99. Springer, Berlin.

Koska, I. and Stegmann, H. (2001) Revitalisierung eines Quellmoorkomplexes am Serbitz-Oberlauf. In: *Landschaftsökologische Moorkunde* (eds. M. Succow. and H. Joosten), pp. 509–17. Schweizerbart'sche Verlagsbuchhandlung, Stuttgart.

Kotowski, W.H. (2002) Fen communities: ecological mechanisms and conservation strategies. PhD thesis, University of Groningen.

Kotowski, W., van Diggelen, R. and Kleinke, J. (1998) Behaviour of wetland plant species along a moisture gradient in two geographically distant areas. *Acta Botanica Neerlandica* 47, 337–49.

Krautzer, B., Parente, G., Spatz G. *et al.* (2003) *Seed Propagation of Indigenous Species and their Use for Restoration of Eroded Areas in the Alps.* Final report CT98-4024, BAL Gumpenstein, Irdning.

Krebs, C.J. (2001) *Ecology: The Experimental Analysis of Distribution and Abundance*, 5th edn. Benjamin Cummings, San Francisco.

Krebs, C.J., Sinclair, A.R.E., Boonstra, R. *et al.* (1999) Community dynamics of vertebrate herbivores: how can we untangle the web? In: *Herbivores: Between Plants and Predators* (eds. H. Olff, V.K. Brown and R.H. Drent), pp. 447–73. Blackwell Science, Oxford.

Kreutzer, K., Deschu, E. and Hösl, G. (1986) Vergleichende Untersuchungen über den Einfluß von Fichte (*Picea abies* Karst.) und Buche (*Fagus sylvatica* L.) auf die Sickerwasserqualität. *Forstwissentschaftliches Centralblatt* 105, 364–71.

Kristensen, P. and Hansen, H.O. (eds.) (1994) *European Rivers and Lakes: Assessment of their Environmental State.* European Environment Agency, Copenhagen.

Kruse, K. and Sharp, D.E. (2002) *Central Flywayharvest and Population Survey Data Book.* Office of Migratory Bird Management. U.S. Fish and Wildlife Service, Denver.

Ksontini, M., Louguet, P., Laffray, D. and Rejeb, M.N. (1998) Comparaison des effets de la contrainte hydrique sur la croissance, la conductance stomatique et la photosynthèse de jeunes plants de chênes méditerranées (*Quercus suber, Q. faginea, Q. coccifera*) en Tunisie. *Annales des Sciences Forestières* 55, 477–95.

Kufel, L., Prejs, A. and Rybak, J.I. (eds.) (1997) Shallow lakes '95 – Trophic cascades in shallow freshwater and brackish lakes. *Hydrobiologia* 342/343.

Kuhn, W. and Kleyer, M. (2000) A statistical habitat model for the Blue Winged Grasshopper (*Oedipoda caerulescens*) considering the habitat connectivity. *Zeitschrift für Ökologie und Naturschutz* 8, 207–18.

Kuijt, J. (1969) *The Biology of Parasitic Flowering Plants.* University of California Press, Berkeley.

Kuiters, A.T. (1990) Role of phenolic substances from decomposing forest litter in plant-soil interactions. *Acta Botanica Neerlandica* 39, 329–48.

Kulczynski, S. (1949) Peat bogs of Polesie. *Mémoires de l'Académie Polonaise des Sciences et des Lettres. Série B, Sciences naturelles* 15, 1–356.

Küster, H. (1997) The role of farming in the postglacial expansion of beech and hornbeam in the oak woodlands of central Europe. *The Holocene* 7, 239–42.

Kwak, M.M., Velterop, O. and van Andel, J. (1998) Pollen and gene flow in fragmented habitats. *Applied Vegetation Science* 1, 37–54.

Lamers, L.P.M., Tomassen, H.B.M. and Roelofs, J.G.M. (1998) Sulphate-induced eutrophication and phytotoxicity in freshwater wetlands. *Environmental Science and Technology* 32, 199–205.

Lamers, L.P.M., Bobbink, R. and Roelofs, J.G.M. (2000) Natural nitrogen filter fails in raised bogs. *Global Change Biology* 6, 583–6.

Lamers, L.P.M., Smolders, A.J.P. and Roelofs, J.G.M. (2002) The restoration of fens in the Netherlands. *Hydrobiologia* 478, 107–30.

Lammens, E.H.R.R. (1999) The central role of fish in lake restoration and management. *Hydrobiologia* 396, 191–8.

Lammens, E.H.R.R., Gulati, R.D., Meijer, L.M. and van Donk, E. (1990) The first biomanipulation conference: a synthesis. *Hydrobiologia* 200/201, 619–27.

Lammens, E.H.R.R., van Nes, E.H. and Mooij, W.M. (2002) Differences in exploitation of the bream in three shallow lake systems and their relation to water quality. *Freshwater Biology* 47, 2435–42.

Lancaster, J. (2000) The ridiculous notion of assessing ecological health and identifying the useful concepts underneath. *Human and Ecological Risk Assessment* 6, 213–22.

Lande, R. (1988) Genetics and demography in biological conservation. *Science* 241, 1455–60.

Landesregierung Mecklenburg-Vorpommern (2000) *Konzept zur Bestandssicherung und zur Entwicklung der Moore in Mecklenburg-Vorpommern.* Umweltministerium Mecklenburg-Vorpommern.

Landis, T.D., Tinus, R.W., McDonald, S.E. and Barnett, J.P. (1989) *The Container Tree Nursery Manual. Agricultural Handbook.* Report 674. USDA Forest Service, Washington, DC.

Lang, G. (1994) *Quartäre Vegetationsgeschichte Europas.* G. Fischer, Jena, Stuttgart.

Langlois, E., Bonis, A. and Bouzillé, J.B. (2001) The response of *Puccinellia maritima* (Huds.) Parl. to burial: a key to understanding its role in salt-marsh dynamics? *Journal of Vegetation Science* 12, 289–97.

Langton, R.W. and Robinson, W.E. (1990) Faunal associations on scallop grounds in the western Gulf of Maine USA. *Journal of Experimental Marine Biology and Ecology* 144, 157–71.

Lavorel, S. and Cramer, W. (eds.) (1999) Plant functional types and disturbance dynamics. *Journal of Vegetation Science* (Special Feature) 10, 603–730.

Lavorel, S. and Garnier, E. (2001) Aardvarck to Zyzyxia: functional groups across kingdoms. *New Phytologist* 149, 360–4.

Lavorel, S., McIntyre, S., Landsberg, J. and Forbes, T.D.A. (1997) Plant functional classifications: from general groups to specific groups based on response to disturbance. *Trends in Ecology and Evolution* 12, 474–8.

Law, R. (1999) Theoretical aspects of community assembly. In: *Advanced Ecological Theory* (ed. J. McGlade), pp. 143–71. Blackwell Science, Oxford.

Law, R. and Morton, R.D. (1996) Permanence and the assembly of ecological communities. *Ecology* 77, 762–75.

Lawler, S.P., Armesto, J.J. and Kareiva, P. (2001) How relevant to conservation are studies linking biodiversity and ecosystem functioning? In: *The Functional Consequences of Biodiversity: Empirical Progress and Theoretical Extensions* (eds. A.P. Kinzig, S.W. Pacala and D. Tilman), pp. 294–313. Princeton University Press, Princeton.

Lawton, J.H. (1997) The role of species in ecosystems: aspects of ecological complexity and biological diversity. In: *Biodiversity: An Ecological Perspective* (eds. T. Abe, S.A. Levin and M. Higashi), pp. 215–28. Springer Verlag, New York.

Lazzaro, X. (1987) A review of planktivorous fishes: their evolution, feeding behaviours, selectivities, and impacts. *Hydrobiologia* 146, 97–167.

Le Houérou, H.N. (1984) Rain-use efficiency: a unifying concept. *Journal of Arid Environments* 7, 213–47.

Le Houérou, H.N. (2000) Restoration and rehabilitation of arid and semiarid Mediterranean ecosystems in North Africa and West Asia: a review. *Arid Soil Research and Rehabilitation* 14, 3–14.

Leendertse, P.C., Roozen, A.J.M. and Rozema, J. (1997) Long-term changes (1953–1990) in the salt marsh vegetation at the Boschplaat on Terschelling in relation to sedimentation and flooding. *Plant Ecology* 132, 49–58.

Leirós, M.C., Gil-Sotres, F., Trasar-Cepeda, M.C. *et al.* (1996) Soil recovery at the Meirama opencast lignite mine in northwest Spain: a comparison of the effectiveness of cattle slurry and inorganic fertilizer. *Water, Air and Soil Pollution* 91, 109–24.

Lenihan, H.S. (1999) Physical-biological coupling on oyster reefs: how habitat structure influences individual performance. *Ecological Monographs* 69, 251–75.

Lenihan, H.S. and Micheli, F. (2001) Soft-sediment communities. In: *Marine Community Ecology* (eds. M.D. Bertness and S.D. Gaines), pp. 253–88. Sinauer Publishers, Sunderland, MA.

Lenihan, H.S. and Peterson, C.H. (1998) How habitat degradation through fishery disturbance enhances impacts of hypoxia on oyster reefs. *Ecological Applications* 8, 128–40.

Leonard, L.A., Hine, A.C. and Luther, M.E. (1995) Superficial sediment transport and deposition processes in a *Juncus roemerianus* marsh, West-Central Florida. *Journal of Coastal Research* 11, 322–36.

Leuschner, Chr. (1998) Mechanismen der Konkurrenzüberlegenheit der Rotbuche. *Berichte der Reinhold-Tüxen-Geselschaft* 10, 5–18, Hannover.

Levin, D.A., Francisco-Ortega, J. and Jansen, R.K. (1996) Hybridization and the extinction of rare plant species. *Conservation Biology* 10, 10–16.

Levin, S.A. (2001) Immune systems and ecosystems. *Conservation Ecology* 5, 17.

Levine, J.M. and D'Antonio, C.M. (1999) Elton revisited: a review of evidence linking diversity and invasibility. *Oikos* 87, 15–26.

Levine, J., Brewer, S. and Bertness, M.D. (1998) Nutrient competition and plant zonation in a New England salt marsh. *Journal of Ecology* 86, 285–92.

Levins, R. (1969) Some demographic and genetic consequences of environmental heterogeneity for biological control. *Bulletin of the Entomological Society of America* 15, 237–40.

Lewandowski, J., Schauser, I. and Hupfer, M. (2003) Long term effects of phosphorus precipitations with alum in hypereutrophic Lake Süsser See (Germany). *Water Research* 37, 3194–204.

Light, A. (2003) 'Faking Nature' revisited. In: *The Beauty Around Us: Environmental Aesthetics in the Scenic Landscape and Beyond* (eds. D. Michelfelder and B. Wilcox) SUNY Press, Albany, NY.

Light, A. and Higgs, E.S. (1996) The politics of ecological restoration. *Environmental Ethics* 18, 227–47.

Likens, G.E., Bormann, F.H. and Johnson, N.M. (1972) Acid rain. *Environment* 14(2), 33–40.

Limpens, J., Berendse, F. and Klees, H. (2003) N-deposition affects N availability in interstitial water, growth of Sphagnum and invasion of vascular plants in bog vegetation. *New Phytologist* 157, 339–47.

Lindeman, R.L. (1942) The trophic-dynamic aspects of ecology. *Ecology* 23, 399–418.

Little, C. (2000) *The Biology of Soft Shores and Estuaries.* Oxford University Press, Oxford.

Lomolino, M.V. and Channell, R. (1995) Splendid isolation: patterns of range collapse in endangered mammals. *Journal of Mammalogy* 76, 335–47.

Lomolino, M.V. and Channell, R. (1998) Range collapse, re-introductions, and biogeographic guidelines for conservation. *Conservation Biology* 12, 481–4.

Londo, G. (2002) Is *Rhytidiadelphus squarrosus* (Hedw.) Warnst. increasing in The Netherlands? *Lindbergia* 27, 63–70.

Looijen, R.C. and van Andel, J. (1999) Ecological communities: conceptual problems and definitions. *Perspectives in Plant Ecology, Evolution and Systematics* 2, 210–22.

Looijen, R.C. and van Andel, J. (2002) A reply to Thomas Parker's critique of the community concept. *Perspectives in Plant Ecology, Evolution and Systematics* 5, 131–7.

Loreau, M., Naeem, S. and Inchausti, P. (eds.) (2002) *Biodiversity and Ecosystem Functioning.* Oxford University Press, Oxford.

Losvik, M.H. and Austad, I. (2002) Species introduction through seeds from an old, species-rich meadow: effects of management. *Applied Vegetation Science* 5, 185–94.

Loucks, O.L. (1990) Land-water interactions. In: *Wetlands and Shallow Continental Water Bodies*, vol. 1: *Natural and Human Relationships* (ed. B.C. Patten), pp. 243–58. SPB Academic Publishing, The Hague.

Loucks, O.L. (1992) Prediction tools for rehabilitating linkages between land and wetland ecosystems. In: *Ecosystem Rehabilitation*, vol. 2: *Ecosystem Analysis and Synthesis* (ed. M.K. Wali), pp. 297–308, SPB Academic Publishing, The Hague.

Lovejoy, T.E. (1997) Biodiversity: what is it? In: *Biodiversity II. Understanding and Protecting our Biological Resources* (eds. M.L. Reaka-Kudla, D.E. Wilson and E.O. Wilson). pp. 7–14. Joseph Henry Press, Washington, DC.

Lovelock, J.E. (1991) *Gaia: The Practical Science of Planetary Medicine.* Gaia Books, London.

Lovich, J.E. and Bainbridge, D. (1999) Anthropogenic degradation of the Southern California desert ecosystem and prospects for natural recovery and restoration. *Environmental Management* 24, 309–26.

Lucassen, E., Bobbink, R., Oonk, M.M.M. *et al.* (1999) The effect of liming and reacidification on the growth of *Juncus bulbosus*: a mesocosm experiment. *Aquatic Botany* 64, 95–103.

Lucassen, E.C.H.E.T., Smolders, A.J.P., Van de Crommenacker, J. and Roelofs, J.G.M. (2004) Effects of stagnating sulphate-rich water on the mobility of phosphorus in freshwater wetlands: a field experiment. *Archiv für Hydrobiologie* **160**, 117–31.

Ludwig, J.A. and Tongway, D.J. (1995) Spatial organization and its function in semi-arid woodlands, Australia. *Landscape Ecology* **10**, 51–63.

Ludwig, J.A. and Tongway, D.J. (1996) Rehabilitation of semiarid landscapes in Australia. II. Restoring vegetation patches. *Restoration Ecology* **4**, 398–406.

Lugo, A. and Helmer, E. (2004) Emerging forests on abandoned land: Puerto Rico's new forests. *Forest Ecology and Management* **190**, 145–61.

Lundberg, P., Ranta, E. and Kaitala, V. (2000) Species loss leads to community closure. *Ecology Letters* **3**, 465–8.

Lütke-Twenhöven, F. (1992) Untersuchungen zur Wirkung stickstoffhaltiger Niederschläge auf die Vegetation von Hochmooren. *Mitteilungen der Arbeitsgemeinschaft Geobotanik in Schleswig-Holstein und Hamburg* **44**, 1–172.

Lynch, M. and Shapiro, J. (1981) Predation, enrichment and phytoplankton structure. *Limnology and Oceanography* **26**, 86–102.

MacArthur, R.H. (1955) Flucutations of animal populations and a measure of community stability. *Ecology* **36**, 533–6.

MacArthur, R.H. and Wilson, E.O. (1967) *The Theory of Island Biogeography*. Princeton University Press, Princeton.

Mack, R.N., Simberloff, D., Lonsdale, W.M. *et al.* (2000) Biotic invasions: causes, epidemiology, global consequences, and control. *Ecological Applications* **10**, 689–710.

Macklin, M.J. and Lewin, J. (1997) Channel, floodplain and drainage basin response to environmental change. In: *Applied Fluvial Geomorphology for River Engineering and Management* (eds. C.R. Thorne, R.D. Hey and M.D. Newson). Wiley, Chichester.

Madsen, T.R., Shine, R., Olsson, M. and Wittzel, H. (1999) Restoration of an inbred adder population. *Nature* **402**, 34–5.

Maestre, F.T. (2003) Small-scale spatial patterns of two soil lichens in semi-arid Mediterranean steppe. *Lichenologist* **35**, 71–81.

Maestre, F.T., Bautista, S., Cortina, J. and Bellot, J. (2001) Potential for using facilitation by grasses to establish shrubs on a semi-arid degraded steppe. *Ecological Applications* **11**, 1641–55.

Maestre, F.T., Huesca, M.T., Zaady, E. *et al.* (2002) Infiltration, penetration resistance and microphytic crust composition in contrasted microsites within a Mediterranean semi-arid steppe. *Soil Biology and Biochemistry* **34**, 895–8.

Maestre, F.T., Cortina, J., Bautista, S. and Bellot, J. (2003a) Does *Pinus halepensis* facilitate the establishment of shrubs in Mediterranean semi-arid afforestations? *Forest Ecology and Management* **176**, 147–60.

Maestre, F.T., Cortina, J., Bautista, S. *et al.* (2003b) Small-scale environmental heterogeneity and spatiotemporal dynamics of seedling establishment in a semi-arid degraded ecosystem. *Ecosystems* **6**, 630–43.

Mageau, M.T., Costanza, R. and Ulanowicz, R.E. (1995) The development and initial testing of a quantitative assessment of ecosystem health. *Ecosystem Health* **1**, 201–13.

Markart, G., Kohl, B. and Zanetti, P. (1997) Oberflächenabfluß bei Starkregen: Abflußbildung auf Wals-, Weide- und Feuchtflächen (am beispiel des oberen Einzuggebietes der Schesa-Bürsenberg, Voralberg). *Centralblatt für das gesamte Forstwesen* **114**, 123–44.

Marris, C., Wynne, B., Simmons, P. and Weldon, S. (2001) *Public Perceptions of Agricultural Biotechnologies in Europe* (PABE), final report. IEPPP, Lancaster.

Marrs, R.H. (1993) Soil fertility and nature conservation in Europe: theoretical considerations and practical management solutions. *Advances in Ecological Research* **24**, 241–300.

Marrs, R.H. and Le Duc, M.G. (2000) Factors controlling vegetation change in long-term experiments designed to restore heathland in Breckland, UK. *Applied Vegetation Science* **3**, 135–46.

Marsh, G.P. (1871) *Man and Nature: Or Physical Geography as Modified by Human Action*. Scribner, New York.

Marsh, T., Black, A., Acreman, M. and Elliott, C. (2000) River flows. In: *The Hydrology of the UK* (ed. M. Acreman), pp. 101–33. Routledge, London.

Mason, C.F. (1996) *Biology of Freshwater Pollution*, 3rd edn. Longman Group, London.

Mate, B.R. (1972) Sea otter transplants to Oregon in 1970 and 1971. *Proceedings of the 8th Annual Conference on Biological Sonar and Diving Mammals*, pp. 47–53. Stanford Research Institute Press, Stanford.

Mattsson, A. (1997) Predicting field performance using seedling quality assessment. *New Forests* **13**, 227–52.

Matus, G., Verhagen, R., Bekker, R.M. and Grootjans, A.P. (2003) Restoration of Cirsio dissecti-Molinietum in the Netherlands; can we rely on the soil seed banks? *Applied Vegetation Science* **6**, 73–84.

Mayer, Ph., Abs, C. and Fischer, A. (2004) Colonisation by vascular plants after soil disturbance in the

Bavarian Forest: key factors and relevance for forest dynamics. *Forest Ecology and Management* 188, 279–89.

McCallum, H., Timmers, P. and Hoyle, S. (1995) Modelling the impact of predation on reintroductions of bridled nailtail wallabies. *Wildlife Research* 22, 63–171.

McCann, K.S. (2000) The diversity-stability debate. *Nature* 405, 228–33.

McClanahan, T.R. and Wolfe, R.W. (1993) Accelerating forest succession in a fragmented landscape: the role of birds and perches. *Conservation Biology* 7, 279–88.

McDonald, A.W. (1992) Succession in a three-year old flood-meadow near Oxford. *Journal of Applied Ecology* 29, 345–52.

McDonald, A.W. (1993) The relationship between seed bank and sown-seeds in a re-created flood-meadow after one year. *Journal of Vegetation Science* 4, 395–400.

McDonald, A.W. (2001) Succession during the re-creation of a flood-meadow 1985–1999. *Applied Vegetation Science* 4, 167–76.

McDonald, A.W., Bakker, J.P. and Vegelin, K. (1996) Seed bank classification and its importance for the restoration of species-rich flood-meadows. *Journal of Vegetation Science* 7, 157–64.

McFadyen, R.C. and Skarratt, B. (1996) Potential distribution of *Chromolaena odorata* (siam weed) in Australia, Africa and Oceania. *Agriculture, Ecosystems and Environment* 59, 89–96.

McIntyre, S. and Lavorel, S. (2001) Livestock grazing in subtropical pastures: steps in the analysis of attribute response and plant functional types. *Journal of Ecology* 89, 209–26.

McMichael, A.J., Butler, C.D. and Folke, C. (2003) New visions for addressing sustainability. *Science* 302, 1919–20.

McNaughton, S.J. (1976) Serengeti migratory wildebeest: facilitation of energy flow by grazing. *Science* 191, 92–4.

McNaughton, S.J. (1984) Grazing lawns: animals in herds, plant form and coevolution. *American Naturalist* 124, 863–86.

McNaughton, S.J. (1985) Ecology of a grazing ecosystem: the Serengeti. *Ecological Monographs* 55, 259–94.

McNeil, J. (1992) *Mountains of the Mediterranean World: an Environmental History.* Cambridge University Press, Cambridge.

McQueen, D.J., Post, J.R. and Mills, E.L. (1986) Trophic relationships in freshwater pelagic ecosystems. *Canadian Journal of Fisheries and Aquatic Sciences* 43, 1571–81.

Meadows, D., Meadows, H., Meadows, D.L. *et al.* (1972) *The Limits to Growth.* Potomac Associates, New York.

Meffe, G.K. (1995) Genetic and ecological guidelines for species reintroduction programs: application to Great Lakes fishes. *Journal of Great Lakes Research* 21, 3–9.

Meijer, M.L. (2000) Biomanipulation in the Netherlands: 15 years of experience. PhD thesis, Wageningen University.

Meijer, M.L., De Haan, M.W., Breukelaar, A.W. and Buiteneld, H. (1990) Is reduction of the benthivorous fish an important cause of high transparency following biomanipulation in shallow lakes? *Hydrobiologia* 200/201, 303–15.

Meijer, M.L., Jeppesen, E., van Donk, E. *et al.* (1994a) Long-term responses to fish-stock reduction in small shallow lakes: interpretation of five-year results of four biomanipulation cases in the Netherlands and Denmark. *Hydrobiologia* 275/276, 457–66.

Meijer, M.L., van Nes, E.H., Lammens, E.H.R.R. *et al.* (1994b) The effects of fish stock reduction: the consequences of a drastic fish stock reduction in the large and shallow lake Wolderwijd, The Netherlands: can we understand what happened? *Hydrobiologia* 275/276, 31–42.

Meijer, M.L., De Bois, I., Scheffer, M. *et al.* (1999) Biomanipulation in shallow lakes in The Netherlands: an evaluation of 18 case studies. *Hydrobiologia* 408/409, 13–30.

Mellado, J. (1989) S.O.S. Souss: Argan forest destruction in Morocco. *Oryx* 23, 87–93.

Mendelssohn, I.A. (1979) Nitrogen metabolism in the height forms of *Spartina alterniflora* in North Carolina. *Ecology* 60, 574–84.

Menge, B.A. and Sutherland, J.P. (1976) Species diversity gradients: synthesis of the roles of predation, competition and temporal heterogeneity. *American Naturalist* 110, 351–69.

Menge, B.A. and Sutherland, J.P. (1987) Community regulation: variation in disturbance, competition and predation in relation to environmental stress and recruitment. *American Naturalist* 130, 730–57.

Menge, B.A., Berlow, E.L., Blanchette, C.A. *et al.* (1994) The keystone species concept: variation in interaction strength in a rocky intertidal habitat. *Ecological Monographs* 64, 249–86.

Mensink, B.P., Everaarts, J.M., Kralt, H. *et al.* (1996) Tributyltin exposure in early life stages induces the development of male sexual characteristics in the common whelk, *Buccinum undatum. Marine Environmental Research* 42, 151–4.

Menzel, A. and Fabian, P. (1999) Growing season extended in Europe. *Nature* **397**, 659.

Mesón, M. and Montoya, M. (1993) *Selvicultura Mediterránea*. Mundi-Prensa, Madrid.

Middelkoop, H. and van Haselen, C.O.G. (1999) *Twice a River: Rhine and Meuse in the Netherlands*. National Institute for Inland Water Management and Waste Water Treatment (RIZA), Lelystad.

Middleton, B. (1999) *Wetland Restoration, Flood Pulsing and Disturbance Dynamics*. Wiley, New York.

Middleton, B. (2000) Hydrochory, seed banks, and regeneration dynamics along the landscape boundaries of a forested wetland. *Plant Ecology* **146**, 169–84.

Millán, M.M. (1998) Procesos atmosféricos en el Mediterráneo. Implicaciones en la gestión de nuestros bosques. In: *Problemas sanitarios en los sistemas forestales: de los espacios protegidos a los cultivos de especies de crecimiento rápido* (ed. R. Montoya), pp. 263–90. Coleccion Técnica. Ministerio de Medio Ambiente. O.A. Parques Naturales, Madrid.

Milton, S.J. (2003) 'Emerging ecosystems': a washing-stone for ecologists, economists and sociologists? *South African Journal of Science* **99**, 404–6.

Milton, S.J., Dean, W.R.J. and Richardson, D.M. (2003) Economic incentives for restoring natural capital: trends in southern African rangelands. *Frontiers in Ecology and the Environment* **1**, 247–54.

Mitchell, R.J., Marrs, R.H., Le Duc, M.G. and Auld, M.H.D. (1998) A study of the restoration of heathland on successional sites: changes in vegetation and soil chemical properties. *Journal of Applied Ecology* **36**, 770–83.

Mitchell, R.J., Auld, M.H.D., Le Duc, M.G. and Marrs, R.H. (2000) Ecosystem stability and resilience: a review of their relevance for the conservation management of lowland heaths. *Perspectives in Plant Ecology, Evolution and Systematics* **3**, 142–60.

Mitlacher, K., Poschlod, P., Rosén, E. and Bakker, J.P. (2002) Restoration of wooded meadows – comparative analysis along a chronosequence on Öland (Sweden). *Applied Vegetation Science* **5**, 63–74.

Mitsch, W.J. and Jørgenson, S.E. (eds.) (1989) *Ecological Engineering – An Introduction to Ecotechnology*. Wiley, New York.

Moen, A. (1990) The plant cover of the boreal uplands of central Norway I. Vegetation geology of Solendet Nature Reserve, haymaking fens and birch woodlands. *Gunneria* **63**, 1–451.

Möller, A. (1921) *Dauerwaldwirtschaft*. Springer, Berlin.

Montalvo, A.M., Williams, S.I., Rice, K.J. *et al.* (1997) Restoration biology: a population biology perspective. *Restoration Ecology* **5**, 277–90.

Montanarella, L. (2002) The protection of arid and semi-arid soils in Europe. In: *Sustainable Use and Management of Soils in Arid and Semiarid Regions* (eds. A. Faz, R. Ortiz and A.R. Mermut), pp. 148–61. Quaderna Ed., Cartagena.

Moog, D., Poschlod, P., Kahmen, S. and Schreiber, K.F. (2002) Comparison of species composition between different grassland management treatments after 25 years. *Applied Vegetation Science* **5**, 99–106.

Moora, M. and Zobel, M. (1996) Effect of arbuscular mycorrhiza on inter- and intraspecific competition of two grassland species. *Oecologia* **108**, 79–84.

Moore, P.D. (2002) The future of cool temperate bogs. *Environmental Conservation* **29**, 3–20.

Mortimer, S.R., Hollier, J.A. and Brown, V.K. (1998) Interactions between plant and insect diversity in the restoration of lowland calcareous grasslands in southern Britain. *Applied Vegetation Science* **1**, 101–14.

Mortimer, S.R., van der Putten, W.H. and Brown, V.K. (1999) Insect and nematode herbivory below ground: interactions and role in vegetation succession. In: *Herbivores: Between Plants and Predators* (eds. H. Olff, V.K. Brown and R.H. Drent), pp. 205–38. Blackwell Science, Oxford.

Mosandl, R. and Kleinert, A., (1998) Development of oaks (*Quercus petraea* (Matt.) Liebl.) emerged from bird-dispersed seeds under old-growth pine (*Pinus sylvestris* L.) stands. *Forest Ecology and Management* **106**, 35–44.

Mosimann, T. (1984) Das Stabilitätspotential alpiner Geoökosysteme gegenüber Bodenstörungen durch Schi-Pistenbau. *Verhandlungen der Gesellschaft für Ökologie (Bern)* **Bd. XII**, 167–76.

Moss, B. (1990) Engineering and biological approaches to the restoration from eutrophication of shallow lakes in which aquatic plant communities are important components. *Hydrobiologia* **200/201**, 367–77.

Moss, B. (1998) Shallow lakes, biomanipulation and eutrophication. *Scope News Letter* **29**, 1–45.

Moss, B. (2001) *The Broads: The People's Wetlands*. Harper Collins, London.

Moss, B., Balls, H., Irvine, K. and Stansfield, J. (1986) Restoration of two lowland lakes by isolation from nutrient-rich water sources with and without removal of sediment. *Journal of Applied Ecology* **28**, 586–602.

Mössmer, R. and Fischer, A. (1999) Waldentwicklung nach Sturmwurf im Universitätswald Landshut. *Forstliche Forschungsberichte (München)* **176**, 70–81.

Mouissie, A.M. (2004) Seed dispersal by large herbivores. PhD thesis, University of Groningen.

Müller, A. (2003) A flower in full blossom? Ecological economics at the crossroads between normal and post-normal science. *Ecological Economics* **45**, 19–27.

Muller, S., Dutoit, T., Alard, D. and Grevilliot, F. (1998) Restoration and rehabilitation of species–rich grassland ecosystems in France: a review. *Restoration Ecology* **6**, 94–101.

Müller-Motzfeld, G. (1997) Renaturierung eines Überflutungsgrünlandes an der Ostseeküste. *Schriftenreihe für Landschaftspflege und Naturschutz* **54**, 239–63.

Myers, N., Mittermeier, R.A., Mittermeier, C.G. *et al.* (2000) Biodiversity hotspots for conservation priorities. *Nature* **403**, 853–8.

Navarro, R. and Martínez, A. (1997) Las marras producidas por ausencia de cuidados culturales. *Cuadernos de la SECF* **4**, 43–57.

Navas, A., Machín, J. and Navas, B. (1999) Use of biosolids to restore the natural vegetation on degraded soils in the badlands of Zaragoza (NE Spain). *Bioresource Technology* **69**, 199–205.

Naveh, Z. (1990) Ancient man's impact on the Mediterranean landscape in Israel: ecological and evolutionary perspectives. In: *Man's Role in Shaping of the Eastern Mediterranean Landscape* (eds. S. Bottema, G. Entjes-Nieborg and W. van Zeist), pp. 43–50. Balkema, Rotterdam.

Naveh, Z. (1991) Mediterranean uplands as anthropogenic perturbation-dependent systems and their dynamic conservation management. In: *Terrestrial and Aquatic Ecosystems. Perturbation and Recovery* (ed. O. Ravera), pp. 545–56. Ellis Horwood, New York.

Naveh, Z. (2000) What is holistic landscape ecology? A conceptual introduction. *Landscape and Urban Planning* **50**, 7–26.

Ne'eman, G. (1997) Regeneration of natural pine forest. Review of work done after the 1989 fire in Mount Carmel, Israel. *International Journal of Wildland Fire* **7**, 295–306.

Nehls, G. and Thiel, M. (1993) Large-scale distribution patterns of the mussel *Mytilus edulis* in the Wadden Sea of Schleswig-Holstein: do storms structure the ecosystem? *Netherlands Journal of Sea Research* **31**, 181–7.

Nei, M., Maruyama, T. and Chakraborty, R. (1975) The bottleneck effect and genetic variability in populations. *Evolution* **29**, 1–10.

NFNA (1999) *The Skjern River Restoration Project.* National Forest and Nature Agency, Danish Ministry of Environment and Energy, Copenhagen.

Nienhaus, K. and Bayer, A.K. (2001) Highlights in mining research. In: *Presentation at the 1st European Mineral Resources Day of EmiReC*, 22 March 2001, Brussels.

Nienhuis, P.H. and Gulati, R.D. (eds.) (2002) Ecological restoration of aquatic and semi-aquatic ecosystems in the Netherlands: an introduction. *Hydrobiologia* **478**, 1–6.

Nienhuis, P.H., Buijse, A.D., Leuven, R.S.E.W. *et al.* (2002) Ecological rehabilitaton of the lowland basin of the river Rhine (NW Europe). *Hydrobiologia* **478**, 53–72.

Niering, W.A. (1997a) Human-dominated ecosystems and the role of restoration ecology. *Restoration Ecology* **5**, 273–4.

Niering, W.A. (1997b) Tidal wetlands restoration and creation along the east coast of North America. In: *Restoration Ecology and Sustainable Development* (eds. K.M. Urbanska, N.R. Webb and P.J. Edwards), pp. 259–85. Cambridge University Press, Cambridge.

Nijland, H.J. and Cals, M.J.R. (2001) River restoration in Europe: practical approaches. *Proceedings of the 2000 River Restoration Conference*. RIZA, Wageningen.

Nilsson, C., Gardfjell, M. and Grelsson, G. (1991) Importance of hydrochory in structuring plant communities along rivers. *Canadian Journal of Botany* **69**, 2631–3.

Nilsson, M.-C. (1994) Separation of allelopathy and resource competition by the boreal dwarf shrub *Empetrum hermaphroditum* Hagerup. *Oecologia* **98**, 1–7.

Nilsson, M.-C., Gallet, C. and Wallstedt, A. (1998) Temporal variability of phenolics and batatasin-III in *Empetrum hermaphroditum* leaves over an eight-year period: interpretations of ecological function. *Oikos* **81**, 6–16.

Nixon, S.W. (1995) Coastal marine eutrophication: a definition, social causes, and future concerns. *Ophelia* **41**, 199–219.

Noakes, D.L. and Curry, R.A. (1995) Lessons to be learned from attempts to restore *Salvelinus* species other than *S. namaycush*: a review of reproductive behavior. *Journal of Great Lakes Research* **21**, 54–64.

Noordhuis, R.H., Reeders, H.H. and bij de Vaate, A. (1992) Filtering rate and pseudofaeces production in zebra mussels and their application in water quality management. *Limnologie Aktuell* **4**, 101–14.

Noy-Meir, I. (1973) Desert ecosystems: environment and producers. *Annual Review of Ecology and Systematics* **4**, 25–52.

Noy-Meir, I. (1975) Stability of grazing systems: an application of predator-prey graphs. *Journal of Ecology* **63**, 459–81.

NRC (1992) *Committee on 'Restoration of Aquatic Ecosystems-Science, Technology and Public Policy'*. National Research Council USA, National Academy Press, Washington, DC.

NRC (1999) *Our Common Journey: A Transition toward Sustainability*. National Research Council USA, National Academy Press, Washington, DC.

Nunes, M.A., Catarino, F. and Pinto, E. (1989) Strategies for acclimation to seasonal drought in *Ceratonia siliqua* leaves. *Physiologia Plantarum* **77**, 150–6.

Nunney, L. and Campbell, K.A. (1993) Assessing minimum viable population size: demography meets population genetics. *Trends in Ecology and Evolution* **8**, 234–9.

ÖAG (2000) *Richtlinie für standortgerechte Begrünungen* [Guidelines for Site Specific Restoration], ÖAG (Austrian Grassland Federation), BAL Gumpenstein, Irdning.

Odum, E.P. (1953) *Fundamentals of Ecology*, 1st edn. W.B. Saunders Company, Philadelphia.

Odum, E.P. (1971) *Fundamentals of Ecology*, 3rd edn. W.B. Saunders Company, Philadelphia.

Odum, E.P. (1996) *Ecology: a Bridge between Science and Ssociety*. Sinauer, Sunderland, MA.

Odum, E.P., Finn, J.T. and Franz, E.H. (1979) Perturbation theory and the subsidy-stress gradient. *BioScience* **29**, 349–52.

Odum, H.T. (1983) *Systems Ecology: An Introduction*. Wiley, New York.

O'Grady, M.F. (2001) Salmoid riverine habitat restoration in the Republic of Ireland. In: *River Restoration in Europe: Practical Approaches* (eds. H.J. Nijland and M.J.R. Cals), pp. 237–43. Proceedings of the 2000 River Restoration Conference. RIZA, Wageningen.

Oguto-Ohwayo, R. (1999) Nile perch in Lake Victoria: the balance between benefits and negative impacts of aliens. In: *Invasive Species and Biodiversity Management* (eds. O.T. Sandlund, P.J. Schei and Å. Viken), pp. 47–64. Kluwer Academic Publishers, Dordrecht.

Oksanen, L., Fretwell, S.D., Arruda, J. and Niemelä, P. (1981) Exploitation ecosystems in gradients of primary productivity. *American Naturalist* **118**, 240–61.

Olde Venterink, H., Pieterse, N.M., Belgers, J.D.M. *et al.* (2002) N, P, and K budgets along nutrient availability-productivity gradients in wetlands. *Ecological Applications* **12**, 1010–26.

Olde Venterink, H., Wiegman, F., van der Lee, G.E.M. and Vermaat, J.E. (2003) Role of active floodplains for nutrient retention in the river Rhine. *Journal of Environmental Quality* **32**, 1430–5.

Olem, H. and Flock, G. (eds.) (1990) *Lake and Reservoir Restoration Guidance Manual*, 2nd edn. EPA 440/4-90-006. Environmental Protection Agency, Washington, DC.

Olff, H., de Leeuw, J., Bakker, J.P. *et al.* (1997) Vegetation succession and herbivory in a salt marsh: changes induced by sea-level rise and silt deposition along an elevational gradient. *Journal of Ecology* **85**, 799–814.

Olff, H., Vera, F.W.M., Bokdam, J. *et al.* (1999) Shifting mosaics in grazed woodlands driven by the alternation of plant facilitation and competition. *Plant Biology* **1**, 127–37.

Olff, H., Ritchie, M.E. and Prins, H.H.T. (2002) Global environmental controls of diversity in large herbivores. *Nature* **415**, 901–4.

Oliet, J.A. and Artero, F. (1993) *Estudio del desarrollo y la supervivencia en zonas áridas del repoblado protegido mediante tubos protectores*. Ponencias y Comunicaciones del I Congreso Forestal Español. vol. II, pp. 415–20. Lourizán.

Oomes, M.J.M., Olff, H. and Altena, H.J. (1996) Effects of vegetation management and raising the water table on nutrient dynamics and vegetation change in a wet grassland. *Journal of Applied Ecology* **33**, 576–88.

Oost, A.P. and de Boer, P.L. (1994) Sedimentology and development of barrier islands, ebb-tidal deltas, inlets and backbarrier areas of the Dutch Wadden Sea. *Senckenbergiana Maritima* **24**, 65–115.

Opdam, P. (1991) Metapopulation theory and habitat fragmentation: a review of holarctic breeding bird studies. *Landscape Ecology* **5**, 93–106.

Ormerod, S.J. (2003) Restoration in applied ecology: editor's introduction. *Journal of Applied Ecology* **40**, 44–50.

Orson, R.A. (1999) A palaeoecological assessment of *Phragmites australis* in New England tidal marshes: changes in plant community structure during the last few millennia. *Biological Invasions* **1**, 149–58.

Ozenda, P. (1988) *Die Vegetation der Alpen im europäischen Gebirgsraum*. Gustav Fischer Verlag, Stuttgart.

Ozinga, W.A., van Andel, J. and McDonnell-Alexander, M.P. (1997) Nutritional soil heterogeneity and mycorrhiza as determinants of plant species diversity. *Acta Botanica Neerlandica* **46**, 237–54.

Pace, M.L., Cole, J.J., Carpenter, S.R. and Kitchell, J.F. (1999) Trophic cascades revealed in diverse ecosystems. *Trends in Ecology and Evolution* **14**, 483–8.

Page, D.S., Ozbal, C.C. and Lanphear, M.E. (1996) Concentration of butyltin species in sediments associated with shipyard activity. *Environmental Pollution* 91, 237–43.

Pahl-Wostl, C. (1995) *The Dynamic Nature of Ecosystems: Chaos and Order Entwined.* Wiley, Chichester.

Pahl-Wostl, C. and Ulanowicz, R. (1993) Quantification of species as functional units within an ecological network. *Ecological Modelling* 66, 65–79.

Paine, R.T. (1966) Food web complexity and species diversity. *American Naturalist* 100, 65–75.

Paine, R.T. (1980) Food webs: linkage, interaction strength and community infrastructure. *Journal of Animal Ecology* 49, 667–85.

Pakeman, R.J., Thwaites, R.H., Le Duc, M.G. and Marrs, R.H. (2000) Vegetation re-establishment on land previously subject to control of *Pteridium aquilinum* by herbicide. *Applied Vegetation Science* 3, 95–104.

Pakeman, R.J., Thwaites, R.H., Le Duc, M.G. and Marrs, R.H. (2002) The effects of cutting and herbicide treatment on *Pteridium aquilinum* treatment. *Applied Vegetation Science* 5, 203–12.

Palik, B.J., Goebel, P.C., Kirkman, L.K. and West, L. (2000) Using landscape hierarchies to guide restoration of disturbed ecosystems. *Ecological Applications* 10, 189–202.

Palmer, M.A., Allan, J.D. and Butman, C.A. (1996) Dispersal as a regional process affecting the local dynamics of marine and stream benthic invertebrates. *Trends in Ecology and Evolution* 11, 322–6.

Palmer, M.A., Ambrose, R.F. and Poff, N.L. (1997) Ecological theory and community restoration ecology. *Restoration Ecology* 5, 291–300.

Palmer, M., Bernhardt, E., Chornesky, E. *et al.* (2004) Ecology for a crowded planet. *Science* 304, 1251–2.

Palmer, M.W. (1992) The coexistence of species in fractal landscapes. *American Naturalist* 139, 375–97.

Parry, M.L. (ed.) (2000) *Assessment of Potential Effects and Adaptations for Climate Change in Europe: Summary and Conclusions.* Jackson Environmental Institute, University of East Anglia, Norwich.

Pärtel, M., Zobel, M., Zobel, K. and van der Maarel, E. (1996) The species pool and its relation to species richness: evidence from Estonian plant communities. *Oikos* 75, 111–17.

Pastor, J., Aber, J.D. and McClaugherty, C.A. (1984) Aboveground production and N and P cycling along a nitrogen mineralization gradient on Blackhawk Island, Wisconsin. *Ecology* 65, 256–68.

Paterson, D.M. (1997) Biological mediation of sediment erodibility: ecology and physical dynamics. In: *Cohesive Sediments* (eds. N. Burt, R. Parker, and J. Watts), pp. 215–29. Wiley, Chichester.

Pausas, J.G. (2004) Changes in fire and climate in the eastern Iberian Peninsula (Mediterranean basin). *Climatic Change* 63, 337–50.

Pausas, J. and Vallejo, V.R. (1999) The role of fire in European Mediterranean ecosystems. In: *Remote Sensing of Large Wildfires in the European Mediterranean Basin* (ed. E. Chuvieco), pp. 3–16. Springer, Berlin.

Pausas, J.G., Carbó, E., Caturla, R.N. *et al.* (1999) Post-fire regeneration patterns in the Eastern Iberian Peninsula. *Acta Oecologica* 20, 499–508.

Pausas, J.G., Rusch, G.M. and Leps, J. (eds.) (2003) Plant functional types in relation to disturbance and land use. *Journal of Vegetation Science* (special feature) 14, 305–416.

Pearce, D.W. (1993) *Economic Values and the Natural World.* Earthscan Publications, London.

Pennings, S.C. and Callaway, R.M. (1996) Impact of a parasitic plant on the structure and dynamics of salt marsh vegetation. *Ecology* 77, 1410–19.

Pennings, S.C. and Callaway, R.M. (2002) Parasitic plants: parallels and contrasts with herbivores. *Oecologia* 131, 479–89.

Pennycuick, C.J. (1972) Soaring behaviour and performance of some East African birds, observed from a motor glide. *Ibis* 114, 178–218.

Peñuelas, J.L., Ocaña, L., Domínguez, S. and Renilla, I. (1997) Experiencias sobre control de la competencia herbácea en repoblaciones de terrenos agrícolas abandonados. Resultados de tres años en campo. *Cuadernos de la SECF* 4, 119–26.

Peratoner, G. (2003) *Organic Seed Propagation of Alpine Species and their Use in Ecological Restoration of Ski Runs in Mountain Regions.* Dissertation, Kassel University Press.

Perrow, M.R. and Davy, A.J. (eds.) (2002a) *Handbook of Ecological Restoration*, vol. 1: *Principles of Restoration.* Cambridge University Press, Cambridge.

Perrow, M.R. and Davy, A.J. (eds.) (2002b) *Handbook of Ecological Restoration*, vol. 2: *Restoration in Practice.* Cambridge University Press, Cambridge.

Perrow, M.R., Meijer, M.-L., Dawidowicz, P. and Coops, H. (1997) Biomanipulation in shallow lakes: state of the art. *Hydrobiologia* 342/343, 355–65.

Persson, L. (1999) Trophic cascades: abiding heterogeneity and the trophic level concept at the end of the road. *Oikos* 85, 385–97.

Persson, T.S. (1995) *Management of Roadside Verges: Vegetation Changes and Species Diversity.* Sveriges Lantbrucksuniversitet Institutionen for Ekologi och Miljovard, Report 82, Uppsala.

Peterson, C.H., Summerson, H.C. and Fegley, S.R. (1987) Ecological consequences of mechanical harvesting of clams. *Fishery Bulletin* **85**, 281–98.

Peterson, S.A. (1981) *Sediment Removal as a Lake Restoration Technique.* Corvallis Environmental Research Laboratory, USEPA. EPA-600/3-81-013, Corvallis.

Petts, G. (1984) *Impounded Rivers: Perspectives for Ecological Management.* Wiley, Chichester.

Petts, G., Heathcote, J. and Martin, D. (2002) *Urban Rivers: Our Inheritance and Future.* IWA Publishing, London.

Pfadenhauer, J. and Klötzli, F. (1996) Restoration experiments in middle European wet terrestrial ecosystems: an overview. *Vegetatio* **126**, 101–15.

Pfadenhauer, J. and Grootjans, A.P. (1999) Wetland restoration in Central Europe: aims and methods. *Applied Vegetation Science* **2**, 95–106.

Pfadenhauer, J., Höper, H., Borkowski, B. *et al.* (2001) Entwicklung planzenartreichen Niedermoorgrünlands. In: *Ökosystemmanagement für Niedermoore: Strategien und Verfahren zur Renaturierung* (eds. R. Kratz and J. Pfadenhauer), pp. 134–55. Ulmer, Stuttgart.

Phillips, G.L., Jackson, R., Bennet, C. and Chilvers, A. (1994) The importance of sediment phosphorus release in the restoration of very shallow lakes (the Norfolk Broads, England) and implications for biomanipulation. *Hydrobiologia* **275/276**, 445–56.

Pickett, S.T.A. and Parker, V.T. (1994) Avoiding the old pitfalls: opportunities in a new discipline. *Restoration Ecology* **2**, 75–9.

Pickett, S.T.A., Kolasa, J., Armesto, J.J. and Collins, S.L. (1989) The ecological concept of disturbance and its expression at various hierarchical levels. *Oikos* **54**, 129–36.

Pielou, E.C. (1979) *Biogeography.* Wiley, New York.

Piersma, T. and Koolhaas, A. (1997) *Shorebirds, Shellfish(eries) and Sediments around Griend, Western Wadden Sea, 1988–1996. Single Large-scale Exploitative Events Lead to Long-term Changes of the Intertidal Birds-benthos Community.* NIOZ-report 1997-7, Texel.

Piersma, T., Koolhaas, A., Dekinga, A., Beukema, J.J., Dekker, R. and Essink, K. (2001) Long-term indirect effects of mechanical cockle-dredging on intertidal bivalve stocks in the Wadden Sea. *Journal of Applied Ecology* **38**, 976–90.

Pietsch, W. (1998) Sukzession der Vegetation im NSG 'Insel im Senftenberger See' (1970–1996). *Berichte Institut für Landschafts- und Pflanzenökologie Universität Hohenheim,* Beiheft 5, 59–68.

Pimm, S.L. and Lawton, J.H. (1978) On feeding on more than one level. *Nature* **275**, 542–4.

Pimm, S.L., Lawton, J.H. and Cohen, J.E. (1991) Food web patterns and their consequences. *Nature* **350**, 669–74.

Pinkham, R. (2000) *Daylighting: New Life for Buried Streams.* Rocky Mountain Institute, Old Snowmass, CO.

Polis, G.A. and Strong, D.R. (1996) Food web complexity and community dynamics. *American Naturalist* **147**, 813–46.

Polis, G.A., Anderson, W.B. and Holt, R.D. (1997) Toward an integration of landscape and food web ecology: the dynamics of spatially subsidized food webs. *Annual Review of Ecology and Systematics* **28**, 289–316.

Polis, G.A., Sears, A.L.W., Huxel, G.R., Strong, D.R. and Maron, J. (2000) When is a trophic cascade a trophic cascade? *Trends in Ecology and Evolution* **15**, 473–5.

Pomarol, M. (1994) Releasing Montagu's harrier (*Circus pygargus*) by the method of hacking. *Journal of Raptor Research* **28**, 19–22.

Poschlod, P. (1990) *Vegetationsentwicklung in abgetorften Hochmooren des bayerischen Alpenvorlandes unter besonderer Berücksichtigung standortskundlicher und populationsbiologischer Faktoren.* Dissertationes Botanicae 152. Cramer, Berlin.

Poschlod, P. and Bonn, S. (1998) Changing dispersal processes in the central European landscape since the last ice age: an explanation for the actual decrease of plant species richness in different habitats? *Acta Botanica Neerlandica* **47**, 27–44.

Poschlod, P. and WallisDeVries, M.F. (2002) The historical and socioeconomic perspective of calcareous grasslands: lessons from the distant and recent past. *Biological Conservation* **104**, 361–76.

Poschlod, P., Bakker, J.P., Bonn, S. and Fischer, S. (1996) Dispersal of plants in fragmented landscapes. In: *Species Survival in Fragmented Landscapes* (eds. J. Settele, C.R. Margules, P. Poschlod and K. Henle), pp. 123–7. Kluwer Academic Publishers, Dordrecht.

Poschlod, P., Kiefer, S., Tränkle, U., Fischer, S. and Bonn, S. (1998) Plant species richness in calcareous grasslands as affected by dispersability in space and time. *Applied Vegetation Science* **1**, 75–90.

Postel, S. and Richter, B. (2003) *Rivers for Life: Managing Water for People and Nature.* Island Press, Washington, DC.

Pott, R. and Hüppe, J. (1991) *Die Hudelandschaften Nordwestdeutschlands.* Westfälisches Museum für Naturkunde, Münster.

PPG25 (2001) *Regions Planning Policy Guidance Notes on Development and Flood Risk.* DTLR, London.

Prach, K., Jenik, J. and Large, A.R.G. (1996) *Floodplain Ecology and Management: The Luznice River in the Trebon Biospere Reserve, Central Europe.* SPB Academic Publishing, Amsterdam.

Prescott-Allen, R., Munro, D.A. and Holdgate, M.W. (1991) *Caring for the Earth: A Strategy for Sustainable Living.* International Union for Conservation of Nature and Natural Resources, United Nations Environment Programme, World Wildlife Fund. IUCN, Gland.

Preston, F.W. (1962) The canonical distribution of commonness and rarity. Part I. *Ecology* 43, 185–215.

Price, P.W., Westoby, M., Rice, B. *et al.* (1986) Parasite mediation in ecological interactions. *Annual Review of Ecology and Systematics* 17, 487–505.

Prins, H.H.T. and Douglas-Hamilton, I. (1989) Stability in a multi-species assemblage of large herbivores in East Africa. *Oecologia* 83, 392–400.

Prins, H.H.T. and van der Jeugd, H.P. (1993) Herbivore population crashes and woodland structure in East Africa. *Journal of Ecology* 81, 305–14.

Prins, H.H.T. and Olff, H. (1998) Species-richness of African grazer assemblages: towards a functional explanation. In: *Dynamics of Tropical Communities* (eds. D.M. Newbury, H.H.T. Prins and N.D. Brown), pp. 449–90. Blackwell Science, Oxford.

Prugh, T., Costanza, R. and Daly, H. (2003) *The Local Politics of Global Sustainability.* Island Press, Washington, DC.

Pugh, C.E., Hossner, L.R. and Dixon, J.B. (1984) Oxidation rate of pyrite as affected by surface area, morphology, oxygen concentration, and autotrophic bacteria. *Soil Science* 137, 309–14.

Pugnaire, F.I., Haase, P. and Puigdefábregas, J. (1996) Facilitation between higher plant species in a semiarid environment. *Ecology* 77, 1420–6.

Puhe, J. and Ulrich, B. (2000) *Global and Human Impacts on Forest Ecosystems.* Ecological Studies Series, vol. 143, Springer, Berlin.

Puigdefábregas, J. and Mendizabal, T. (1998) Perspectives on desertification: Western Mediterranean. *Journal of Arid Environments* 39, 209–24.

Pulliam, H.R. (1988) Sources, sinks, and population regulation. *American Naturalist* 132, 652–61.

Puurman, E. and Ratas, U. (1995) Problems of conservation and management of the west Estonian seashore meadows. In: *Directions in European Coastal Management* (eds. M.G. Healy and J.P. Doodey), pp. 345–50. Samara Publishing, Cardigan.

Pye, K. (2000) Saltmarsh erosion in southeast England: mechanisms, causes and implications. In: *British Saltmarshes* (eds. B.R. Sherwood, B.G. Gardiner and T. Harris), pp. 359–96. Linnean Society of London, London.

Pyšek, P. (1995) Recent trends in studies on plant invasions (1974–1994). In: *Plant Invasions: General Aspects and Special Problems* (eds. P. Pyšek, K. Prach, M. Rejmánek and M. Wade). pp. 223–36. SPB Academic Publishing, Amsterdam.

Pyšek, P., Jarosik, V. and Kučera, T. (2002) Patterns of invasion in temperate nature reserves. *Biological Conservation* 104, 13–24.

Pywell, R.F., Bullock, J.M., Roy, D.B. *et al.* (2003) Plant traits as predictors of performance in ecological restoration. *Journal of Applied Ecology* 40, 65–77.

Quézel, P., Médail, F., Loisel, R. and Barbero, M. (1999) Biodiversity and conservation of forest species in the mediterranean basin. *Unasylva* 197, 21–8.

Rabinowitz, D. (1981) Seven forms of rarity. In: *The Biological Aspects of Rare Plant Conservation* (ed. H. Synge), pp. 205–17. Wiley, Chichester.

Raffaelli, D. and Hawkins, S. (1996) *Intertidal Ecology.* Chapman and Hall, London.

Rapport, D.J. (1995) Ecosystem health: an emerging integrative science. In: *Evaluating and Monitoring the Health of Large-Scale Ecosystems* (eds. D.J. Rapport, C.L. Gaudet and P. Calow), pp. 3–31. Springer-Verlag, Berlin.

Rapport, D.J., Costanza, R. and McMichael, A.J. (1998) Assessing ecosystem health. *Trends in Ecology and Evolution* 13, 397–402.

Rast, W. and Holland, M. (1988) Eutrophication of lakes and reservoirs – a framework for making management decisions. *Ambio* 17, 2–12.

Rast, W. and Thornton, J.A. (1996) Trends in eutrophication research and control. *Hydrological Processes* 10, 295–313.

Raunkiaer, C. (1934) *The Life Forms of Plants and Statistical Plant Geography.* Oxford University Press, Oxford.

Raven, P.J., Holmes, N.T.H., Dawson, F.H. *et al.* (1998) *River Habitat Quality: The Physical Character of Rivers and Streams in the UK and Isle of Man.* Environment Agency, Bristol.

Reed, D.J. (1988) Sediment dynamics and deposition in a retreating coastal salt marsh. *Estuarine, Coastal and Shelf Science* **26**, 67–79.

Read, D.J. (1991) Mycorrhizas in ecosystems. *Experientia* **47**, 376–91.

Reeders, H.H. and bij de Vaate, A. (1990) Zebra mussels (*Dreissena polymorpha*): a new perspective for water quality management. *Hydrobiologia* **200/201**, 437–50.

Reeders, H.H. and bij de Vaate, A. (1992) Bioprocessing of polluted suspended matter from the water column by the zebra mussel (*Dreissena polymorpha* Pall.). *Hydrobiologia* **239**, 53–63.

Reise, K. (1982) Long-term changes in the macrobenthic invertebrate fauna of the Wadden Sea: are polychaetes about to take over? *Netherlands Journal of Sea Research* **16**, 29–36.

Reise, K. (1985) *Tidal Flat Ecology. An Experimental Approach to Species Interactions.* Springer, Berlin.

Reise, K., Gollasch, S. and Wolff, W.J. (1999) Introduced marine species of the North Sea coasts. *Helgoländer Meeresuntersuchungen* **52**, 219–34.

Reisigl, H. and Keller, R. (1987) *Alpenpflanzen im Lebensraum; Alpine Rasen, Schut- und Felsvegetation.* Gustav Fischer Verlag, Stuttgart.

Rejmánek, M. (1999) Invasive plant species and invasible ecosystems. In: *Invasive Species and Biodiversity Management* (eds. O.T. Sandlund, P.J. Schei and Å. Viken), pp. 79–102. Kluwer Academic Publishers, Dordrecht.

Rejmánek, M. and Rosén, E. (1992) Influence of colonizing shrubs on species area relationships in alvar plant communities. *Journal of Vegetation Science* **3**, 625–30.

Reynolds, C.S. (1994) The ecological basis for the successful biomanipulation of aquatic communities. *Archiv für Hydrobiologie* **130**, 1–34.

Rice, E.L. (1974) *Allelopathy.* Academic Press, New York.

Richards, C.M. (2000) Inbreeding depression and genetic rescue in a plant metapopulation. *American Naturalist* **155**, 383–94.

Richert, M., Dietrich, O., Koppisch, D. and Roth, S. (2000) The influence of rewetting on vegetation development and decomposition in a degraded fen. *Restoration Ecology* **8**, 185–96.

Richter, B.S. and Stutz, J.C. (2002) Mycorrhizal inoculation of Big Sacaton: implications for grassland restoration of abandoned agricultural fields. *Restoration Ecology* **10**, 607–16.

Rickers, J.R., Queen, L.P. and Arthaud, G.J. (1995) A proximity-based approach to assessing habitat. *Landscape Ecology* **10**, 309–21.

Riemann, B. and Hoffmann, E. (1991) Ecological consequences of dredging and bottom trawling in the Limfjord, Denmark. *Marine Ecology: Progress Series* **69**, 171–8.

Rieseberg, L.H. (1991) Hybridization in rare plants: insights from case studies in *Cercocarpus* and *Helianthus*. In: *Genetics and Conservation of Rare Plants* (eds. D.A. Falk and K.E. Holsinger), pp. 171–81. Oxford University Press, Oxford.

Rietkerk, M. and van de Koppel, J. (1997) Alternate stable states and threshold effects in semi-arid grazing systems. *Oikos* **79**, 69–76.

Riley, A.L. (1998) *Restoring Streams in Cities: A Guide for Planners, Policymakers and Citizens.* Island Press, Washington, DC.

Risk, M.J. and Moffat, J.S. (1977) Sedimentological significance of fecal pellets of *Macoma balthica* in the Minas Basin, Bay of Fundy. *Journal of Sedimentary Petrology* **47**, 1425–36.

Ritchie, M.E. and Olff, H. (1999a) Herbivore diversity and plant dynamics: compensatory and additive effects. In: *Herbivores: Between Plants and Predators* (eds. H. Olff, V.K. Brown and R.H. Drent), pp. 175–204. Blackwell Science, Oxford.

Ritchie, M.E. and Olff, H. (1999b) Spatial scaling laws yield a synthetic theory of biodiversity. *Nature* **400**, 557–60.

RIZA (1996) *Landscape Planning of the River Rhine in the Netherlands: Towards a Balance in River Management.* RIZA, Lelystad.

Roberts, C.M. (1997) Ecological advice for the global fisheries crisis. *Trends in Ecology and Evolution* **12**, 35–8.

Robertson, D.P. and Hull, R.B. (2001) Beyond biology: toward a more public ecology for conservation. *Conservation Biology* **15**, 970–9.

Robichaud, P.R., Bayers, J.L., Neary, D.G. (2000) *Evaluating the Effectiveness of Postfire Rehabilitation Treatments.* General Technical Report RMRS-GTR-63. Rocky Mountain Research Station, USDA, Ford Collins, CO.

Röder, H., Fischer, A. and Klöck, W. (1996) Waldentwicklung auf Quasi-Dauerflächen im Luzulo-Fagetum der Buntsandsteinrhön (Forstamt Mittelsinn). *Forstwissenschaftliches Centralblatt* **115**, 321–35.

Rodwell, J.S. (1992) (ed.) *British Plant Communities.* vol. 3: *Grasslands and Montane Communities.* Cambridge University Press, Cambridge.

Roe, E. (1998) *Taking Complexity Seriously: Policy Analysis, Triangulation and Sustainable Development.* Kluwer Academic Publishers, Boston.

Roelofs, J.G.M. (1991) Inlet of alkaline river water into peaty lowlands; effects on water quality and *Stratiotes aloides* stands. *Aquatic Botany* **39**, 267–93.

Roelofs, J.G.M., Brandrud, T.E. and Smolders, A.J.P. (1994) Massive expansion of *Juncus bulbosus* L. after liming of acidified SW Norwegian lakes. *Aquatic Botany* **48**, 187–202.

Roelofs, J.G.M., Brouwer, E. and Bobbink, R. (2002) Restoration of aquatic macrophyte vegetation in acidified and eutrophicated soft water wetlands in the Netherlands. *Hydrobiologia* **478**, 171–80.

Rogers, D.I., Piersma, T., Lavaleye, M., Pearson, G.B. and de Goeij, P. (2003) *Life Along Land's Edge: Wildlife on the Shores of Roebuck Bay, Broome.* Department of Conservation and Land Management, Crawley, WA.

Röhle, H. (1995) Zum Wachstum der Fichte auf Hochleistungsstandorten in Südbayern. *Mitteilungen der Staatsforstverwaltung Bayerns* **48**, München.

Roldán, A., Querejeta, I., Albaladejo, J. and Castillo, V. (1996) Survival and growth of *Pinus halepensis* Miller seedlings in a semi-arid environment after forest soil transfer, terracing and organic amendments. *Annales des Sciences Forestières* **53**, 1099–112.

Roman, C.T., Garvine, R.W. and Portnoy, J.W. (1995) Hydrologic modeling as a predictive basis for ecological restoration of salt marshes. *Environmental Management* **19**, 559–66.

Ronen, D. and Magaritz, M. (1991) Groundwater quality as affected by managerial decisions in agricultural areas: effect of land development and irrigation with sewage effluents. In: *Hydrological Basis of Ecologically Sound Management of Soil and Groundwater* (eds. H.P. Nachtnebel and K. Kovar), pp. 153–62. International Association of Hydrological Sciences publication no. 202. IAHS Press, Wallingford.

Root, R. (1967) The niche exploitation pattern of the blue-grey gnatcatcher. *Ecological Monographs* **37**, 317–50.

Roozen, A.J.M. and Westhoff, V. (1985) A study on long-term salt marsh succession using permanent plots. *Vegetatio* **61**, 23–32.

Rosén, E. and Bakker, J.P. (2005) Effects of agri-environment schemes on scrub clearance, livestock grazing and plant diversity in a low-intensity farming system on Öland, Sweden. *Basic and Applied Ecology* **6**, in press.

Rosenzweig, M.L. (1995) *Species Diversity in Space and Time.* Cambridge University Press, Cambridge.

Rosenzweig, M. (2003) *Win-Win Ecology. How the Earth's Species can Survive in the Midst of Human Enterprise.* Oxford University Press, Oxford.

Rosenzweig, M.L. and MacArthur, R.H. (1963) Graphical representation and stability conditions of predator-prey interactions. *American Naturalist* **97**, 209–23.

Rosenzweig, M.L. and Abramsky, Z. (1993) How are diversity and productivity related? In: *Species Diversity in Ecological Communities* (eds. R.E. Ricklefs and D. Schluter), pp. 52–76. University of Chicago Press, Chicago.

Rosgen, D. (1996) *Applied River Morphology.* Wildland Hydrology, Pagosa Springs, CO.

Roth, S., Seeger, T., Poschlod, P. *et al.* (2001) Etablierung von Röhrichten und Seggenrieden. In: *Ökosystemmanagement für Niedermoore; Strategien und Verfahren zur Renaturierung* (eds. R. Kratz and J. Pfadenhauer), pp. 125–34. Ulmer Verlag, Stuttgart.

Rothschild, B.J., Ault, J.S., Goulletquer, P. and Héral, M.. (1994) Decline of the Chesapeake Bay oyster population: a century of habitat destruction and overfishing. *Marine Ecology: Progress Series* **111**, 29–39.

RRC (1999) *Manual of River Restoration Techniques.* River Restoration Centre, Bedford.

RRC (2002) *Update for the Manual of River Restoration Techniques.* River Restoration Centre, Bedford.

Rumpel, C., Kögel-Knabner, I. and Hüttl, R.F. (1999) Organic matter composition and degree of humification in lignite-rich mine soils under a chronosequence of pine. *Plant and Soil* **213**, 161–8.

Rutherford, I.D., Jerie, K. and Marsh, N. (2000) *A Rehabilitation Manual for Australian Scheme*, vols. 1 and 2, Cooperative Research Centre for Catchment Hydrology, Canberra. Land and Water Research and Development Corporation, Clayton.

Ryding, S.O. (1981) Reversibility of man-introduced eutrophication: experience of a lake recovery study in Sweden. *Internationale Revue der gesamten Hydrobiologie* **64**, 449–503.

Ryding, S.O. and Rast, W. (eds.) (1989) *The Control of Eutrophication of Lakes and Reservoirs.* UNESCO, Man and the Biosphere Series, vol. 1. The Parthenon Publishing Group, London.

Sakai, A.K., Allendorf, F.W., Holt, J.S. *et al.* (2001) The population biology of invasive species. *Annual Review of Ecology and Systematics* **32**, 305–32.

Sala, O.E., Chapin, III, F.S., Armesto, J.J. *et al.* (2000) Biodiversity: Global biodiversity scenarios for the year 2001. *Science* **287**, 1770–4.

Samson, F. and Knopf, F. (1994) Prairie conservation in North America. *BioScience* **44**, 418–21.

Sanderson, E.W., Jaiteh, M., Levy, M.A. *et al.* (2002) The human footprint and the last of the wild. *BioScience* **52**, 891–904.

Saris, F., Vergeer, J.W., Hustings, F. and van Turnhout, C. (2002) Volkstelling onder broedvogels biedt gevarieerd beeld. *De Levende Natuur* 103, 196–205.

Sas, H., Ahlgren, I., Bernardt, H. *et al.* (1989) *Lake Restoration by Reduction of Nutrient Loading: Expectations, Experiences, Extrapolations.* Akademia Verlag Richarz. St. Augustin, Germany.

Sasser, C.E. and Gosselink, J.G. (1984) Vegetation and primary production in a floating freshwater marsh in Louisiana. *Aquatic Botany* 20, 245–55.

Saunders, D.A., Hobbs, R.J. and Ehrlich, P. (eds.) (1993) *Nature Conservation 3: Reconstruction of Fragmented Ecosystems.* Surrey Beatty and Sons, Chipping Norton.

Sax, D.F. and Gaines, S.D. (2003) Species diversity: from global decreases to local increases. *Trends in Ecology and Evolution* 18, 561–6.

Schaaf, W., Wilden, R. and Gast, M. (1998) Soil solution composition and element cycling as indicators of ecosystem development along a chronosequence of post-lignite mining sites in Lusatia/Germany. In: *Land Reclamation – Achieving Sustainable Benefits* (eds. H.R. Fox, H.M. Moore and A.D. McIntosh). pp. 241–7. A.A. Balkema, Rotterdam.

Schaaf, W., Gast, M., Wilden, R., Scherzer, J., Blechschmidt, R. and Hüttl, R.F. (1999) Temporal and spatial development of soil solution chemistry and element budgets in different mine soils of the Lusatian lignite mining area. *Plant and Soil* 213, 169–79.

Schaaf, W., Neumann, C. and Hüttl, R.F. (2001) Actual cation exchange capacity in lignite containing pyritic mine spoils. *Journal of Plant Nutrition and Soil Science* 164, 77–8.

Schama, S. (1995) *Landscape and Memory.* Harper Collins Publishers, London.

Scheffer, F. and Schachtschabel, P. (1992) Lehrbuch der Bodenkunde, 13th edn. Enke Verlag, Stuttgart.

Scheffer, M., Carpenter, S., Foley, J.A. *et al.* (2001) Catastrophic shifts in ecosystems. *Nature* 413, 591–6.

Scheffer, M., Hosper, S.H., Meijer, M.-L. *et al.* (1993) Alternative equilibria in shallow lakes. *Trends in Ecology and Evolution* 8, 275–9.

Scheiner, S.M. and Berrigan, D. (1998) The genetics of phenotypic plasticity. VIII. The cost of plasticity in *Daphnia pulex. Evolution* 52, 368–78.

Scheu, S., Theenhaus, A. and Jones, T.H. (1999) Links between the detritivore and herbivore system: effects of earthworms and Collembola on plant growth and aphid development. *Oecologia* 119, 541–51.

Schiefer, J. (1981) Bracheversuche in Baden-Würtemberg. *Beihefte zu den Veröffentlichungen für Naturschutz und Landschaftspflege in Baden-Würtemberg* 22, 1–325.

Schindler, D.W. (1974) Experimental lakes area: whole lakes experiments in eutrophication. *Journal of the Fisheries Research Board of Canada* 31, 937–53.

Schindler, D.W. (1988) Effects of acid rain on freshwater ecosystems. *Science* 239, 149–57.

Schindler, D.W. (1997) Liming to restore acidified lakes and streams: a typical approach to restoring damaged ecosystems? *Restoration Ecology* 5, 1–6.

Schlesinger, W.H. and Pilmanis, A.M. (1998) Plant-soil interactions in deserts. *Biogeochemistry* 42, 169–87.

Schmidt, P.A. (1998) Potential natural vegetation as an objective of close-to-nature forest management? *Forstwissenschaftliches Centralblatt* 117, 193–237.

Schmidt, W. (1993) Sukzession und Sukzessionslenkung auf Brachflächen. Neue Erkenntnisse aus einem Dauerflächenversuch. *Scripta Geobotanica (Göttingen)* 20, 65–104.

Schmidt, W. (1995) Einfluss der Wiedervernässung auf physikalische Eigenschaften des Moorkörpers der Frieländer Grossen Wiese. *Zeitschrift für Kulturtechnik und Landentwicklung* 36, 107–12.

Schmitthüsen, J. (1963) Der wissenschaftliche Landschaftsbegriff. *Mitteilungen der floristisch – soziologischen Arbeitsgemeinschaft* 10, 9–19.

Schönenberger, W., Fischer, A. and Innes, J. (eds.) (2002) Vivian's legacy in Switzerland: impact of windthrow on forest dynamics. *Forest Snow and Landscape Research* 77, 1–224.

Schot, P.P., Dekker, S.C. and Poot, A. (2004) The dynamic form of rainwater lenses in drained fens. *Journal of Hydrology* 293, 74–84.

Schrautzer, J., Asshoff, M. and Müller, F. (1996) Restoration strategies for wet grasslands in northern Germany. *Ecological Engineering* 7, 255–78.

Schreiber, K.F. (1997) Grundzüge der Sukzession in 20-jährigen Grünlandbrachen in Baden-Württemberg. *Forstwissenschaftliches Centralblatt* 116, 243–58.

Schröder, B. and Richter, O. (2000) Are habitat models transferable in space and time? *Zeitschrift für Ökologie und Naturschutz* 8, 195–205.

Schulze, E.D. and Mooney, H.A. (eds.) (1993) *Biodiversity and Ecosystem Function.* Ecological Studies, vol. 99. Springer, Berlin.

Schulze, E.-D., Oren, R. and Lange, O.L. (1989) Processes leading to forest decline: a synthesis. *Ecological Studies* 77, 459–68.

Schumm, S.A. (1979) Geomorphic thresholds: the concept and its applications. *Transactions of the Institute of British Geographers. New Series* 4, 485–515.

Schumm, S.A. and Litchy, R.W. (1965) Time, space and causality. *American Journal of Science* 263, 110–19.

Seddon, P.J. and Soorae, P.S. (1999) Guidelines for sub-specific substitutions in wildlife restoration projects. *Conservation Biology* 13, 177–84.

SER (2002) *The SER Primer on Ecological Restoration.* Society for Ecological Restoration, Science and Policy Working Group. www.ser.org.

Seva, J.P., Valdecantos, A., Vilagrosa, A. *et al.* (2000) Seedling morphology and survival in some Mediterranean tree and shrub species. In: *Mediterranean Desertification. Research Results and Policy Implications*, vol. 2, EC Report EUR 19303 (eds. P. Balabanis, D. Peter, A. Ghazi and M. Tsogas), pp. 397–406. Brussels.

Severinghaus, C.W. and Darrow, R.W. (1976) Failure of elk to survive in the Adirondacks. *New York Fish and Game Journal* 23, 98–9.

Shapiro, J. and Wright, D.I. (1984) Lake restoration by biomanipulation: Round Lake, Minnesota, the first two years. *Freshwater Biology* 14, 371–83.

Shapiro, J., LaMarra, V. and Lynch, M. (1975) Biomanipulation: an ecosystem approach to lake restoration. In: *Water Quality Management through Biological Control* (eds. P.L. Brezonik and J.L. Fox), pp. 85–96. University of Gainesville, Gainesville, FL.

Shepherd, P.C.F., Partridge, V.A. and Hicklin, P.W. (1995) *Changes in Sediment Types and Invertebrate Fauna in the Intertidal Mudflats of the Bay of Fundy between 1977 and 1994.* Technical Report No. 237. Canadian Wildlife Service, Sackville.

Short, J., Bradshaw, S.D., Giles, J. *et al.* (1992) Reintroduction of macropods (Marsupialia, Macropodoidea) in Australia: a review. *Biological Conservation* 62, 189–204.

Shugart, H.H. (1998) *Terrestrial Ecosystems in Changing Environments.* Cambridge University Press, Cambridge.

Sierdsema, H. and Bonte, D. (2002) Duinstruwelen en samenstelling broedvogelbevolking: meer vogels, minder kwaliteit. *De Levende Natuur* 103, 88–93.

Sinclair, A.R.E. (1995) Equilibria in plant-herbivore interactions. In: *Serengeti II, Dynamics, Management and Conservation of an Ecosystem* (eds. A.R.A. Sinclair and P. Arcese), pp. 91–113. Chicago University Press, Chicago.

Singer, F.J., Moses, M.E., Bellew, S. and Sloan, W. (2000) Correlates to colonizations of new patches by translocated populations of bighorn sheep. *Restoration Ecology* 8 (suppl. 4), 66–74.

Sirin, A.A. and Minaeva, T. (eds.) (2001) *Peatlands of Russia: Towards an Analysis of Sectoral Information.* Geos Publishing House, Moscow.

Sjörs, H. (1948) Myrvegetation I Bergslagen. *Acta Phytogeographica Suecica* 21, 1–299.

Skoglund, S.J. (1990) Seed dispersing agents in two regularly flooded river sites. *Canadian Journal of Botany* 68, 754–60.

Smith, O.J., Hamback, P.A. and Beckerman, A.P. (2000) Trophic cascades in terrestrial systems: a review of the effects of carnivore removal on plants. *American Naturalist* 155, 141–53.

Smith, R.M. and Stamey, W.L. (1965) Determining the range of tolerable erosion. *Soil Science* 100, 414–24.

Smits, A.J.M., Cals, M.J.R. and Drost, H.J. (2001) In: *River Restoration in Europe: Practical Approaches* (eds. H.J. Nijland and M.J.R. Cals). Proceedings of the 2000 River Restoration Conference, pp. 41–9. RIZA, Wageningen.

Smolders, A.J.P. and Roelofs, J.G.M. (1993) Sulphate mediated iron limitation and eutrophication in aquatic ecosystems. *Aquatic Botany* 46, 247–53.

Smolders, A.J.P., Nijboer, R.C. and Roelofs, J.G.M. (1995) Prevention of sulphide accumulation and phosphate mobilization by the addition of iron(II)chloride to a reduced sediment: an enclosure experiment. *Freshwater Biology* 34, 559–68.

Smolders, A.J.P., Tomassen, H.B.M., Pijnappel, H.W. *et al.* (2001) Substrate-derived CO_2 is important in the development of *Sphagnum* spp. *New Phytologist* 152, 325–32.

Søndergaard, M.E., Jeppesen, E., Mortensen, E. *et al.* (1990) Phytoplankton biomass reduction after planktivorous fish reduction in a shallow, eutrophic lake: a combined effect of reduced internal P-loading and increased zooplankton grazing. *Hydrobiologia* 200, 229–40.

Sort, X. and Alcañiz, J.M. (1996) Contribution of sewage sludge to erosion control in the rehabilitation of limestone quarries. *Land Degradation and Development* 7, 69–76.

Soulé, M.E. (1980) Thresholds for survival: maintaining fitness and evolutionary potential. In: *Conservation Biology: An Evolutionary-Ecological Perspective* (eds. M.E. Soulé and B.A. Wilcox), pp. 111–24. Sinauer, Sunderland, MA.

Soulé, M.E. (1987) *Viable Populations for Conservation.* Cambridge University Press, Cambridge.

Soulé, M.E. and Simberloff, D.S. (1986) What do genetics and ecology tell us about the design of nature reserves? *Biological Conservation* 35, 19–40.

Soulé, M.E. and Terborgh, J. (1999) Conserving nature at regional and continental scales: a scientific program for North America. *BioScience* 49, 809–17.

SOVON Vogelonderzoek Nederland (2002) *Atlas van de Nederlandse broedvogels, 1998-2000: verspreiding, aantallen, verandering.* Nederlandse Fauna 5. KNNV, Utrecht.

Sperber, G. (1968) Die Reichswälder bei Nürnberg. Aus der Geschichte des ältesten Kunstforstes. *Mitteilungen Aus der Staatsforstverwaltung Bayerns*, vol. 37. Frankenverlag, Nürnberg.

Spieker, H. and Hansen, J. (2002) *Central Europe: Conifer to Broadleaf Transformation: Conversion of Secondary Norway Spruce Forests on Sites Naturally Dominated by Broadleaves?* In: Proceedings of the IUFRO conference on Restoration of Boreal and Temperate Forests, Denmark. *Reports* 11, 50–1.

Splichtinger, N., Wenig, M., James, P. *et al.* (2001) Satellite detection of a continental-scale plume of nitrogen oxides from boreal forest fires. *Geophysical Research Letters* 28, 4579–82.

Stace, C.A. (1975) *Hybridization and the Flora of the British Isles.* Academic Press, London.

Stace, C.A. (1991) *New Flora of the British Isles.* Cambridge University Press, Cambridge.

Stacey, P.B., Johnson, V.A. and Taper, M.L. (1997) Migration within metapopulations: the impact upon local population dynamics. In: *Metapopulation Biology: Ecology, Genetics and Evolution* (eds. I. Hanski and M.E. Gilpin), pp. 267–91. Academic Press, San Diego.

Stahl, J. (2001) *Limits to the Co-occurrence of Avian Herbivores.* PhD Thesis, University of Groningen.

Stahr, A. (1996) Zur Genese und Dynamik von Blattanbrüchen auf Almen in den nördlichen Kalkalpen. *Geöökodynamik* 17, 217–48.

Stanley-Price, M.R. (1989) *Animal Re-introductions: The Arabian Oryx in Oman.* Cambridge University Press, Cambridge.

Steenwerth, K.L., Jackson, L.E., Caldéron, F.J. *et al.* (2002) Soil microbial community composition and land use history in cultivated and grassland ecosystems of Coastal California. *Soil Biology and Biochemistry* 34, 1559–611.

Steffan-Dewenter, I. and Tscharntke, T. (1999) Effects of habitat isolation on pollinator communities and seed set. *Oecologia* 78, 550–8.

Steffan-Dewenter, I. and Tscharntke, T. (2002) Insect communities and biotic interactions on fragmented calcareous grasslands – a mini review. *Biological Conservation* 104, 275–84.

Steinberg, C.E.W. and Wright, R.F. (eds.) (1994) *Acidification of Freshwater Ecosystems: Implications for Future.* Wiley, Chichester.

Stevens, D.R. and Goodson, N.J. (1993) Assessing effects of removals for transplanting on a high-elevation bighorn sheep population. *Conservation Biology* 7, 908–15.

Stewart, O.C. (1956) Fire as the first great force employed by man. In: *Man's Role in Changing the Face of the Earth* (ed. W.L. Thomas), pp. 115–33. University of Chicago, Chicago.

Stock, M. and Hofeditz, F. (2000) Der Einfluss des Salzwiesen-Managements auf die Nutzung des Habitates durch Nonnen- und Ringelgänse. In: Die Salzwiesen der Hamburger Hallig (eds. M. Stock and K. Kiehl). *Schriftenreihe des Nationalparks Schleswig-Holsteinisches Wattenmeer* 11, 43–55.

Stock, M. and Hofeditz, F. (2002) Einfluss des Salzwiesen-Managements auf Habitatnutzung und Bestandsentwicklung von Nonnengänsen *Branta leucopsis* im Wattenmeer. *Vogelwelt* 123, 265–82.

Stocking, M.A. and Elwell, H.A. (1976) Vegetation and erosion: a review. *Scottish Geographical Magazine* 92, 4–16.

Stockwell, C.A., Mulvey, M. and Vinyard, G.L. (1996) Translocations and the preservation of allelic diversity. *Conservation Biology* 10, 1133–41.

Stone, P.B. (1992) *The State of the World's Mountains.* Zed Books, London.

Strasser, M. (1999) *Mya arenaria:* an ancient invader of the North Sea coast. *Helgoländer Meeresuntersuchungen* 52, 309–24.

Strauss, S.Y. (1994) Levels of herbivory and parasitism in host hybrid zones. *Trends in Ecology and Evolution* 9, 209–14.

Strijker, D. (2000) Ruimtelijke verschuivingen in de EU landbouw 1950–1982. PhD thesis, University of Amsterdam.

Stroh, M., Storm, C., Zehm, A. and Schwabe, A. (2002) Restorative grazing as a tool for directed succession with diaspore inoculation: the model of sand ecosystems. *Phytocoenologia* 32, 595–625.

Strykstra, R.J. (2000) Reintroduction of plant species: shifting settings. PhD thesis, University of Groningen.

Strykstra, R.J., Bekker, R.M. and Verweij, G.L. (1996a) Establishment of *Rhinanthus angustifolius* in a successional hayfield after seed dispersal by mowing machinery. *Acta Botanica Neerlandica* 45, 557–62.

Strykstra, R.J., Verweij, G.L. and Bakker, J.P. (1996b) Seed dispersal by mowing machinery in a Dutch brook valley system. *Acta Botanica Neerlandica* 46, 387–401.

Strykstra, R.J., Bekker, R.M. and Bakker, J.P. (1998) Assessment of dispersule availability: its practical use in restoration management. *Acta Botanica Neerlandica* 47, 57–70.

Strykstra, R.J., Bekker, R.M. and van Andel, J. (2002) Dispersal and life span spectra in plant communities: a key to safe site dynamics, species coexistence and conservation. *Ecography* 25, 145–60.

Strzyszcz, Z. (1996) Recultivation and landscaping in areas after brown-coal mining in middle-east European countries. *Water, Air and Soil Pollution* 91, 145–57.

Stumm, W. and Morgan, J.J. (1981) *Aquatic Chemistry: An Introduction Emphasizing Chemical Equilibria in Natural Waters*, 2nd edn. Wiley, New York.

Stuyfzand, P.J. (1989) Hydrology and water quality aspects of Rhine bank groundwater in The Netherlands. *Journal of Hydrology* 106, 341–63.

Suc, J.P. (1984) Origin and evolution of the Mediterranean vegetation and climate in Europe. *Nature* 307, 429–32.

Succow, M. (1982) Topische und chorische Naturraumtypen der Moore. In: *Naturräumliche Grundlagen der Landnutzung* (eds. D. Kopp, K.-D. Jäger and M. Succow), pp. 138–83. Akademie Verlag, Berlin.

Succow, M. (1988) *Landschaftsökologische Moorkunde: am Beispiel der Moore der DDR*. Gustav Fischer, Jena.

Succow, M. and Joosten, H. (eds.) (2001) *Landschaftsökologische Moorkunde*, 2nd edn. Schweizerbart'sche Verlagsbuchhandlung, Stuttgart.

Suding, K.N., Gross, K.L. and Houseman, G.R. (2004) Alternative states and positive feedbacks in restoration ecology. *Trends in Ecology and Evolution* 19, 46–53.

Sukopp, H. (1972) Wandel von Flora und Vegetation in Mitteleuropa unter dem Einfluß des Menschen. *Berichte über Landwirtschaft der Landnutzung am Beispiel des Tieflandes der DDR* 50, 112–39.

Sukopp, H. and Trepl, L. (1987) Extinction and naturalization of plant species as related to ecosystem structure and function. *Ecological Studies* 61, 245–76.

Sutherland, W.J. (2002) Restoring a sustainable countryside. *Trends in Ecology and Evolution* 17, 148–50.

Svenning, J.C. (2002) A review of natural vegetation openness in north-western Europe. *Biological Conservation* 104, 133–48.

Svensson, J.E., Henrikson, L., Larsson, S. and Wilander, A. (1995) Liming strategies and effects: the Lake Gårdsjön case study. In: *Liming of Acidified Surface Waters: A Swedish Synthesis* (eds. L. Henrikson and Y.-W. Brodin), pp. 309–25. Springer, Berlin.

Swaine, M.D. and Whitmore, T.C. (1988) On the definition of ecological species groups in tropical rain forest. *Vegetatio* 75, 81–6.

Swart, J.A.A., van der Windt, H.J. and Keulartz, J. (2001) Valuation of nature in conservation and restoration. *Restoration Ecology* 9, 230–8.

SWS (2000) *Position Paper on the Definition of Wetland Restoration*. Society of Wetlands Scientists, McLean, VA.

Sykora, K.V., van der Krogt, G. and Rademakers, J. (1990) Vegetation change on embankments in the south-western part of the Netherlands under the influence of different management practices (in particular sheep grazing). *Biological Conservation* 52, 49–81.

Tabacchi, E., Planty-Tabacchi, A.M. and Decamps, H. (1990) Continuity and discontinuity of the riparian vegetation along a fluvial corridor. *Landscape Ecology* 5, 9–20.

Tackenberg, O., Poschlod, P. and Bonn, S. (2003) Assessment of wind dispersal potential in plant species. *Ecological Monographs* 73, 191–205.

Tallowin, J.R.B. and Smith, R.E.N. (2001) Restoration of a Cirsio-Molinietum fen meadow on an agricultural improved pasture. *Restoration Ecology* 9, 167–78.

Tamis, W.L.M. and van 't Zelfde, M. (2003) KilometerhokFrequentieKlassen, een nieuwe zeldzaamheidsschaal voor de Nederlandse flora. *Gorteria* 29, 57–83.

Tanner, J.E. and Hughes, T.P. (1994) Species coexistence, keystone species, and succession: a sensitivity analysis. *Ecology* 75, 2204–19.

Tansley, A.G. (1935) The use and abuse of vegetational concepts and terms. *Ecology* 16, 284–307.

Tappeiner, U. (1996) Ökologie des alpinen Rasens, Grenzen der Begrünung. *Rasen-Turf-Gazon* 27, 36–40.

Tasser, E., Newesely, C., Höller, P. *et al.* (1999) Potential risks through land-use changes. In: *Land-Use Changes in European Mountain Ecosystems* (eds. A. Cernuska, U. Tappeiner and N. Bayfield), pp. 218–24. Blackwell Science, Oxford.

Tasser, E., Mader, M. and Tappeiner, U. (2003) Effects of land use in alpine grasslands on the probability of landslides. *Basic and Applied Ecology* 4, 271–80.

Taylor, P.D., Fahrig, L., Henein, K. and Merriam, G. (1993) Connectivity is a vital element of landscape structure. *Oikos* 68, 571–2.

Temmerman, S. (2003) Sedimentation on tidal marshes in the Scheldt estuary: a field and numerical modelling study. PhD thesis, Catholic University of Leuven.

Temperton, V.M., Hobbs, R.J., Nuttle, T. and Halle, S. (eds.) (2004) *Assembly Rules and Restoration Ecology*. Island Press, Washington, DC.

ter Borg, S.J. (1985) Population biology and habitat relations of some hemiparasitic Scrophulariaceae. In: *The Population Structure of Vegetation* (ed. J. White), pp. 463–87. Junk Publishers, Dordrecht.

ter Heerdt, G.N.J., Schutter, A. and Bakker, J.P. (1997) Kiemkrachtig heidezaad in de bodem van ontgonnen heidevelden. *De Levende Natuur* 98, 142–6.

Thirgood, J.V. (1981) *Man and the Mediterranean Forest.* Academic Press, New York.

Thistle, D. (1981) Natural physical disturbances and communities of marine soft bottoms. *Marine Ecology: Progress Series* 6, 223–8.

Thompson, K., Bakker, J.P. and Bekker, R.M. (1997) *The Soil Seed Banks of North West Europe: Methodology, Density and Longevity.* Cambridge University Press, Cambridge.

Thompson, K., Bakker, J.P., Bekker, R.M. and Hodgson, J.G. (1998) Ecological correlates of seed persistence in soil in the north-west European flora. *Journal of Ecology* 86, 163–9.

Thorne, C.R., Hey R.D. and Newson, M.D. (eds.) (1997) *Applied Fluvial Geomorphology for River Engineering and Management.* John Wiley and Sons, Chichester.

Thornes, J. (1987) Erosional equilibria under grazing. In: *Conceptual Issues in Environmental Archaeology* (eds. J.L. Bintliff, D.A. Davidson and E.G. Grant), pp. 193–210. Elsevier, New York.

Thornes, J.B. and Brandt, J. (1994) Erosion-vegetation competition in a stochastic environment undergoing climatic change. In: *Environmental Change in Drylands: Biogeographical and Geomorphological Perspectives* (eds. A.C. Millington and K. Pye), pp. 205–20. Wiley, London.

Thrall, P.H., Biere, A. and Uyenoyama, M.K. (1995) Frequency-dependent disease transmission and the dynamics of the *Silene-Ustilago* host-pathogen system. *American Naturalist* 145, 43–62.

Tilman, D. (1982) *Resource Competition and Community Structure.* Princeton University Press, Princeton.

Tilman, D. (1988) *Plant Strategies and the Dynamics and Structure of Plant Communities.* Princeton University Press, Princeton.

Tilman, D. (1996) Biodiversity: population versus ecosystem stability. *Ecology* 77, 350–63.

Tilman, D. and Pacala, S. (1993) The maintenance of species richness in plant communities. In: *Species Diversity in Ecological Communities* (eds. R.E. Ricklefs and D. Schluter), pp. 13–25. University of Chicago Press, Chicago.

Timms, R.M. and Moss, B. (1984) Prevention of growth of potentially dense phytoplankton populations by zooplankton grazing, in the presence of zooplanktivorous fish in a shallow wetland ecosystem. *Limnology and Oceanography* 29, 472–86.

Tischendorf, L. and Fahrig, L. (2000) How should we measure landscape connectivity? *Landscape Ecology* 15, 633–41.

Tischew, S., Mahn, E.-G. and Schmiedeknecht, A. (1995) Von der Natur lernen: Rekultivierung von Bergbaufolgelandschaften. *Landschaftsarchitektur* 4, 160–4.

Tomassen, H.B.M., Smolders, A.J.P., Limpens, J. *et al.* (2004a) Expansion of invasive species on ombrotrophic bogs; desiccation or high N-deposition? *Journal of Applied Ecology* 41, 139–50.

Tomassen, H.B.M., Smolders, A.J.P., Lamers, L.P.M. and Roelofs, J.G.M. (2004b) Development of floating rafts after the rewetting of cut-over bogs; the importance of peat quality. *Biogeochemistry* 71, 69–87.

Tongway, D.J. and Hindley, N. (1995) *Manual for Soil Condition Assessment of Tropical Grasslands.* CSIRO. Division of Wildlife and Ecology, Canberra.

Tongway, D.J. and Ludwig, J.A. (1996) Rehabilitation of semiarid landscapes in Australia I. Restoring productive soil patches. *Restoration Ecology* 4, 388–97.

Tongway, D. and Ludwig, J. (2002) Australian semi-arid lands and savannas. In: *Handbook of Ecological Restoration.* vol. 2: *Restoration in Practice* (eds. M.R. Perrow and A.J. Davy), pp. 486–502. Cambridge University Press, Cambridge.

Török, K., Szili-Kovács, T., Halassy, M. *et al.* (2000) Immobilization of soil nitrogen as a possible method for the restoration of sandy grassland. *Applied Vegetation Science* 3, 7–14.

Toth, J. (1963) A theoretical analysis of groundwater flow in small drainage basins. *Journal of Geophysical Research* 68, 4795–812.

Trabaud, L. (1998) Man and fire: impacts on Mediterranean vegetation. In: *Mediterranean-type Shrublands.* (eds. F. Di Castri, D.W. Goodall and R. Specht), pp. 523–38. Elsevier, Amsterdam.

Troll, C. (1939) Luftbildplan und ökologische Bodenforschung (Aerial photography and ecological studies of the earth). *Zeitschrift der Gesellschaft für Erdkunde*, Berlin, Heft 7/8.

Tscharntke, T. (1992) Fragmentation of *Phragmites* habitats, minimum viable population size, habitat suitability, and local extinction of moths, midges, flies, aphids, and birds. *Conservation Biology* 6, 530–6.

Tubbs, C.R. (1977) Wildfowl and waders in Langstone Harbour. *British Birds* 70, 177–99.

Turner, M.G., Romme, W.H., Gardner, R.H. *et al.* (1993) A revised concept of landscape equilibrium: disturbance and stability on scaled landscapes. *Landscape Ecology* 8, 213–27.

Turner, R.E., Swenson, E.M. and Milan, C.S. (2000) Organic and inorganic contributions to vertical accretion in salt marsh sediments. In: *Concepts and Controversies in Tidal Marsh Ecology* (eds. M.P. Weinstein and D.A. Kreeger), pp. 583–96. Kluwer Academic Publishers, Dordrecht.

Tüxen, J. (1983) Die Schutzwürtigkeit der niedersachsischen Kleinstmoore im Hinblick auf ihre Vegetation. *Tüxenia* 3, 423–35.

Tüxen, R. (1954) Pflanzengesellschaften und Grundwasser-Ganglinien. *Angewandte Pflanzensoziologie* 8, 64–98.

Tüxen, R. (1956) Die heutige potentielle natürliche Vegetation als Gegenstand der Vegetationskartierung. *Angewandte Pflanzensoziologie* 13, 5–42.

Ulanowicz, R.E. (1997) *Ecology, the Ascendent Perspective.* Columbia University Press, New York.

Ulrich, B. (1986) Natural and anthropogenic components of soil acidification. *Zeitschrift für Pflanzenernährung und Bodenkunde* 149, 702–17.

UN (1982) *World Charter for Nature.* Great Assembly of the United Nations, 28 October 1982. United Nations, New York.

UN (1992a) *Report of the United Nations Conference on Environment and Development. Annex 1. Rio Declaration on Environment and Development.* Great Assembly of the United Nations, 12 August 1992. United Nations, New York.

UN (1992b) *Agenda 21.* UN Department of Economic and Social Affairs. Division of Sustainable Development, New York.

UN (1994) *Protocol on Further Reduction of Sulphur Emissions.* Oslo, 1994, United Nations, Geneva.

UNCCD (1994) *Adoption of the United Nations Convention to Combat Desertification*, Paris 17 June 1994. UNCCD, Paris.

UNCCD (2001) *Proceedings of the Sixth Ministerial Meeting of UNCCD, Annex IV: Regional Implementation Annex for the Northern Mediterranean.* Ancona. UNCCD, Bonn.

UNFCCC (1994) *Framework Convention on Climate Change* (convention text), 21 March 1994, Bonn. UNFCCC.

UNFCCC (1997) *Kyoto Protocol to the UNFCCC*, 11 December 1997, Kyoto. UNFCCC.

Urbanska, K. (1986) High altitude revegetation research in Switzerland: problems and perspectives. *Veröffentlichungen des Geobotanischen Institutes ETH* 87, 155–67.

Urbanska, K. (1997) Restoration ecology research above the timberline: colonization of safety islands on a machine-graded alpine ski run. *Biodiversity and Conservation* 6, 1655–70.

Valdecantos, A. (2001) Aplicación de fertilizantes orgánicos e inorgánicos en la repoblación de zonas forestales degradadas de la Comunidad Valenciana. PhD thesis, University of Alicante.

Valdecantos, A., Vilagrosa, A., Seva, J.P. *et al.* (1996) Mycorhization et application du compost urbain pour l'amélioration de la survie et de la croissance des semis de *Pinus halepensis* en milieu semiaride. *Cahiers Options Méditerranéennes* 20, 87–104.

Valdecantos, A., Cortina, J., Fuentes, D. *et al.* (2002) Use of biosolids for reforestation in the Region of Valencia (E Spain): first results of a pilot project. *Bioprocessing of Solid Waste and Sludge* 1, 1–6.

Vallejo, R.V. (ed.) (1996) *La restauración de la cubierta vegetal en la Comunidad Valenciana.* Centro de Estudios Ambientales del Mediterráneo, Valencia.

Vallejo, R.V. and Alloza, J.A. (1998) The restoration of burned lands: the case of eastern Spain. In: *Large Forest Fires* (ed. J.M. Moreno), pp. 91–108. Backhuys, Leiden.

Vallejo, R.V. and Alloza, J.A. (2003) I + D aplicado a la gestión forestal y la lucha contra la desertificación: la experiencia del CEAM en la Comunidad Valenciana. *Ecosistemas* 2004/1.

Vallejo, R.V., Cortina, J., Ferran, A. *et al.* (1998) Sobre els trets distintius dels sòls mediterranis. *Acta Botanica Barcelonesa* 45, 603–32.

Vallejo, R., Bautista, S. and Cortina, J. (1999) Restoration for soil protection after disturbances. In: *Life and Environment in the Mediterranean* (ed. L. Trabaud), pp. 301–43. WIT Press, Southampton.

van Andel, J. (1998a) Two approaches towards the relationship between plant species diversity and ecosystem functioning. *Applied Vegetation Science* 1, 9–14.

van Andel, J. (1998b) Intraspecific variability in the context of ecological restoration projects. *Perspectives in Plant Ecology, Evolution and Systemtics* 1, 221–37.

van Andel, J., Bakker, J.P. and Snaydon, R.W. (eds.) (1987) *Disturbance in Grasslands: Causes, Effects and Processes.* Junk Publishers, Dordrecht.

van Andel, J., Bakker, J.P. and Grootjans, A.P. (1993) Mechanisms of vegetation succession: a review of concepts and perspectives. *Acta Botanica Neerlandica* 42, 413–33.

van Breemen, N. and Buurman, P. (2002) *Soil Formation*, 2nd edn. Kluwer Academic Publishers, Dordrecht.

van de Kam, J., Ens, B., Piersma, T. and Zwarts, L. (2004) *Shorebirds: An Illustrated Behavioural Ecology.* KNNV Publishers, Utrecht.

van de Koppel, J., Huisman, J., van der Wal, R. and Olff, H. (1996) Patterns of herbivory along a productivity gradient: an empirical and theoretical investigation. *Ecology* 77, 736–45.

van de Koppel, J., Rietkerk, M. and Weissing, F.J. (1997) Catastrophic vegetation shifts and soil degradation in terrestrial grazing systems. *Trends in Ecology and Evolution* 12, 352–6.

van de Koppel, J., Rietkerk, M., van Langevelde, F. *et al.* (2002) Spatial heterogeneity and irreversible vegetation change in semiarid grazing systems. *American Naturalist* 159, 209–18.

van den Berg, L.J.L., Vergeer, P. and Roelofs, J.G.M. (2003) Heathland restoration in The Netherlands: effects on germination of *Arnica montana*. *Applied Vegetation Science* 6, 117–24.

van den Berg, M.S., Scheffer, M., van Nes, E. and Coops, H. (1999) Dynamics and stability of *Chara* sp and *Potamogeton pectinatus* in a shallow lake changing in eutrophication level. *Hydrobiologia* 409, 335–42.

van den Bergh, J. and Verbruggen, H. (1999) An evaluation of the 'ecological footprint': reply to Wackernagel and Ferguson. *Ecological Economics* 31, 319–21.

van der Does, J., Verstraelen, P., Boers, P. *et al.* (1992) Lake restoration with and without dredging of phosphorus-enriched upper sediment layers. *Hydrobiologia* 233, 197–210.

van der Heijden, M.G.A., Boller, T., Wiemken, A. and Sanders, I.R. (1998) Different arbuscular mycorrhizal fungal species are potential determinants of plant community structure. *Ecology* 79, 2082–91.

van der Maarel, E. (1990) Ecotones and ecoclines are different. *Journal of Vegetation Science* 1, 135–8.

van der Meijden, E. and van der Veen-van Wijk (1997) Tritrophic metapopulation dynamics. In: *Metapopulation Biology: Ecology, Genetics and Evolution* (eds. I. Hanski and M.E. Gilpin), pp. 387–405. Academic Press, San Diego.

van der Meijden, E. and Klinkhamer, P.G.L. (2000) Conflicting interests of plants and the natural enemies of herbivores. *Oikos* 89, 202–8.

van der Meijden, E., Klinkhamer, P.G.L., de Jong, T.J. and van Wijk, C.A.M. (1992) Meta-population dynamics of biennial plants: how to exploit temporary habitats. *Acta Botanica Neerlandica* 41, 249–70.

van der Meulen, F. and Salman, A.H.P.M. (1996) Management of Mediterranean coastal dunes. *Ocean and Coastal Management* 30, 177–95.

van der Putten, W.H., van Dijk, C. and Peters, B.A.M. (1993) Plant-specific soil-borne diseases contribute to succession in foredune vegetation. *Nature* 362, 53–6.

van der Putten, W.H., Vet, L.E.M., Harvey, J.A. and Wäckers, F.L. (2001) Linking above- and below-ground multitrophic interactions of plants, herbivores, pathogens, and their antagonists. *Trends in Ecology and Evolution* 16, 547–54.

van der Wal, R., Egas, M., van der Veen, A. and Bakker, J. (2000a) Effects of resource competition and herbivory on plant performance along a natural productivity gradient. *Journal of Ecology* 88, 317–30.

van der Wal, R., van Wieren, S., van Wijnen, H. *et al.* (2000b) On facilitation between herbivores: how Brent Geese profit from Brown Hares. *Ecology* 81, 969–80.

van der Woude, B.J., Pegtel, D.M. and Bakker, J.P. (1994) Nutrient limitation after long-term fertilizer application in cut grasslands. *Journal of Applied Ecology* 31, 405–12.

van Diggelen, R. (1998) Moving gradients: assessing restoration prospects of degraded brook valleys. PhD thesis, University of Groningen.

van Diggelen, R., Grootjans, A.P. and Burkunk, R. (1994) Assessing restoration perspectives of disturbed brook valleys: the Gorecht area, The Netherlands. *Restoration Ecology* 2, 87–96.

van Diggelen, R., Grootjans, A.P. and Wierda, A. (1995) Hydro-ecological landscape analysis: a tool for wetland restoration. *Zeitschrift für Kulturtechnik und Landentwicklung* 36, 125–31.

van Diggelen, R., Molenaar, W.J. and Kooijman, A.M. (1996) Vegetation succession in a floating mire in relation to management and hydrology. *Journal of Vegetation Science* 7, 809–20.

van Diggelen, R., Grootjans, A.P. and Harris, J.A. (2001) Ecological restoration: state of the art or state of the science? *Restoration Ecology* 9, 115–18.

van Diggelen, R., Sijtsma, F., Strijker, D. *et al.* (2005) Searching indicators for the relationship between biodiversity changes and economic developments at the regional scale. *Basic and Applied Ecology*, in press.

van Donk, E. and van de Bund, W.J. (2002) Impact of submerged macrophytes including charyophytes on phyto- and zooplankton communities: allelopathy versus other mechanisms. *Aquatic Botany* 72, 261–74.

van Donk E., Gulati, R.D., Iedema, A. and Meulemans, J.T. (1993) Macrophyte-related shifts in the nitrogen and phosphorus contents of the different trophic levels in a biomanipulated shallow lake. *Hydrobiologia* 251, 19–26.

van Dorp, D., van den Hoek, W.P.M. and Daleboudt, C. (1996) Seed dispersal capacity of six perennial grassland species measured in a wind tunnel at varying wind speed and height. *Canadian Journal of Botany* 74, 1956–63.

van Duren, I.C., Grootjans, A.P., Strijkstra, R.J. *et al.* (1998) A multidiciplinary evaluation of restoration measures in a degraded fen meadow (Cirsio-Molinietum). *Applied Vegetation Science* 1, 115–30.

van Eeten, M.J.G. and Roe, E. (2002) *Ecology, Engineering and Management: Reconciling Ecosystem Rehabilitation and Service Reliability.* Oxford University Press, New York.

van Eerden, M.R. (ed.) (2000) *Pechora Delta: Structure and Dynamics of the Pechora Delta Ecosystems (1995–1999).* RIZA report 200.037, Institute for Inland Water Management and Waste Water Treatment. RIZA, Lelystad.

van Gemerden, B.S., Olff, H., Parren, M.P.E. and Bongers, F. (2003) The pristine rain forest? Remnants of historical human impacts on current tree species composition and diversity. *Journal of Biogeography* 30, 1–10.

van Impe, J. (1985) Estuarine pollution as a probable cause of increase of estuarine birds. *Marine Pollution Bulletin* 16, 271–6.

van Katwijk, M.M. and Hermus, D.C.R. (2000) Effects of water dynamics on Zostera marina: transplantation experiments in the intertidal Dutch Wadden Sea. *Marine Ecology: Progress Series* 208, 107–18.

van Katwijk, M.M., Schmitz, G.H.W., Hanssen, L.S.A.M. and den Hartog, C. (1998) Suitability of *Zostera marina* populations for transplantation to the Wadden Sea as determined by a mesocosm shading experiment. *Aquatic Botany* 60, 283–305.

van Katwijk, M.M., Hermus, D.C.R., de Jong, D.J., Asmus, R.M. and de Jonge, V.N. (2000) Habitat suitability of the Wadden Sea for restoration of *Zostera marina* beds. *Helgoländer Marine Research* 54, 117–28.

van Langevelde, F., van der Knaap, W.G.M. and Claasen, G.D.H. (1998) Comparing connectivity in landscape networks. *Environment and Planning* B 25, 849–63.

van Liere, L. and Gulati, R.D. (eds.) (1992) Restoration and recovery of shallow eutrophic lake ecosystems in the Netherlands. *Hydrobiologia* 233.

van Liere, L. and Janse, J.H. (1992) Restoration and resilience to recovery of the Lake Loosdrecht ecosystem in relation to its phosphorus flow. *Hydrobiologia* 233, 95–104.

van Mechelen, L., Groenemans, R. and van Ranst, E. (1997) *Forest Soil Conditions in Europe: Results of a Large Scale Soil Survey.* Forest Soil Coordinating Centre, University of Gent.

van Nes, E.H. (2002) Controlling complexity in individual-based models of aquatic vegetation and fish communities. PhD thesis, Wageningen University.

van Ommering, G., van Halder, I., van Swaay, C.A.M. and Wynhoff, I. (1995) *Bedreigde en kwetsbare dagvlinders in Nederland: toelichting op de Rode Lijst.* IKC-rapport 18. IKC-Natuurbeheer, Wageningen.

van Rooij, S.A.M. and Drost, H.J. (1996) *Het Lauwersmeergebied: 25 jaar onderzoek ten dienste van natuurontwikkeling en beheer.* Flevobericht 387. RIZA, Lelystad.

van Rooij, S.A.M. and Groen, K.P. (1996) *De oevergebieden van het Volkerak-Zoommeer: ontwikkeling van abiotisch milieu en vegetatie sinds 1987.* Flevobericht 393. RIZA, Lelystad.

van Swaay, C.A.M. (2002) The importance of calcareous grasslands for butterflies in Europe. *Biological Conservation* 104, 315–18.

van Til, M. and van Mourik, J. (2001) Vegetatieherstel in vochtige duinvalleien. Een analyse van de effecten van beheer bij regeneratie in de AWD. *Landschap* 18, 161–72.

van Tooren, B.F., During, H.J. and Lensink, M.J.G. (1987) The influence of the bryophyte layer on the microclimate in chalk grasslands. *Abstracta Botanica* 9, 219–30.

van Wieren, S.E. (1998) Effects of large herbivores upon the animal community. In: *Grazing and Conservation Management* (eds. M.F. WallisDeVries, J.P. Bakker and S.E. van Wieren), pp. 185–214. Kluwer Academic Publishers, Dordrecht.

van Wijnen, H.J. (1999) Nitrogen dynamics and vegetation succession in salt marshes. PhD thesis, University of Groningen.

van Wijnen, H.J. and Bakker, J.P. (1999) Nitrogen and phosphorus limitation in a coastal barrier salt marsh: the implications for vegetation succession. *Journal of Ecology* 87, 265–72.

van Wijnen, H.J. and Bakker, J.P. (2001) Long-term surface elevation changes in salt marshes: a prediction of marsh response to future sea-level rise. *Estuarine, Coastal and Shelf Science* 52, 381–90.

van Wirdum, G. (1995) The regeneration of fens in abandoned peat pits below sea level in the Netherlands. In: *Restoration of Temperate Wetlands* (eds. B.D. Wheeler, S.C. Shaw, W.J. Fojt and R.A. Robertson), pp. 251–72. Wiley, Chichester.

van Wirdum, G., den Held, A.J. and Schmitz, M. (1992) Terrestrializing fen vegetation in former turbaries in the Netherlands. In: *Fens and Bogs in the Netherlands: Vegetation, History, Nutrient Dynamics and Conserva-*

tion (ed. J.T.A. Verhoeven), pp. 323–60. Kluwer Academic Publishers, Dordrecht.

van Zanten, B.O. (1993) Historisch overzicht van het onderzoek naar de mogelijkheden voor lange afstand verspreiding bij mossen. *Buxbaumiella* 27, 31–4.

Vasander, H. (ed.) (1996) *Peatlands of Finland*. Finnish Peatland Society, Helsinki.

Veit, H. (2002) *Die Alpen: Geoökologie and Landschaftsentwicklung*. Ulmer, Stuttgart.

Vepsäläinen, K., Savolainen, R., Tiainen, J. and Vilén, J. (2000) Successional changes of ant assemblages: from virgin and ditched bogs to forests. *Annales Zoologici Fennici* 37, 135–49.

Vera, F.W.M. (2000) *Grazing Ecology and Forest History*. CABI International, Oxford.

Verboom, J., Schotman, A., Opdam, P. and Metz, A.J. (1991) European nuthatch metapopulations in a fragmented agricultural landscape. *Oikos* 61, 149–56.

Verdú, M. and García-Fayos, P. (1996) Nucleation processes in a Mediterranean bird-dispersed plant. *Functional Ecology* 10, 275–80.

Vergeer, P., Rengelink, R., Copal, A. and Ouborg, N.J. (2003) The interacting effects of genetic variation, habitat quality and population size on performance of *Succisa pratensis*. *Journal of Ecology* 91, 18–26.

Verhagen, R., Klooker, J., Bakker, J.P. and van Diggelen, R. (2001) Restoration success of low-production plant communities on former agricultural soils after top-soil removal. *Applied Vegetation Science* 4, 75–82.

Verhagen, R., van Diggelen, R. and Bakker, J.P. (2003) *Natuurontwikkeling op minerale gronden: veranderingen in de vegetatie en abiotische omstandigheden gedurende de eerste tien jaar na ontgronden*. Report, Laboratory of Plant Ecology, University of Groningen and It Fryske Gea, Olterterp.

Verhagen, R., van Diggelen, R. and Bakker, J.P. (2004) Ontgronden van voormalige landbouwgronden: welk resultaat na tien jaar voor de vegetatie? *De Levende Natuur* 105, 44–50.

Verhoeven, J.T.A., Koerselman, W. and Meuleman, A.F.M. (1996) Nitrogen- or phosphorus-limited growth in herbaceous, wet vegetation: relation with atmospheric inputs and management regimes. *Trends in Ecology and Evolution* 11, 25–40.

Verkaar, H.J. (1990) Corridors as a tool for plant species conservation. In: *Species Dispersal in Agricultural Habitats* (eds. R.H.G. Bunce *et al.*), pp. 82–97. Belhaven Press, London.

Vermeer, J.G. and Joosten, H. (1992) Conservation and management of bog and fen reserves in the Nether-lands. In: *Fens and Bogs in the Netherlands: Vegetation, History, Nutrient Dynamics and Conservation* (ed. J.T.H. Verhoeven), pp. 433–78. Kluwer Academic Publishers, Dordrecht.

Versteegh, M., Piersma, T. and Olff, H. (2004) Biodiversity in the Dutch Wadden Sea: possible consequences of ignoring the evidence on ecological effects of mobile fishing gear. *De Levende Natuur* 105, 6–9.

Verwey, J. (1952) On the ecology of distribution of cockle and mussel in the Dutch Waddensea, their role in sedimentation and the source of their food supply, with a short review of the feeding behaviour of bivalve mollusks. *Archives Neerlandaises de Zoologie* 10, 171–239.

Vesey-Fitzgerald, D.F. (1960) Grazing succession among East African game mammals. *Journal of Mammalogy* 41, 161–72.

Viessman, W. and Hammer, M.J. (1993) *Water Supply and Pollution Control*, 5th edn. Harper Collins College Publishers, New York.

Vilagrosa, A., Cortina, J., Gil, E. and Bellot, J. (2003) Suitability of drought-preconditioning techniques in Mediterranean climate. *Restoration Ecology* 11, 208–16.

Villar, P., Peñuelas, J.L. and Carrasco, I. (2000) *Influencia del endurecimiento por estrés hídrico y la fertilización en algunos parámetros funcionales relacionados con la calidad de la planta de Pinus pinea*. 1er Simposio sobre el Pino piñonero (*Pinus pinea*), pp. 211–18, Valladolid.

Visser, P.M., Ibelings, B.W., van der Veer, B. *et al.* (1996) Artificial mixing prevents nuisance blooms of the cyanobacterium *Microcystis* in Lake Nieuwe Meer, the Netherlands. *Freshwater Biology* 36, 436–50.

Vivian-Smith, G. and Stiles, E.W. (1994) Dispersal of salt marsh seeds on the feet and feathers of waterfowl. *Wetlands* 14, 316–19.

Vogel, S. (2003) The nature of artifacts. *Environmental Ethics* 25, 149–68.

Vogl, R.J. (1980) The ecological factors that produce perturbation-dependent ecosystems. In: *The Recovery Process in Damaged Ecosystems* (ed. Cairns, Jr, J.), pp. 63–94. Ann Arbor Publisher, Ann Arbor.

Voigt, H.J. (1980) *Hydrogeochemie. Eine Einführung in die Beschaffenheitsentwicklung des Grundwassers*. Urania Verlag, Leipzig.

Vollenweider, R.A. (1968) *Scientific Fundamentals of Eutrophication of Lakes and Flowing Waters, with Particular Reference to Nitrogen and Phosphorus as Factors in Eutrophication*. OECD Report DAS/CSI/68.27. OECD, Paris.

Vollenweider, R.A. (1987) Scientific concepts and methodologies pertinent to lake research and lake restoration. *Schweizerische Zeitschrift für Hydrologie* **49**, 129–47.

Vollenweider, R.A. and Kerekes, J. (1981) *OECD Eutrophication Programme*. Synthesis Report. OECD, Paris.

von Carlowitz, Hans Carl (1713) *Sylvicultura oeconomica oder hauswirthliche Nachricht und naturmäßige Anweisung zur wilden Baum-Zucht*. J.F. Braun, Leipzig [reprint Freiberg 2000].

von Zezschwitz, E. (1985) Qualitätsänderung des Waldhumus. *Forstwissenschaftliches Centralblatt* **104**, 205–20.

Vos, M., Berrocal, S.M., Kramaouna, F., Hemerik, L. and Vet, L.E.M. (2001) Plant-mediated indirect effects and the persistence of parasitoid-herbivore communities. *Ecology Letters* **4**, 38–45.

Wackernagel, M. and Rees, W. (1996) *Our Ecological Footprint: Reducing Human Impact on the Earth*. New Society Publishers, Gabriola Island, BC.

Wackernagel, M., Schulz, N.B., Deumling, D. *et al.* (2002) Tracking the ecological overshoot of the human economy. *Proceedings of the National Academy of Sciences USA* **99**, 9266–71.

Wade, K.R., Ormerod, S.J. and Gee, A.S. (1989) Classification and ordination of macroinvertebrate assemblages to predict stream acidity in upland Wales. *Hydrobiologia* **171**, 59–78.

Waide, R.B., Willig, M.R., Steiner, C.F. *et al.* (1999) The relationship between productivity and species richness. *Annual Review of Ecology and Systematics* **30**, 257–300.

Wainwright, J. (1994) Anthropogenic factors in the degradation of semi-arid regions: a prehistoric case study in southern France. In: *Environmental History in Drylands: Biogeographical and Geomorphological Perspectives* (eds. A.C. Millington and K. Pye), pp. 285–304. Wiley, New York.

Walentowski, H. and Gulder, H.-J. (2001) Es wachst zusammen was zusammen gehört. Die neue Karte der natürlichen Waldzusammensetzung Bayerns. Bayerische Landesanstalt für Wald und Forstwirtschaft. *LWF aktuell* **31**, 49 pp.+map, Freising.

Wali, M.K. (ed.) (1992) *Ecosystem Rehabilitation*, vol. 1: *Policy Issues*; vol. 2: *Ecosystem Analysis and Synthesis*. SPB, The Hague.

Walker, B.H. (1992) Biological diversity and ecological redundancy. *Conservation Biology* **6**, 18–23.

Walker, B., Carpenter, S., Anderies, J. *et al.* (2002) Resilience management in Social-ecological Systems: a working hypothesis for a participatory approach. *Conservation Ecology* **6**, 14.

Walker, L.R. and Del Moral, R. (eds.) (2003) *Primary Succession and Ecosystem Rehabilitation*. Cambridge University Press, Cambridge.

WallisDeVries, M.F., Poschlod, P. and Willems, J.H. (2002) Challenges for the conservation of calcareous grasslands in northwestern Europe: integrating the requirements of flora and fauna. *Biological Conservation* **104**, 265–73.

Walz, N. and Nixdorf, B. (eds.) (1999) Shallow lakes, 1998: trophic interactions in shallow water and brackish water bodies. *Hydrobiologia* **408/409**.

Ward, D., Holmes, N. and José, P. (1993) *The New Rivers and Wildlife Handbook*. Royal Society for the Protection of Birds, Sandy.

Ward, L.G., Kemp, W.M. and Buynton, W.R. (1984) The influence of waves and seagrass communities on suspended particulates in an estuarine embayment. *Marine Geology* **59**, 85–103.

Wardle, D.A., Bonner, K.I., Barker, G.M. *et al.* (1999) Plant removals in perennial grassland: vegetation dynamics, decomposers, soil biodiversity, and ecosystem properties. *Ecological Monographs* **69**, 535–68.

Warwick, R.M. and Uncles, R.J. (1980) Distribution of benthic macrofauna associations in the Bristol Channel in relation to tidal stress. *Marine Ecology: Progress Series* **3**, 97–103.

Wassen, M.J., Barendregt, A., Schot, P.P. and Beltman, B. (1990) Dependency of local mesotrophic fens on a regional groundwater flow system in a poldered river plain in the Netherlands. *Landscape Ecology* **5**, 21–38.

Wassen, M.J., van Diggelen, R., Verhoeven, J.T.A. and Wolejko, L. (1996) A comparison of fens in natural and artificial landscapes. *Vegetatio* **126**, 5–26.

Watling, L. and Norse, E.A. (1998) Disturbance of the seabed by mobile fishing gear: a comparison to forest clearcutting. *Conservation Biology* **12**, 1180–97.

WCED (1987) *Our Common Future: The Brundtland Report*, World Commission on Environment and Development. Oxford University Press, New York.

WCMC (1992) *Global Biodiversity: Status of the Earth's Living Resources*, World Conservation Monitoring Centre. Chapman and Hall, London.

Webb, N.R. (1997) The development of criteria for ecological restoration. In: *Restoration Ecology and*

Sustainable Development (eds. K.M. Urbanska, N.R. Webb and P.J. Edwards), pp. 133–58. Cambridge University Press, Cambridge.

Weber, E. (2000) Querschnittsaufgabe 'Punkt zu Fläche', In: *Ökologisches Entwicklungspotential der Bergbaufolgelandschaften* (eds. R.F. Hüttl, E. Weber and D. Klem), pp. 357–62. Teubner Stuttgart, Leipzig.

Weber, N. (1998) Afforestation in Europe: experiences and future possibilities. In: *Carbon Dioxide Mitigation in Forestry and Wood Industry* (eds. G.H. Kohlmaier, M. Weber and R.A. Houghton), pp. 153–65. Springer, Berlin.

Wedin, D.A. (1992) Biodiversity conservation in Europe and North America: grasslands a common challenge. *Restoration and Management Notes* 10, 137–43.

Weiher, E. and Keddy, P. (eds.) (1999) *Ecological Assembly Rules: Perspectives, Advances, Retreats.* Cambridge University Press, Cambridge.

Werkgroep Florakartering Drenthe (1999) *Atlas van de Drentse Flora.* Schuyt & Co, Haarlem.

Werritty, A., Black, A.R., Duck, R.W. *et al.* (2002) *Climate Change: Flooding Occurrences Review.* Report to Central Research Unit, Scottish Executive, Edinburgh.

Wesstrom, I. and Steen, E. (1993) Recover of vegetation after soil conservation measures in mountain areas of central Tunisia. *Ecologia Mediterranea* 19, 99–109.

Westhoff, V. (1952) The management of nature reserves in densely populated countries considered from a botanical viewpoint. In: *Proceedings and Papers of the Technical Meeting of the International Union for the Protection of Nature*, pp. 77–82. The Hague.

Westhoff, V. (1983) Man's attitude towards vegetation. In: *Man's Impact on Vegetation* (eds. W. Holzner, M.J.A. Werger and I. Ikusima), pp. 7–24. Junk, The Hague.

Westhoff, V. and Sykora, K.V. (1979) A study of the influence of desalination on the Juncetum gerardii. *Acta Botanica Neerlandica* 28, 505–12.

Wheeler, B.D. and Shaw, S.C. (1995a) Plants as hydrologists? An assessment of the value of plants as indicators of water conditions in fens. In: *Hydrology and Hydrochemistry of British Wetlands* (eds. R.M.J. Hughes and S.L. Heathwaite), pp. 63–82. Wiley, New York.

Wheeler, B.D. and Shaw, S.C. (1995b) *Restoration of Damaged Wetlands.* HMSO, London.

Wheeler, B.D. and Proctor, M.C.F. (2000) Ecological gradients, subdivisions and terminology of north-west European mires. *Journal of Ecology* 88, 187–203.

Wheeler, B.D., Money, R.P. and Shaw, S.C. (2002) Freshwater wetlands. In: *Handbook of Ecological Restoration*, vol. 2: *Restoration in Practice* (eds. M.R. Perrow and A.J. Davy), pp. 325–54. Cambridge University Press, Cambridge.

Whelan, R.J. (1986) Seed dispersal in relation to fire. In: *Seed Dispersal* (ed. D.R. Murray), pp. 237–71. Academic Press, Sydney.

Whisenant, S. (2002) Terrestrial ecosystems. In: *Handbook of Ecological Restoration*, vol. 1. (eds. M. Perrow and A. Davy), pp. 83–105. Cambridge University Press, Cambridge.

Whisenant, S.G. (1999) *Repairing Damaged Wildlands. A Process-oriented Landscape-scale Approach.* Cambridge University Press, Cambridge.

White, P.S. and Walker, J.L. (1997) Approximating nature's variation: selecting and using reference information in restoration ecology. *Restoration Ecology* 5, 338–49.

White, P.S. and Jentsch, A. (2001) The search for generality in studies of disturbance and ecosystem dynamics. In: *Progress in Botany* (eds. K. Esser, U. Lüttge, J.W. Kadereit and J.W.Z. Beischlag), pp. 399–450. Springer, Berlin.

Whittaker, R.H. (1965) Dominance and diversity in plant communities. *Science* 147, 250–60.

Whittaker, R.H. (1975) *Communities and Ecosystems*, 2nd edn. MacMillan Publishing, New York.

Wichtman, W. and Succow, M. (2001) Nachwachsenden Rohstoffe. In: *Ökosystemmanagement für Niedermoore; Strategien und Verfahren zur Renaturierung* (eds. R. Kratz, and J. Pfadenhauer), pp. 177–84. Ulmer, Stuttgart.

Wiegleb, G. and Felinks, B. (2001) Primary succession in post-mining landscapes of Lower Lusatia: chance or necessity? *Ecological Engineering* 17, 199–217.

Wiens, J.A. (1976) Population responses to patchy environments. *Annual Review of Ecology and Systematics* 7, 81–120.

Wiens, J.A. (1992) What is landscape ecology, really? *Landscape Ecology* 7, 149–50.

Wiens, J.A., Crawford, C.S. and Gocz, J.R. (1985) Boundary dynamics: a conceptual framework for studying landscape ecosystems. *Oikos* 45, 421–7.

Wilcove, D.S. (1987) Recall to the wild: wolf reintroduction in Europe and North America. *Trends in Ecology and Evolution* 2, 146–7.

Wild, A. and Florineth, F. (1999) Untersuchung von Begrünungsmethoden über der Waldgrenze. *Rasen-Turf-Gazon* 30, 4–13.

Wilmanns, O. and Bogenrieder, A. (1986) Veränderungen der Buchenwälder des Kaiserstuhls im Laufe von vier Jahrzehnten und ihre Interpretation. Pflanzensoziologische Tabellen als Dokumente. *Abhandlungen, Münster* **48**, 55–80.

Wilson, E.O. (1994) *The Diversity of Life.* Penguin Books, London.

Wilson, E.O. (2002) *The Future of Life.* Knopf, New York.

Wilson, H.M., Gibson, M.T. and O'Sullivan, P.E. (1993) Analysis of current policies and alternative strategies for the reduction of nutrient loads on eutrophicated lakes: the example of Slapton Ley, Devon. *Aquatic Conservation: Marine and Freshwater Ecosystems* **3**, 239–51.

Wilson, J.B. and Gitay, H. (1995) Limitations to species coexistence: evidence for competition from field observations, using a patch model. *Journal of Vegetation Science* **6**, 369–76.

Winterhalder K., Clewell A.F. and Aronson, J. (2004) Values and science in ecological restoration: a response to Davis and Slobodkin. *Restoration Ecology* **12**, 4–7.

With, K.A., Gardner, R.H. and Turner, M.G. (1997) Landscape connectivity and population distributions in heterogeneous environments. *Oikos* **78**, 151–69.

Wittig, R. (1998) Vegetationskundliche Bewertung der Buchenwälder auf den Rekultivierungsflächen des Braunkohletagebaugebietes Ville. In: *Braunkohletagebauten und Rekultivierung* (ed. W. Pflug), pp. 256–68. Springer, Berlin.

Wittmann, H. and Rücker, T. (1999) *Rekultivierung von Hochlagen*, Laufener Seminarbeitrag, Bayerische Akademie für Naturschutz und Landschaftspflege, Laufen, pp. 69–78.

Wolejko, L., Aggenbach, C., van Diggelen, R. and Grootjans, A.P. (1994) Vegetation and hydrology in a spring mire complex in western Pomerania, Poland. *Proceedings of the Royal Netherlands Academy of Arts and Sciences* **97**, 219–45.

Wolff, B. and Riek, W. (1997) *Deutscher Waldbodenbericht 1996: Ergebnisse der bundesweiten Bodenzustandserhebung im Wald von 1987–1993 (BZE).* Bundesministerium für Ernährung, Landwirtschaft und Forsten (BMELF), Bonn.

Wolf, C.M., Griffith, B., Reed, C. and Temple, S.A. (1996) Avian and mammalian translocations: update and reanalysis of 1987 survey data. *Conservation Biology* **10**, 1142–54.

Wolfert, H.P. (2001) *Geomorphological Change and River Rehabilitation: Case Studies on Lowland Fluvial Systems in the Netherlands*, Alterra, Wageningen.

Wolfram, Ch., Hörcher, U., Lorenzen, D. *et al.* (2000) Vegetation succession in a salt-water lagoon in the polder Beltringharder Koog, German Wadden Sea. In: *Vegetation Science in Retrospect and Perspective* (eds. P.S. White, L. Mucina, J. Lepš and E. van der Maarel), Proceedings 41st IAVS Symposium Uppsala 1998, pp. 42–6. Opulus Press, Uppsala.

Wöllecke, J. (2001) Charakterisierung der Mykorrhizazönosen zweier Kiefernforste unterschiedlicher Trophie. *Cottbuser Schriften* **17**.

Wolters, H.A., Platteeuw, M. and Schoor, M.M. (eds.) (2001) *Guidelines for Rehabilitation and Management of Floodplains: Ecology and Safety Combined.* NRC, Delft.

Wolters, M. and Bakker, J.P. (2002) Soil seed bank and driftline compositionalong a successional gradient on a temperate salt marsh. *Applied Vegetation Science* **5**, 55–62.

Wolters, M., Garbutt, A. and Bakker, J.P. (2005) Recreation of intertidal habitat; evaluating the success of de-embankments in north-west Europe. *Biological Conservation* **123**, 249–68.

Woodward, F.I. and Cramer, W. (1996) Plant functional types and climate changes: introduction. *Journal of Vegetation Science* **7**, 306–8.

Wootton, J.T. (1994) The nature and consequences of indirect effects in ecological communities. *Annual Review of Ecology and Systematics* **25**, 443–66.

Worster, D. (1977) *Nature's Economy: The Roots of Ecology.* Cambridge University Press, Cambridge.

Wright, R.F. (1985) Liming and reacidification of Hovvatan, a chronically acidified lake in southernmost Norway. *Canadian Journal of Fisheries and Aquatic Sciences* **42**, 1103–13.

Wu, J. and Loucks, O.L. (1995) From balance of nature to hierarchical patch dynamics: a paradigm shift in ecology. *Quarterly Review of Biology* **70**, 439–66.

Wyant, J.G., Meganck, R.A. and Ham, S.H. (1995) A planning and decision-making framework for ecological restoration. *Environmental Management* **19**, 789–96.

Yates, C.J. and Hobbs, R.J. (1997) Woodland restoration in the western Australian wheatbelt: a conceptual framework using a state and transition model. *Restoration Ecology* **5**, 28–35.

Young, A., Boyle, T. and Brown, T. (1996) The population genetic consequences of habitat fragmentation for plants. *Trends in Ecology and Evolution* 11, 413–18.

Young, T.P. (2000) Restoration ecology and conservation biology. *Biological Conservation* 92, 73–83.

Zabel, J. and Tscharntke, T. (1998) Does fragmentation of *Urtica* habitats affect phytophagous and predatory insects differentially? *Oecologia* 116, 419–25.

Zackrisson, O., Nilsson, M.-C. and Wardle, D.A. (1996) Key ecological function of charcoal from wildfire in Boreal forest. *Oikos* 77, 10–19.

Zadoks, J.C. (1987) The function of plant pathogenic fungi in natural communities. In: *Disturbance in Grasslands: Causes, Effects and Processes* (eds. J. van Andel, J.P. Bakker and R.W. Snaydon), pp. 201–7. Junk Publishers, Dordrecht.

Zavaleta, E.S., Hobbs, R.J. and Mooney, H.A. (2001) Viewing invasive species removal in a whole-ecosystem context. *Trends in Ecology and Evolution* 16, 454–9.

Zedler, J.B. and Callaway, J.C. (1999) Tracking wetland restoration: do mitigation sites follow desired trajectories? *Restoration Ecology* 7, 69–73.

Zedler, J.B. and Lindig-Cisneros, R. (2000) Functional equivalency of restored and natural salt marshes. In: *Concepts and Controversies in Tidal Marsh Ecology* (eds. M.P. Weinstein and D.A. Kreeger), pp. 565–82. Kluwer Academic Publishers, Dordrecht.

Zink, T.A. and Allen, M.F. (1998) The effects of organic amendments on the restoration of a disturbed coastal sage scrub habitat. *Restoration Ecology* 6, 52–8.

Zobel, M. (1997) The relative role of species pools in determining plant species richness: an alternative explanation of coexistence? *Trends in Ecology and Evolution* 12, 266–9.

Zobel, M., van der Maarel, E. and Dupré, C. (1998) Species pool: the concept, its determination and significance for community restoration. *Applied Vegetation Science* 1, 55–66.

Zöckler, C. (2000) *Wise Use of Floodplains: Review of River Restoration Projects in a Number of European Countries*. WWF European Freshwater Programme, Cambridge.

Zohary, D. and Hopf, M. (1993) *Domestication of Plants in the Old World*. Clarendon, Oxford.

Internet websites

Internet websites can be of great use in the exploration of information. But one should be aware of the fact that the content of a website is not always consistent in time. The following websites may be useful and have been referred to in this book. Their accessibility was confirmed on 16 August 2004.

International organizations

United Nations (www.un.org)

For Agenda 21 search on *economic and social development*, then on *sustainable development* and then on *documents*.

- Food and Agricultural Organizations of the United Nations (FAO), www.fao.org
- United Nations Convention to Combat Desertification (UNCCD), www.unccd.int
- United Nations Economic Commission for Europe, www.unece.org

For the Protocol on Further Reduction of Sulphur Emissions, search on *environment*.

- United Nations Educational, Scientific and Cultural Organization (UNESCO), www.unesco.org
  UNESCO Man and Biosphere, www.unesco.org/mab
- United Nations Framework Convention on Climate Change (UNFCCC), www.unfccc.org

Other international organizations

- Commission Internationale pour la Protection des Alpes (CIPRA), www.cipra.org
- Convention of Biological Diversity (CBD), www.biodiv.org
- Intergovernmental Panel on Climate Change (IPCC), www.ipcc.ch
- International Council for the Exploration of the Sea (ICES), www.ices.dk
- International Union for the Conservation of Nature (IUCN), http://iucn.org
- Organisation for Economic Co-operation and Development (OECD), www.oecd.org
- Society for Ecological Restoration International (SERI), www.ser.org
- Society of Wetlands Scientists (SWS), www.sws.org

For the Position Paper on the Definition of Wetland Restoration, search on *wetland concerns*.

- World Conservation Monitoring Centre (WCMC), www.wcmc.org.uk
- World Wildlife Fund (WWF), www.wwf.org. On this website the WWF divisions of different countries can be found easily.

Europe

- European Union, www.europa.eu.int

Austria

- Bundesministerium für Land- und Forstwirtschaft, Umwelt und Wasserwirtschaft, Republik Österreich (Federal Ministry of Agriculture, Forestry, Environment and Water Management), www.bal.bmlf.gv.at

Denmark

- Miljøministeriet, Miljøstyrelsen (Danish Environmental Protection Agency), www.mst.dk
- Miljøministeriet, Skov- og Naturstyrelsen (Danish Forest and Nature Agency), www.sns.dk
- Miljøministeriet (Danish Ministry of the Environment), www.mem.dk

Germany

- Bundesamt für Naturschutz (The German Federal Agency for Nature Conservation), www.bfn.de
- Bundesministerium für Umwelt, Naturschutz und Reaktorsicherheit, Umweltbundesamt (Federal Environmental Agency), www.umweltbundesamt.de

Italy

- Centro Italiano per la Riqualificazione Fluviale (Italian Centre for River Restoration; CIRF), www.cirf.org

The Netherlands

- Vogelonderzoek Nederland (Dutch Centre for Field Ornithology; SOVON), www.sovon.nl
- Ministerie van Landbouw, Natuur en Voedselkwaliteit (Dutch Ministry for Agriculture, Nature and Food Quality), www.minlnv.nl/international
- Ministerie van Verkeer en Waterstaat, Rijksinstituut voor Integraal Zoetwaterbeheer en Afvalwaterbehandeling (National Institute for Inland Water Management and Waste Water Treatment; RIZA), www.riza.nl

- Ministerie van Verkeer en Waterstaat, Rijksinstituut voor Kust en Zee (National Institute for Coastal and Marine Management; RIKZ), www.rikz.nl
- Rijksinstituut voor Volksgezondheid en Milieu (National Institute for Public Health and the Environment; RIVM), www.rivm.nl

For RIVM environmental data, http://arch.rivm.nl/environmentaldata

Switzerland

- Bundesamt für Umwelt, Wald und Landschaft (Swiss Agency for the Environment, Forests and Landscape; SAEFL), www.umwelt-schweiz.ch

UK

- Department for Environment, Food and Rural Affairs (DEFRA), www.defra.gov.uk
- River Restoration Centre, www.therrc.co.uk

Canada

- Canadian Wildlife Service, www.cws-scf.ec.gc.ca

USA

- Everglades, www.evergladesplan.org
- Federal Interagency Stream Restoration Working Group (FISRWG), www.nrcs.usda.gov
- United States Department of Agriculture, www.usda.gov
- USDA Forest Service, www.usda.gov (Agencies and Offices)
- United States Environmental Protection Agency (EPA), www.epa.gov
- United States Fish and Wildlife Service, www.fws.gov

Index

Page references to figures are in *italic*, to tables are in **bold**

10.30